# TRENDS IN MATHEMATICS

*Trends in Mathematics* is a series devoted to the publication of volumes arising from conferences and lecture series focusing on a particular topic from any area of mathematics. Its aim is to make current developments available to the community as rapidly as possible without compromise to quality and to archive these for reference.

Proposals for volumes can be sent to the Mathematics Editor at either

Birkhäuser Verlag
P.O. Box 133
CH-4010 Basel
Switzerland

or

Birkhäuser Boston Inc.
675 Massachusetts Avenue
Cambridge, MA 02139
USA

Material submitted for publication must be screened and prepared as follows:

All contributions should undergo a reviewing process similar to that carried out by journals and be checked for correct use of language which, as a rule, is English. Articles without proofs, or which do not contain any significantly new results, should be rejected. High quality survey papers, however, are welcome.

We expect the organizers to deliver manuscripts in a form that is essentially ready for direct reproduction. Any version of TEX is acceptable, but the entire collection of files must be in one particular dialect of TEX and unified according to simple instructions available from Birkhäuser.

Furthermore, in order to guarantee the timely appearance of the proceedings it is essential that the final version of the entire material be submitted no later than one year after the conference. The total number of pages should not exceed 350. The first-mentioned author of each article will receive 25 free offprints. To the participants of the congress the book will be offered at a special rate.

# Groups and Geometries

## Siena Conference, September 1996

Lino di Martino
William M. Kantor
Guglielmo Lunardon
Antonio Pasini
Maria Clara Tamburini
Editors

Springer Basel AG

Editors' address:

Antonio Pasini
Department of Mathematics
University of Siena
Via del Capitano 15
53100 Siena

1991 Mathematical Subject Classification 20–06, 51–06, 05–06

A CIP catalogue record for this book is available from
the Library of Congress, Washington D.C., USA

Deutsche Bibliothek Cataloging-in-Publication Data

**Groups and geometries** : Siena conference, September 1996/Antonio Pasini ed.
- Basel ; Boston ; Berlin : Birkhäuser, 1998
(Trends in Mathematics)
ISBN 978-3-0348-9785-3


Originally published by Birkhäuser Verlag in 1998
Softcover reprint of the hardcover 1st edition 1998

Printed on acid-free paper produced of chlorine-free pulp. TCF ∞

ISBN 978-3-0348-9785-3 ISBN 978-3-0348-8819-6 (eBook)
DOI 10.1007/978-3-0348-8819-6

9 8 7 6 5 4 3 2 1

# Contents

# Contents

# Preface

On September 1–7, 1996 a conference on Groups and Geometries took place in lovely Siena, Italy. It brought together experts and interested mathematicians from numerous countries. The scientific program centered around invited expository lectures; there also were shorter research announcements, including talks by younger researchers.

The conference concerned a broad range of topics in group theory and geometry, with emphasis on recent results and open problems. Special attention was drawn to the interplay between group-theoretic methods and geometric and combinatorial ones.

Expanded versions of many of the talks appear in these Proceedings. This volume is intended to provide a stimulating collection of themes for a broad range of algebraists and geometers. Among those themes, represented within the conference or these Proceedings, are aspects of the following:

1. the classification of finite simple groups,
2. the structure and properties of groups of Lie type over finite and algebraically closed fields of finite characteristic,
3. buildings, and the geometry of projective and polar spaces, and
4. geometries of sporadic simple groups.

We are grateful to the authors for their efforts in providing us with manuscripts in LaTeX. Barbara Priwitzer and Thomas Hintermann, Mathematics Editors of Birkhäuser, have been very helpful and supportive throughout the preparation of this volume.

The organizing committee consisted of L. Di Martino (Milan), W. M. Kantor (Eugene), G. Lunardon (Naples), A. Pasini (Siena), and M. C. Tamburini (Brescia). The Conference Service of the University of Siena provided valuable assistance for the organization of the Conference. We thank the University of Siena, the Italian National Research Council (C.N.R.), the team of the National Research Project 'Non-commutative Algebra' (supported by the Italian Ministry for University and Scientific Research) and the Bank Monte dei Paschi di Siena for their financial support to the conference. The Universities of Brescia and Naples also gave a precious support to the Conference. We warmly thank them for this.

# Talks

M. Aschbacher, *Quasithin groups.*

L. Bader, *On infinite flocks of quadratic cones.*

B. Baumeister, *A new computerfree existence proof of the third group of Janko.*

A. Bonisoli, *On two transitive ovals in finite projective planes.*

A. Brouwer, *The geometry far away from a point or a chamber.*

F. Buekenhout, *The flag-transitive incidence geometries of small almost simple groups and an application to the Mathieu group* $M_{12}$.

M. Buratti, *Constructions for block designs by difference sets.*

A. Camina, *Can we classify line-transitive linear spaces?*

C. Casolo, *Character degrees of finite groups.*

A. Cohen, *Integral representations of finite groups in algebraic groups.*

B. Cooperstein, *On a connection between the hyperbolic quadric* $Q^+(9,q)$ *and the* $E_{6,1}$*-geometry.*

H. Cuypers, *Local recognition of graphs related to polarities.*

H. Gottschalk, *Geometries for the group* $J_1$.

R. Guralnick, *The first cohomology group and generation of simple groups.*

S. Heiss, *On the* $P_3$ *sequenceability of finite groups.*

J. Hirschfeld, *The geometry of the ternary unitary group.*

C. Huybrechts, *A characterization of the Hall-Janko group* $J_2$ *by a* $c.L^*$*-geometry.*

A. Ivanov, *Flag-transitive extended classical dual polar spaces.*

N. Johnson, *Derivable nets in non derivable planes.*

M. Liebeck, *Regular orbits of linear groups.*

R. Liebler, *Towards a chamber systems representation theory.*

A. Lucchini, *Generating minimally transitive groups.*

G. Malle, *Finite irreducible linear groups with polynomial ring of invariants.*

J. Saxl, *Generating classical groups.*

G. Seitz, *Maximal subgroups of finite exceptional groups.*

E. Shult, *Some aspects of buildings.*

S. Smith, *Application of geometry to group cohomology.*

L. Soicher, *Computing fundamental groups and covers of simplicial complexes.*

A. Shalev, *Hausdorff dimension, pro-p groups and Kac-Moody algebras.*

S. Shpektorov, *A characterization of rank 2 Chevalley groups in even characteristic.*

A. Steinbach, *Generalized quadrangles weakly embeddable in projective spaces.*

B. Stellmacher, *Locally s-transitive groups.*

G. Stroth, *Diagram geometries and sporadic simple groups.*

M.C. Tamburini, *Carter subgroups in classical groups.*

J. Thas, *Embedding of geometries in finite projective spaces: a survey.*

F. Timmesfeld, *Moufang polygons and abstract root-subgroups.*

H. Van Maldeghem, *Spreads and ovoids in finite Moufang hexagons.*

N. Vavilov, *The geometry of tori.*

S. Yoshiara, *The Borel-Tits property for finite groups.*

F. Zara, *Generalized reflection groups.*

# Participants

P. ARROYO JORDA (Valencia, Spain)
M. ARROYO JORDA (Alcoy, Spain)
M. ASCHBACHER (Pasadena, California, USA)
L. BADER (Roma, Italy)
C. BARDINI (Genova, Italy)
C. BARTOLONE (Palermo, Italy)
A. BASILE (Verona, Italy)
B. BAUMEISTER (Halle, Germany)
M. BIANCHI (Milano, Italy)
G. BINI (Roma, Italy)
A. BLUNCK (Darmstadt, Germany)
A. BONISOLI (Potenza, Italy)
L. BROUNS (Gent, Belgium)
A. BROUWER (Eindhoven, The Netherlands)
F. BUEKENHOUT (Bruxelles, Belgium)
M. BURATTI (L Aquila, Italy)
C. CASOLO (Firenze, Italy)
M. CAZZOLA (Milano, Italy)
P. CECCHERINI (Roma, Italy)
D. CHILLAG (Haifa, Israel)
A. COHEN (Eindhoven, The Netherlands)
B. COOPERSTEIN (Santa Cruz, California, USA)
A. COSSIDENTE (Potenza, Italy)
H. CUYPERS (Eindhoven, The Netherlands)
F. DALLA VOLTA (Milano, Italy)
U. DARDANO (Napoli, Italy)
A. DEL FRA (L'Aquila, Italy)
E. DETOMI (Padova, Italy)
L. DI MARTINO (Milano, Italy)
M. ENEA (Palermo, Italy)
G. FALCONE (Palermo, Italy)
P. FRANCINI (Siena, Italy)
D. FROHARDT (Detroit, Michigan, USA)
A. FUKSCHANSKI (Halle, Germany)
N. GAVIOLI (L'Aquila, Italy)
A. GILLIO (Milano, Italy)
H. GOTTSCHALK (Giessen, Germany)
T. GRUNDHÖFER (Würzburg, Germany)
R. GURALNICK (Los Angeles, California, USA)
S. HEISS (Halle, Germany)
M. HERZOG (Tel Aviv, Israel)
J. HIRSCHFELD (Brighton, UK)
D. HUGHES (Torrita di Siena, Italy)
A. HULPKE (Aachen, Germany)
C. HUYBRECHTS (Bruxelles, Belgium)
A. IVANOV (London, UK)
E. JABARA (Mestre, Italy)
V. JHA (Glasgow, UK)
N. JOHNSON (Iowa City, Iowa, USA)
A. JUHASZ (Haifa, Israel)
W. KANTOR (Eugene, Oregon, USA)
O. KING (Newcastle upon Tyne, UK)
G. KORCHMAROS (Potenza, Italy)
R. LAWTHER (Lancaster, UK)
D. LEEMANS (Bruxelles, Belgium)

M. LIEBECK (London, UK)
R. LIEBLER (Fort Collins, Colorado, USA)
P. LONGOBARDI (Napoli, Italy)
H. LÖWE (Braunschweig, Germany)
R. LÖWEN (Braunschweig, Germany)
A. LUCCHINI (Brescia, Italy)
G. LUNARDON (Napoli, Italy)
K. LUX (Aachen, Germany)
M. MAINARDIS (Udine, Italy)
M. MAJ (Napoli, Italy)
A. MAKHNEV (Ekaterinburg, Russia)
G. MALLE (Heidelberg, Germany)
A. MARTINEZ PASTOR (Valencia, Spain)
S. MATTAREI (Padova, Italy)
F. MENEGAZZO (Padova, Italy)
M. MORIGI (Padova, Italy)
C. O'KEEFE (Adelaide, Australia)
G. PARMEGGIANI (Padova, Italy)
D. PASECHNICK (Eindhoven, The Netherlands)
A. PASINI (Siena, Italy)
G. PICA (Napoli, Italy)
P. QUATTROCCHI (Modena, Italy)
G. RINALDI (Modena, Italy)
S. RINAURO (Potenza, Italy)
G. RÖHRLE (Bielefeld, Germany)
L. ROSATI (Firenze, Italy)
D. SAELI (Potenza, Italy)
P. SANCHINI (Brescia, Italy)
L. SERENA (Firenze, Italy)
J. SAXL (Cambridge, UK)
G. SEITZ (Eugene, Oregon, USA)
A. SHALEV (Jerusalem, Israel)
S. SHPEKTOROV (Columbus, Ohio, USA)
E. SHULT (Manhattan, Kansas, USA)
S. SMITH (Chicago, Illinois, USA)
L. SOICHER (London, UK)
A. STEINBACH (Giessen, Germany)
B. STELLMACHER (Kiel, Germany)
G. STROTH (Halle, Germany)
M. TAMBURINI (Brescia, Italy)
J. THAS (Gent, Belgium)
C. TIBILETTI (Milano, Italy)
F. TIMMESFELD (Giessen, Germany)
M. VACCARO (Napoli, Italy)
J. VAN BON (London, UK)
H. VAN MALDEGHEM (Gent, Belgium)
N. VAVILOV (St. Petersburg, Russia)
T. WEIGEL (Freiburg im Br., Germany)
C. WIEDORN (Halle, Germany)
J. WILSON (Birmingham, UK)
S. YOSHIARA (Osaka, Japan)
G. ZACHER (Padova, Italy)
F. ZARA (Amiens, France)
P. ZUCCA (Palermo, Italy)

Trends in Mathematics, © 1998 Birkhäuser Verlag Basel/Switzerland

# Quasithin Groups

Michael G. Aschbacher* and Stephen D. Smith †

**Abstract**

Geoff Mason announced in about 1980 the classifcication of quasithin groups of characteristic 2; but never published this step in the classification of the finite simple groups. In January 1996, the authors began work toward a new and more general classification of quasithin groups; the paper gives an exposition of the approach and considerable progress to date.

The treatment of quasithin groups of characteristic 2 was one of the last steps in the Classification of the finite simple groups. Geoff Mason [Mas80] announced a classification of these groups in about 1980, but never published his work. A few people have a copy of a large manuscript containing his efforts, but because it was distributed slowly, section by section, it was only during the last few years that it was realized that Mason's manuscript is incomplete in various ways. A few years ago the first author wrote up a treatment which begins where Mason's manuscript ends and finishes the problem assuming the results he says he proves. Neither of us have read Mason's manuscript other than superficially, but it appears there are missing lemmas even for the part of the problem the theorems in his manuscript cover. We do believe however that he has seriously addressed the issues involved and that he could turn his manuscript into a proof with enough work. However Mason is now involved with Moonshine and has no interest in completing or publishing his manuscript.

Thus to have a complete proof of the Classification, someone needs to produce a treatment of the quasithin groups of even characteristic. But there are more complications. In their revision of the Classification, Gorenstein, Lyons, and Solomon (GLS) have settled on a weaker definition of "even characteristic" than in the original approach to the Classification. This makes the analysis of quasithin groups of even characteristic more difficult. Also GLS would like to weaken the definition of "quasithin".

In January 1996, we began to work toward a classification of quasithin groups. We have not completed that work yet, but believe we are far enough along to say that our approach is working, we will complete it, and the treatment which

*Partially supported by NSF grants DMS–91-01237 and 96-22843.

†Partially supported by NSA grant MDA 904-93-H-3039.

emerges will be a much simpler treatment than Mason's. Ulrich Meierfrankenfeld and Bernd Stellmacher are also working on the problem, although from a somewhat different point of view.

In this paper we will give you the original definitions of "quasithin" and "even characteristic" and the new definition of "even characteristic" we are working with. We are sticking with the original definition of "quasithin", and will tell you why. We also give some history of the problem and some vague idea of the approach we are taking. The end of the paper consists of a slightly more detailed description of the part of the problem where the generic examples arise.

The quasithin problem is very difficult and complex. There are many possible paths to the goal, and extra thought and care or an extra observation can result in significant simplifications. Thus even if we and/or Meirfrankenfeld and Stellmacher produce a solution to the problem, others should examine those solutions with an eye toward simplification.

The techniques available are technical and difficult to describe in an hour lecture to an audience which is unfamiliar with such techniques. Thus for the most part we will seek to avoid technical details.

Let $G$ be a finite group, $T \in Syl_2(G)$, and $\mathcal{M}$ the set of maximal 2-local subgroups of $G$. Define

$$e(G) = \max\{m_p(M) : M \in \mathcal{M} \text{ and } p \text{ is an odd prime}\}$$

where $m_p(M)$ is the *p-rank* of $M$. If $G$ is a group of Lie type over a field of characteristic 2, then $e(G)$ is a good approximation of the Lie rank of $G$. Define $G$ to be *quasithin* if $e(G) \leq 2$. The definition comes from the N-group paper [Tho68].

Originally a 2-local $H$ of $G$ was said to be of *characteristic 2* if $F^*(H) = O_2(H)$, or equivalently $C_H(O_2(H)) \leq O_2(H)$. Further $G$ was said to be of *characteristic 2 type* if all 2-locals of $G$ were of characteristic 2. Equivalently all members of $\mathcal{M}$ are of characteristic 2. If $G$ is of Lie type over a field of characteristic 2 then $G$ is of characteristic 2 type. The generic example of a simple quasithin group of characteristic 2-type is a group of Lie type over a field of characteristic 2 and Lie rank at most 2, but there are other degenerate examples. I will show you a list of the examples in a while.

We relax the definition of "characteristic 2 type" and say that $G$ is of *even characteristic* if

$$C_M(O_2(M)) \leq O_2(M) \quad \text{for all} \quad M \in \mathcal{M}(T)$$

where for $X \subseteq G$,

$$\mathcal{M}(X) = \{M \in \mathcal{M} : X \subseteq M\}$$

Thus we now require only that 2-locals containing a Sylow 2- subgroup are of characteristic 2 and allow other 2-locals to be nonconstrained. This is in line with the GLS approach, although the definition is somewhat different than theirs. The following example may shed some light on this change of definition.

**Example 1.** Let $L$ be a simple group of Lie type over a field of characteristic 2 and assume either

(a) $G = L\langle t \rangle$ is $L$ extended by an involutory outer automorphism, or

(b) $G = (L_1 \times L_2)\langle t \rangle$, where $L_i \cong L$ is a component of $G$ and $L_1^t = L_2 \neq L_1$ for some involution $t$; i.e., $G$ is the wreath product of $L$ by $\mathbf{Z}_2$.

Then $G$ is of even characteristic but rarely of characteristic 2 type, as $C_G(t)$ usually has a component which is a group of Lie type over a field of characteristic 2. In the context of trying to prove the Classification, groups with the 2- local structure of $G$ will arise, and in the opinion of GLS (and we agree) it is easier to treat such groups using characteristic 2 methods rather than focus on the "semisimple element" $t$.

Let $\mathcal{K}$ denote the list of known simple groups. $G$ is a *$\mathcal{K}$-group* if all composition factors of $G$ are in $\mathcal{K}$. Define $G$ to be a *QTK-group* if $G$ is quasithin of even characteristic and all proper subgroups of $G$ are $\mathcal{K}$-groups. Here is a list (hopefully accurate) of the QTK- groups in $\mathcal{K}$.

## QTK- groups in $\mathcal{K}$

1. Groups of Lie type over fields of characteristic 2 and Lie rank $\leq 2$, but not $U_5(q)$.
2. $L_4(2)$, $L_5(2)$, $Sp_6(2)$, $U_5(4)$.
3. Alternating groups $A_n$, $n \leq 9$.
4. $L_2(p)$, $p$ a Mersenne or Fermat prime, $L_3^\epsilon(3)$, $L_4^\epsilon(3)$, or $G_2(3)$.
5. Mathieu groups, Janko groups, HS, He, or Ru.

We are trying to prove every simple QTK-group is on this list. The generic examples are those in (1).

Groups of even characteristic with $e(G)$ large are handled by looking at local subgroups for odd primes and using signalizer functor methods. There are some special difficulties using this approach when $e(G) = 3$, so GLS have proposed that perhaps groups with $e(G) = 3$ should also be handled using characteristic 2 methods. It is our intuition however (and Aschbacher did the original work for $e(G) = 3$) that the signalizer functor methods on odd primes can give a simpler treatment when $e(G) = 3$. Thus we are sticking to $e(G) \leq 2$, although Meierfrankenfeld and Stellmacher are considering the more general problem.

The methods used in attacking the problem go back to Thompson in the N-group paper [Tho68]. In particular in the N-group paper, Thompson introduced the parameter $e(G)$ and used weak closure, uniqueness theorems, and work of Tutte [Tut59] and Sims [Sim67]. We will return to these techniques in a moment, but first more history. In [Jan72], Janko defined $G$ to be *thin* if $e(G) = 1$; and used Thompson's methods to determine all thin groups of characteristic 2-type in which all 2-locals are solvable. His student Fred Smith extended the solvable 2-local result to quasithin groups in [Smi75]. Aschbacher did the general thin group case in [Asc78] and Mason went a long way toward a complete treatment of the general quasithin case in [Mas80]. There have since been new treatments of portions of the N-group problem due to Stellmacher [Ste] and Gomi and his collaborators [GH92], using extensions of the Tutte-Sims theory which has been come to be known as the *amalgam method.* Aschbacher used some early versions of such extensions due to Glauberman (which eventually expanded to the Glauberman-Niles Theorem [GN83]) in his work on thin groups. Goldschmidt initiated the "modern" amalgam method in [Gol80] and this was extended and the amalgam method modified in [DGS85] by Goldschmidt- Delgado-Stellmacher and in [MS93] by Meierfrankenfeld-Stellmacher.

We now hint at the nature of some of these techniques. Let $G$ be a finite group of even characteristic, $T \in Syl_2(G)$, $M \in \mathcal{M}(T)$, and $V$ nontrivial normal elementary abelian 2-subgroup of $G$. Given a positive integer $n$, define

$$W_n(T,V) = \langle A \leq T : A \leq V^g \text{ and } m(V^g/A) \leq n \text{ for some } g \in G\rangle$$

and

$$C_n(T,V) = C_T(W_n(T,V))$$

Let

$$\mathcal{H}(T) = \{H \leq G : T \leq H \text{ and } O_2(H) \neq 1\}$$

Under suitable hypotheses on the action of $M$ on $V$, for certain $n < m$, and for certain $H \in \mathcal{H}(T)$, one can show

(WC) $$H = \langle N_H(W_n(T,V)), C_H(C_m(T,V))\rangle.$$

Results of this sort and techniques for using them are called *weak closure arguments.* They are used in conjunction with uniqueness theorems. A typical uniqueness theorem says,

(U) $$\mathcal{M}(X) = \{M\}$$

for suitable subgroups $X$ of $M$. eg. if $X = M$ this is trivially true. If $X, Y$ are unique uniqueness groups satisfying (U) with $W_n(T,V) \lhd X$ and $Y \leq C_G(C)$ for some $1 \neq C \leq C_m(T,V)$, then the weak closure theorem (WC) says that $H \leq M$, since (U) says

$$N_G(W_0(T,V)) \leq M \geq C_G(C) \geq C_G(C_m(T,V))$$

In particular if we have shown $\mathcal{M}(T) \neq \{M\}$ and pick $H \in \mathcal{M}(T) \setminus \{M\}$, we would have a contradiction. This gives a hint of how weak closure and uniqueness theorems are used.

Here is a way to produce uniqueness theorems in simple QTK-groups $G$. For $H < G$, define $\mathcal{C}(H)$ to consist of those subnormal subgroups $L$ of $H$ such that $L = L^\infty$ and $L$ is minimal subject to these constraints. Modulo a special class of exceptions that we will ignore here for purposes of exposition, if $L \in \mathcal{C}(H)$ then $L/O_2(L)$ is quasisimple. Let $\mathcal{L}(G,T)$ consist of all subgroups $L$ of $G$ such that $\langle L,T\rangle \in \mathcal{H}(T)$ and $L \in \mathcal{C}(\langle L,T\rangle)$. The set of maximal members of $\mathcal{L}(G,T)$ under inclusion is denoted by $\mathcal{L}^*(G,T)$. It is easy to show

**First Uniqueness Theorem:** If $L \in \mathcal{L}^*(G,T)$ then $\mathcal{M}(\langle L,T\rangle) = \{N_G(\langle L^T\rangle)\}$.

Write $\mathcal{L}_f(G,T)$ for those $L \in \mathcal{L}(G,T)$ such that $L$ does not centralize $Z(O_2(L))$. (These are the *faithful members* of $\mathcal{L}(G,T)$).

If $L \in \mathcal{L}_f^*(G,T)$ and $\{M\} = \mathcal{M}(\langle L,T\rangle)$, then there will be a nontrivial normal elementary abelian 2-subgroup of $M$ not centralized by $L$ with $O_2(L/C_L(V)) = 1$ and we can use $V$ for our weak closure.

Assume we know $\mathcal{M}(T) \neq \{M\}$. Then we let $\mathcal{G}(M,T)$ consist of those $H \in \mathcal{H}(T)$ minimal subject to $H \not\leq M$. There is a parameter $n(H)$ which roughly speaking is $n$ if $F^*(H/O_2(H))$ is of Lie type over $\mathbf{F}_{2^n}$, and $n(H) = 1$ otherwise. The following theorem determines the "generic" QTK-groups.

**Theorem on Generic Groups:** Let $G$ be a simple QTK-group and $T \in Syl_2(G)$. Assume $L \in \mathcal{L}_f^*(G,T)$ with $N_G(L) \in \mathcal{M}(T)$, $L/O_2(L) \cong L_2(2^n)$, and there exists $H \in \mathcal{G}(M,T)$ with $n(H) > 1$. Then either $L$ is a group of Lie type over $\mathbf{F}_{2^n}$ or $\mathbf{F}_{2^{\frac{n}{2}}}$ of Lie rank 2, or $n = 2$ and $L \cong M_{23}$.

Here is a very brief sketch of the proof of the Theorem on Generic Groups. Let $D$ be a Hall $2'$-subgroup of $N_L(T \cap L)$ and $K = H^\infty$. The first and longest step is to show using weak closure arguments and uniqueness theorems that:

**Theorem 1** $D \leq N_G(K)$.

Now it is essentially the case that $K/O_2(K)$ is a rank 1 group of Lie type over a field of even characteristic and $(M \cap K)/O_2(K)$ is a Borel subgroup of $K/O_2(K)$. Let $B$ be a Hall $2'$-subgroup of $K \cap M$. By Theorem 1, the triple $\alpha = (LTB, TBD, KTD)$ is a weak BN-pair in the sense of Delgado-Stellmacher in [DGS85], so by the Main Theorem of [DGS85] (proved using the amalgam method), the amalgam $\alpha$ is isomorphic to that of some rank 2 group $\hat{G}$ of Lie type. We wish to show $G \cong \hat{G}$. First we eliminate a special case:

**Theorem 2** *If $K \notin \mathcal{L}^*(G,T)$ then $G \cong M_{23}$.*

So after this we can assume $K \in \mathcal{L}^*(G,T)$. Let $\tilde{G}$ be the universal completion of the amalgam $\alpha$. Then $G_1 = LTB$ and $G_2 = KTD$ are embedded in $\tilde{G}$ and to simplify notation we regard them as subgroups. Let $L_i = G_i^\infty$, $s_i$ an involution in $N_{L_i}(BD)$, and $W = \langle s_1, s_2\rangle BD$. In $\hat{G}$, the corresponding group $\hat{B}\hat{D}$ is a Cartan group and $\hat{W}/\hat{B}\hat{D}$ is the Weyl group. Let $2m(\alpha) = |\hat{W} : \hat{B}\hat{D}|$ be the order of the Weyl group.

Now $G$ and $\hat{G}$ are images of $\tilde{G}$, and by a result in [Asc93] on the simple connectivity of a certain simplicial complex:

**Theorem 3** *$\hat{G} = \tilde{G}/I$, where $I$ is the normal subgroup generated by $(s_1 s_2)^{m(\alpha)}$.*

Thus to complete the proof of the Theorem on Generic Groups, it remains to show that $G = \langle G_1, G_2 \rangle$ and $s_1 s_2$ has order $m(\alpha)$ in $G$. First we transfer off involutions to insure all involutions in $G$ are fused into $L$. Then the fact that $\mathcal{M}(L_i) = \{G_i\}$ allows us to control centralizers of involutions and the elementary theory of strongly embedded subgroups says $G = \langle G_1, G_2 \rangle$. Then in the case $\hat{G} \cong L_3(q)$, some clever arguments of Suzuki in [Suz65] show $|s_1 s_2| = 3$. In the remaining cases we originally used the Goldschmidt-Alperin Fusion Theorem and the Thompson Order Formula to complete the proof, but as Sergei Shpectorov observed in his talk at this conference, it is easier to show $BD$ is self centralizing in $G$ and use this to argue that $|s_1 s_2| = m(\alpha)$.

## References

[Asc78] M. Aschbacher. Thin finite simple groups. *J. Algebra*, 54:50–112, 1978.

[Asc93] M. Aschbacher. Simple connectivity of $p$-group complexes. *Israel J. Math.*, 82:1–42, 1993.

[DGS85] A. Delgado, D. Goldschmidt, and B. Stellmacher. *Groups and Graphs: New Results and Methods.* Birkhäuser Verlag, Basel, 1985.

[GH92] K. Gomi and M. Hayashi. A pushing up approach to the quasithin finite simple groups with solvable 2-local subgroups. *J. Algebra*, 146:412–426, 1992.

[GN83] G. Glauberman and R. Niles. A pair of characteristic subgroups for pushing up in finite groups. *Proc. London Math. Soc.*, 46:411–453, 1983.

[Gol80] D. Goldschmidt. Automorphisms of trivalent graphs. *Ann. Math.*, 111:377–406, 1980.

[Jan72] Z. Janko. Nonsolvable finite groups all of whose 2-local subgroups are solvable, I. *J. Algebra*, 21:458–517, 1972.

[Mas80] G. Mason. The classification of finite quasithin groups. *(mimeograph typescript)*, (unpublished), around 1980.

[MS93] U. Meierfrankenfeld and B. Stellmacher. Pushing up weak BN-pairs of rank 2. *Commun. in Algebra*, 21:825–934, 1993.

[Sim67] C. Sims. Graphs and finite permutation groups. *Math. Z.*, 95:76–86, 1967.

[Smi75] F. Smith. Finite simple groups all of whose 2-local subgroups are solvable. *J. Algebra*, 34:481–520, 1975.

[Ste] B. Stellmacher. An application of the amalgam method: The 2-local structure of N-groups of characteristic 2 type. *(to appear).*

[Suz65] M. Suzuki. Finite groups in which the centralizer of any element of order 2 is 2-closed. *Ann. Math.*, 82:191–212, 1965.

[Tho68] J. Thompson. Nonsolvable groups all of whose local subgroups are solvable, I. *Bull. Amer. Math. Soc.*, 74:383–437, 1968.

[Tut59] W. Tutte. On the symmetry of cubic graphs. *Canad. J. Math.*, 11:621–624, 1959.

Michael G. Aschbacher
Department of Mathematics
California Institute of Technology
Pasadena California 91125 USA
e-mail: `asch@cco.caltech.edu`

Stephen D. Smith
Department of Mathematics (m/c 249)
University of Illinois at Chicago
851 S. Morgan
Chicago Illinois 60607-7045 USA
e-mail: `smiths@math.uic.edu`

Trends in Mathematics, © 1998 Birkhäuser Verlag Basel/Switzerland

# The Non-canonical Gluings of two Affine Spaces

Barbara Baumeister and Gernot Stroth

**Abstract**

In this paper we determine the flag–transitive non-canonical gluings of two isomorphic desarguesian affine spaces. It turns out that there are fifteen sporadic examples and two infinite series. Moreover, we determine the universal covers of the fifteen sporadic gluings and of the canonical gluing.

## 1 Introduction

Given two geometries each equipped with a suitable parallelism giving rise to the same geometry at infinity, a new geometry can be constructed by gluing the two geometries together along their geometry at infinity. This method of constructing new geometries was introduced by Buekenhout, Huybrechts and Pasini [BHP]. Suppose that the obtained geometry is of rank 3. Then each point is incident to each plane, i.e. $\Gamma$ is *flat*. Conversely, many flat geometries of rank three are gluings – see for instance the classification of flat geometries of type $Af.c^*$, $Af.Af^*$, $c.c^*$ [BDMP, DP, BP]. In [Pas2] Pasini examines the special case of gluing two affine spaces.

Let $(\mathcal{A}_0, ||_0)$ and $(\mathcal{A}_2, ||_2)$ be two affine spaces isomorphic to $AG(n,q)$ and let $\Gamma^\infty = \Gamma_i^\infty$ be the set of parallel classes of $||_i$ $(i = 0, 2)$. Then the gluing $AS(n,q) \circ_\alpha AS(n,q)$ of the points and lines of $\mathcal{A}_0$ and of $\mathcal{A}_2$ via the permutation $\alpha$ of $\Gamma^\infty$ is a geometry of rank three consisting of points, lines and planes defined as follows (see [Pas2]):
The points are the points of $\mathcal{A}_0$, the planes the points of $\mathcal{A}_2$ and the lines the pairs $(l_0, l_2)$ with $l_i$ a line in $\mathcal{A}_i$ such that $l_0^{\infty\alpha} = l_2^\infty$. Any point is incident with any plane and a point $p$ (resp. plane $u$) is incident with a line $(l_0, l_2)$ iff $p$ (resp. $u$) is incident to $l_0$ (resp. $l_2$) in $\mathcal{A}_0$ (resp. $\mathcal{A}_2$).

The gluing is said to be *canonical*, if $\alpha \in P\Gamma L_n(q)$, see [Pas2, 1.3.4]. In this paper we determine all the flag-transitive non-canonical gluings of two isomorphic desarguesian affine spaces $AS(n,q) \circ_\alpha AS(n,q)$.

Let $\Gamma = AS(n,q) \circ_\alpha AS(n,q)$, let $A_i = Aut(\mathcal{A}_i, ||_i)$ and let $K_i^\infty$ be the kernel of the action of $A_i$ on $||_i$, for $i = 0, 2$. Then $A_i^\infty = A_i/K_i^\infty$ is the group induced by $A_i$ on $\Gamma_i^\infty$ $(i = 0, 2)$. Set $X = A_0^\infty \cap A_2^{\infty\alpha}$. Then $Aut(\Gamma) = (K_0^\infty \times K_2^\infty)X$, see Lemma 3.1.

There are fifteen sporadic examples (nine with $n = 2$, two with $n = 4$, two with $n = 6$) and two infinite series. More precisely we prove the following.

**Theorem 1.1** *Let $\Gamma$ be a flag-transitive non-canonical gluing $AS(n, q) \circ_\alpha AS(n, q)$ where $q = p^r$. Then one of the following holds. Conversely each case gives rise to such an example.*

*(1) (i) $n = 2$, $q \in \{5, 7, 11, 23\}$ and $X \cong S_4$;*
*(ii) $n = 2$, $q = 11$ and $X \cong A_4$;*

*(2) $n = 2$, $q \in \{11, 19, 29, 59\}$ and $X \cong A_5$;*

*(3) $n = 4$, $q = 3$, $O_2(X) \cong E_{24}$ and $\mathbb{Z}_5 \leq X/O_2(X) \leq Frob(5:2)$;*

*(4) $n = 6$, $q = 3$ and $X \cong L_2(13)$;*

*(5) $n > 2$ or $n = 2$ and $q \notin \{2, 3, 4, 5, 8, 9\}$; $X = N_{\Gamma L_n(q)}(T)$, with $T$ a Singer cycle of $\Gamma L_n(q)$;*

*(6) $gcd(r \cdot n, (q^n - 1)/(q - 1)) \neq 1$, $X$ is naturally embedded into $\Gamma L_1(q^n)$, but not of type (5).*

*In cases (1), (2) and (3) we may choose $\alpha$ such that $[X, \alpha] = 1$. In case (4) there exists just two examples: one with $[X, \alpha] = 1$ and the other with $X\langle\alpha\rangle \cong PGL_2(13)$.*

Moreover, we determine the universal covers of the 15 exceptional gluings of type (1) – (4) as well as of the canonical gluings $AS(n, q) \circ_\alpha AS(n, q)$. We prove that the gluings in (1), (2) and (3) and in (4) with $X\langle\alpha\rangle \cong PGL_2(13)$ are simply connected (see Propositions (5.4) – (5.6)). In case (4), if $[X, \alpha] = 1$, then the universal cover of $\Gamma$ is a triple cover (see Proposition (5.6)). Let $\Gamma$ be the canonical gluing $\Gamma = AS(n, q) \circ_\alpha AS(n, q)$, then the universal cover is a $q^{\binom{n}{2}}$ (resp. $2^{2^n - n - 1}$) cover, if $q > 2$ (resp. $q = 2$), see Theorem (6.5). After Theorem (6.5) geometric constructions of the universal covers of the canonical gluings are given.

Theorem 1.1 and the determination of certain universal covers solve some of the problems posed by Pasini in [Pas2, (5.4)].

Let $Aut(\Gamma) = (K_0^\infty \times K_2^\infty)X$. Then the group $X$ acts faithfully on the set $\Gamma_0^\infty$ of parallel classes of lines of $\mathcal{A}_0$. Note, that $\Gamma_0^\infty$ is just the set of points of the projective space $PG(n - 1, q)$. Moreover, $Aut(\Gamma)$ acts flag-transitively on $\Gamma$ if and only if $X$ acts transitively on $\Gamma_0^\infty$. In [Pas2, (4.3)] Pasini posed the question, whether a group $X \leq Sym(\Gamma_0^\infty) = Sym((q^n - 1)/(q - 1))$ is transitive on the set of points of $PG(n - 1, q)$ if and only if $X$ contains a Singer-cycle. Since (1) - (4) and (6) of the main theorem do not contain a Singer-cycle, our answer is negative. Pasini also asked the following question. Assume that $X$ contains a Singer-cycle $T$ but not $L_n(q)$. He wondered, whether $X$ is contained in the normalizer of $T$ in $P\Gamma L_n(q)$, if $q$ is large enough. Here, our answer is affirmative.

Our notation is fairly standard, see for instance [Pas1]. For $G = Aut(\Gamma)$ and $\{a_0, a_1, a_2\}$ a maximal flag of $\Gamma$ we denote by $G_i$ the stabilizer of $a_i$ in $G$.

The proof of the main theorem runs as follows. Let $(\Gamma, Aut(\Gamma))$ be a flag-transitive non-canonical gluing $AS(n,q) \circ_\alpha AS(n,q)$. Then $G_0/Z(G_0)$ is doubly transitive of degree $q^n$. Moreover, if $X$ is not isomorphic to a subgroup of $\Gamma L_1(q^n)$, then we prove that we may suppose that $\alpha$ normalizes $X$. By definition $\Gamma$ is a non-canonical gluing if and only if $\alpha$ is in $Sym((q^n-1)/(q-1)) \setminus A_0^\infty$. Hence we just have to determine those doubly transitive permutation groups of degree $q^n$ for which the normalizer of $X$ in $Sym((q^n-1)/(q-1))$ is not contained in $A_0^\infty$.

**Acknowledgement.** We like to thank Antonio Pasini who found a mistake in our first version.

# 2 Preliminaries

**Lemma 2.1** *Let $G$ be a group which acts faithfully and transitively on a set $\Omega$. Then $C_{Sym(\Omega)}(G)$ acts regularly on the set $\Lambda$ of fixed points of $G_\omega$ on $\Omega$ for any $\omega \in \Omega$ and $C_{Sym(\Omega)}(G) \cong N_G(G_\omega)/G_\omega$.*

*Proof:* The representations of $G$ on $\widetilde{\Omega} = \{G_\omega g \mid g \in G\}$ and on $\Omega$ are equivalent: for $\psi : \widetilde{\Omega} \to \Omega$, $G_\omega g \mapsto \omega^g$ holds $x^{\psi g} = x^{g\psi}$ for all $x \in \widetilde{\Omega}$ and all $g \in G$. Therefore $C_{Sym(\Omega)}(G)$ acts regularly on $\Lambda$.

Suppose that $G$ acts regularly on $\Omega$. Then the action of $G$ on $\Omega$ and the action of $G$ on $G$ via right multiplication are equivalent. Moreover,

$$\psi : G \to C_{Sym(\Omega)}(G), \; g \mapsto c_g \text{ with } \omega^{c_g} = \omega^g$$

for all $g \in G$ is an isomorphism. As $C_{Sym(\Omega)}(G)$ and$N_G(G_\omega)/G_\omega$ act regularly on $\Lambda$, we may embed $C_{Sym(\Omega)}(G)$ into $C_{Sym(\Lambda)}(N_G(G_\omega)/G_\omega)$. An application of the above reasoning to $N_G(G_\omega)/G_\omega$ and $\Lambda$ shows that $C_{Sym(\Lambda)}(N_G(G_\omega)/G_\omega) \cong N_G(G_\omega)/G_\omega$. Since $C_{Sym(\Omega)}(G)$ and $N_G(G_\omega)/G_\omega$ are of the same order, the last assertion of the lemma also holds. □

Let $G$ be an affine doubly transitive permutation group of degree $p^n$, $p$ prime. Then $G = O_p(G) : Y$, $V = O_p(G)$ is elementary abelian and $Y$ acts transitively and faithfully on $V^\#$.

In the following, we will use the notation introduced by Hering [He]. Let $L \subseteq Hom(V,V)$ be maximal with respect to the following conditions:

(i) $L$ is normalized by $Y$;

(ii) $L$ contains the identity and $L$ is a field with the addition and multiplication of $Hom(V,V)$.

Then $L$ is a field extension of $GF(p)$. Let $|L| = p^r = q$ and $n$ such that $|V| = q^n$. We say that $Y$ *is naturally embedded into* $\Gamma L_n(q)$. The affine doubly transitive permutation groups are classifyed by Hering and Liebeck [He,Li]. There are four infinite series: $n = 1$ or $H \leq Y \leq \Gamma L_n(q)$, $q = p^r$ with $H \in \{SL_n(q), Sp_n(q), n \geq 4, G_2(q)\}$ (see [He] or [Li]). We will call the remaining affine doubly transitive permutation groups the *exceptional doubly transitive permutation groups.*

**Lemma 2.2** *Let $H = O_p(H) : Y$ be an affine doubly transitive permutation group and suppose that $Y$ is naturally embedded into $G = \Gamma L_n(q)$, $n \geq 2$. Let $Z$ be a subgroup of $G$ which is isomorphic to $Y$ and which acts transitively on $O_p(H)^\#$. Then $Y$ and $Z$ are conjugate in $\Gamma L_n(q)$.*

*Proof:* Set $V = O_p(H)$. As $Z$ acts transitively on $V^\#$, also $V : Z$ is a doubly transitive affine permutation groups of degree $q^n$. Hence $Y$ and $Z$ are one of the groups listed by Hering (see [He] or [Li]).

Let $Z$ be naturally embedded into $\Gamma L_{n_z}(p^{r_z})$. Set $L_y = GF(q)$ and $L_z = GF(p^{r_z})$.

*Case 1.* Let $Y$ be non solvable. Then

$$\begin{aligned} E(Y) \in \{ & SL_n(q);\ Sp_n(q),\ n \geq 4;\ G_2(q),\ n = 6; \\ & SL_2(5),\ n = 2, q = 9, 11, 19, 29, 59; \\ & A_6,\ n = 4, q = 2;\ A_7,\ n = 4, q = 2; \\ & SL_2(13),\ n = 6, q = 3;\ G_2(2)' \cong U_3(3),\ n = 6, q = 2\} \end{aligned}$$

or $Y \cong 2^{1+4}A_5$ or $2^{1+4}Sym(5)$ and $n = 4, q = 3$.

Suppose $E(Y) \cong SL_n(q)$. Then $q = p^{r_z}$ and $n = n_z$. Therefore $L_y^\#$ and $L_z^\#$ are cyclic of order $q - 1$. Moreover, $L_y^\#$ (and also $L_z^\#$) decompose $V$ into $GF(p)$-subspaces of dimension $r$ such that $L_y^\#$ (resp. $L_z^\#$) acts as field multiplication on these subspaces. Since $Y$ (resp. $Z$) acts irreducibly on $V$, these subspaces are isomorphic $GF(q)L_y^\#$ (resp. $GF(q)L_z^\#$)-modules. Therefore $L_y^\#$ and $L_z^\#$ are conjugate in $G$. This implies that $E(C_G(L_y^\#)) = E(Y)$ and $E(C_G(L_z^\#)) = E(Z)$ are conjugate in $G$. Let $E = E(Y)^g = E(Z)$. As $Y \cong Z$, there is a $\phi \in Aut(E)$ such that $Y^{g\phi} = Z$. Since we may assume $\phi \in \Gamma L_n(q)$, the groups $Y$ and $Z$ are conjugate in $G$.

Now let $E(Y) \cong Sp_n(q)$, $(n, q) \neq (4, 2)$. Again $q = p^{r_z}$ and $n = n_z$. $E(Y)$ and $E(Z)$ are contained in subgroups $S, T$ of $G$ isomorphic to $SL_{2n}(q)$. Since $E(Y)$ and $E(Z)$ act transitively on $V$, also $S$ and $T$ act transitively on $V$. Therefore $S$ and $T$ are conjugate in $G$ and we may assume $E(Y), E(Z) \leq S$. As there exist only one symplectic scalar product on the $L$-vectorspace $V$ up to equivalence (see for instance [As]), it follows that $E(Y)$ and $E(Z)$ are conjugate in $G$. As above, we obtain that $Y$ and $Z$ are conjugate in $G$.

Let $E(Y) \cong G_2(q)$ or $G_2(2)'$. Again, $q = p^{r_z}$ and we may assume $E(Y), E(Z) \leq H \leq G$, $H \cong Sp_6(q)$. According to [Kl] $E(Y)$ and $E(Z)$ are conjugate in $N_G(H)$. Since the outer automorphism group of $E(Y)$ is cyclic, the assertion also holds in this case.

So let $E(Y) \cong SL_2(5)$ and $q \in \{9, 11, 19, 29, 59\}$. Again $q = p^{r_z}$. According to [Kl] there are two conjugacy classes of subgroups isomorphic to $SL_2(5)$ in $SL_2(q)$.

Since $PGL_2(q)$ does not contain $S_5$ or $A_5 \times \mathbb{Z}_2$, $E(Y)$ and $E(Z)$ are conjugate in $G$. The fact that $Y \leq E(Y)Z(G)$ and $Z \leq E(Z)Z(G)$ again yields the assertion.

In $L_4(2) \cong A_8$ there is just one conjugacy class of subgroups isomorphic to $A_6$. Therefore the assertion holds for $Y \cong A_6$ and $Y \cong A_7$.

If $Y \cong SL_2(13)$, $q = 3$ and $n = 6$, then $Y \leq H \leq G$, $H \cong Sp_6(3)$. We may assume $Z \leq H$, as well. Now, according to [CCNPW], all subgroups of $Aut(Sp_6(3))$ isomorphic to $SL_2(13)$ are conjugate.

Finally, if $(n, q) = (4, 3)$ and $Y \cong 2^{1+4}A_5$ or $2^{1+4}S_5$, then again, according to [CCNPW], $Y$ and $Z$ are conjugate in $G$. Thus the assertion holds in case 1.

*Case 2.* $Y$ is solvable. Let $n = 2$. Then $Y$ contains a subgroup isomorphic to $SL_2(3)$ [Hu] and $q \in \{3, 5, 7, 11, 23\}$. We have $\Gamma L_1(q^2) \cong \mathbb{Z}_8 : \mathbb{Z}_2$, $\mathbb{Z}_{24} : \mathbb{Z}_2$, $\mathbb{Z}_{48} : \mathbb{Z}_2$, $\mathbb{Z}_{120} : \mathbb{Z}_2$, $\mathbb{Z}_{528} : \mathbb{Z}_2$, respectively. The fact that $|Y|$ does not divide $|\Gamma L_1(q^2)|$ or that $O_3(\Gamma L_1(q^2)) \neq 1$ imply $n_z = 2$. Therefore $Z$ is of the same exceptional type as $Y$ and by [Hu] $Y$ and $Z$ are conjugate in $G$.

If $(n, q) = (4, 3)$ and $Y$ contains a subgroup isomorphic to $2^{1+4}\mathbb{Z}_5$, then, according to [Hu], the assertion holds for $Y$ and $Z$. This proves the lemma. □

**Lemma 2.3** *Let $G = O_p(G) : Y$ be an affine doubly transitive permutation group naturally embedded into $\Gamma L_n(q)$, $q = p^r$, $n \geq 2$ such that $H \leq Y$ with $H \in \{SL_n(q), Sp_n(q), n \geq 4, G_2(q)\}$. Let $m$ and $t$ be natural numbers with $n \mid m$ and $q^n = t^m$ and let $\Gamma^\infty$ be the set of 1-dimensional $GF(t)$-subspaces of $O_p(G)$. Moreover, let $X$ be the group induced by $Y$ on $\Gamma^\infty$. Then $C_{Sym(\Gamma^\infty)}(X) = C_{P\Gamma L_m(t)}(X)$ and $N_{Sym(\Gamma^\infty)}(X) = N_{P\Gamma L_m(t)}(X)$.*

*Proof:* Let $\omega \in \Gamma^\infty$ and let $M = stab_X(\omega)$. Then

$$N_X(M)/M \cong C_{\Gamma L_m(t)}(E(X))/Z(GL_n(q)).$$

Therefore, Lemma 2.1 yields the first part of the assertion.
Since $\Gamma L_n(q)$ embeds into $\Gamma L_m(t)$ also the second part of the assertion follows. □

**Lemma 2.4** *Let $G = O_p(G) : Y$ be an affine exceptional doubly transitive permutation group with $Y \leq \Gamma L_n(q)$, $|O_p(G)| = q^n$. Let $\Gamma^\infty$ be the set of 1-spaces of $V = O_p(G)$ considering $V$ as a natural $GF(q)$-module for $Y$. Then $Y$ induces a group $X$ on $\Gamma^\infty$ as listed in Table 1. Moreover, $X_\omega$, $\omega \in \Gamma^\infty$, and $N_X(X_\omega)/X_\omega$ are as listed in Table 1.*

*Proof:* In order to determine $N_{Sym(\Gamma^\infty)}(X)$ we use the fact that by Lemma 2.1 $C_{Sym(\Gamma^\infty)}(X) \cong N_X(X_\omega)/X_\omega$.

If $n = 2$, then it is easy to calculate $N_X(X_\omega)/X_\omega$ and $N_{Sym(\Gamma^\infty)}(X)$.

If $n = 4$ and $q = 2$, then $|\Gamma^\infty| = 15$, $X \cong A_6$ or $A_7$ and in both cases $C_{Sym(\Gamma^\infty)}(X) = 1$. Since $Sym(7)$ has no permutation representation of degree 15, we obtain $N_{Sym(\Gamma^\infty)}(X) \cong Sym(6)$ or $A_7$, as claimed.

Now let $n = 4$ and $q = 3$. If $Y \cong SL_2(5) * \mathbb{Z}_8$ (resp. $SL_2(5)2 * \mathbb{Z}_8$), then $X \cong A_5 \times \mathbb{Z}_4$ (resp. $Sym(5) \times \mathbb{Z}_4$), $X_\omega \cong Sym(3)$ (resp. $Sym(3) \times \mathbb{Z}_2$) and therefore $C_{Sym(\Gamma^\infty)}(X) = Z(X) \cong \mathbb{Z}_4$ and $N_{Sym(\Gamma^\infty)}(X) = Sym(5) \times \mathbb{Z}_4$.

If $X = O_2(X)N_X(S)$, $S \in Syl_5(X)$, then according to the Theorem of Schur-Zassenhaus $N_{Sym(\Gamma^\infty)}(X) = O_2(X)N_M(N_X(S))$ with $M = N_{Sym(\Gamma^\infty)}(O_2(X))$. If $N_{GL_4(3)}(N_X(S))$ induces the full automorphism group on $N_X(S)$, then

$$N_M(N_X(S)) = C_{Sym(\Gamma^\infty)}(X)N_{GL_4(3)}(N_X(S)).$$

In all cases it is straightforward to determine $N_X(X_\omega)/X_\omega$. Therefore, the assertion holds, if $X = O_2(X)N_X(S)$. If $X/O_2(X) \cong A_5$, then $X_\omega \cong \mathbb{Z}_2A_4$ and therefore $N_X(X_\omega) = X_\omega$. This proves the assertion for $X/O_2(X) \cong A_5$ or $Sym(5)$. Finally, let $n = 6$, $q = 3$ and $X \cong L_2(13)$. Then $X_\omega \cong \mathbb{Z}_3$ which implies the assertion. This proves the lemma. □

**Corollary 2.5** *Let $G = O_p(G) : Y$ be an affine exceptional doubly transitive permutation group with $Y \leq \Gamma L_n(q)$, $|O_p(G)| = q^n$. Let $\Gamma^\infty$ be the set of 1-spaces of $V = O_p(G)$ considering $V$ as a natural $GF(q)$-module for $Y$. Let $X$ be the group induced by $Y$ on $\Gamma^\infty$. If $N_{Sym(\Gamma^\infty)}(X) > N_{\Gamma L_n(q)}(X)$, then one of the following holds.*

*(1) $n = 2$, $q \in \{5, 7, 11, 23\}$ and $A_4 \leq X$;*

*(2) $n = 2$, $q \in \{11, 19, 29, 59\}$ and $X \cong A_5$;*

*(3) $n = 4$, $q = 3$, $O_2(X) \cong 2^4$ and $\mathbb{Z}_5 \leq X/O_2(X) \leq Frob(5:2)$;*

*(4) $n = 6$, $q = 3$ and $X \cong L_2(13)$.*

**Lemma 2.6** *Let $q = p^r$ be a power of a prime $p$. Let $n > 2$ or $n = 2$ and $q \neq 2, 3, 4, 5, 8, 9$. Then $\varphi(q^n - 1/q - 1) > rn$.*

*Proof.* Let $r = 1$, $n = 2$. Then we claim $\varphi(p + 1) > 2$, which is obviously true besides $p = 2, 3$, or 5. Assume next $q^n = 64$. Then $q = 2, 4$, or 8. Hence $q^n - 1/q - 1 = 63, 21$, or 9 and $\varphi(q^n - 1/q - 1) = 36, 12$, or 6. So the conclusion holds in all these cases. Hence we may assume that there is some Zsigmondy prime $t$ dividing $q^n - 1$. In particular $\varphi(q^n - 1/q - 1) \geq t - 1$.

Assume $\varphi(q^n - 1/q - 1) \leq rn$. Then we have $t - 1 \leq rn$. But by the choice of $t$ we have $rn \mid t - 1$. This now shows $t - 1 = rn$ and $\varphi(q^n - 1/q - 1) = t - 1$. Then we get $q^n - 1/q - 1 = t$ or $2t$. As $t$ is odd, we have an integer $u = (t - 1)/2$ and $t \mid p^u + 1$. As $p^{t-1} - 1 = t(q - 1)$ or $2t(q - 1)$, we see that $p^u - 1$ divides $q - 1$ or $2(q - 1)$. So in any case we get $p^u - 1 \mid q - 1$, as $2 \mid p^u + 1$ in case $p$ odd. Then $u \mid r$. But $2u = nr$. This shows $n = 2$ and $u = r$. We now have $p^{(t-1)/2} + 1 = t, (p = 2)$ or $2t, p$ odd. Let $p > 3$. Then $p^{(t-1)/2} > 2^{t-1}$. Hence we get $t \geq 2^{t-2}$. This shows $t = 3$. But then $u = r = 1$, and we are done. So we are left with $p = 2$ and $p = 3$.

Figure 1: Table 1.

| $n$ | $q$ | $X$ | $X_\omega$ | $N_X(X_\omega)/X_\omega$ | $N_{Sym(\Gamma^\infty)}(X)$ |
|---|---|---|---|---|---|
| 2 | 3 | $A_4$ | $\mathbb{Z}_3$ | 1 | $Sym(4)$ |
| | | $Sym(4)$ | $Sym(3)$ | 1 | $Sym(4)$ |
| | 5 | $A_4$ | $\mathbb{Z}_2$ | $\mathbb{Z}_2$ | $Sym(4) \times \mathbb{Z}_2$ |
| | | $Sym(4)$ | $\mathbb{Z}_4$ | $\mathbb{Z}_2$ | $Sym(4) \times \mathbb{Z}_2$ |
| | 7 | $Sym(4)$ | $\mathbb{Z}_3$ | $\mathbb{Z}_2$ | $Sym(4) \times \mathbb{Z}_2$ |
| | 11 | $A_4$ | 1 | $A_4$ | $(A_4 \times A_4)2$ |
| | | $Sym(4)$ | $\mathbb{Z}_2$ | $\mathbb{Z}_2$ | $Sym(4) \times \mathbb{Z}_2$ |
| | 23 | $Sym(4)$ | 1 | $Sym(4)$ | $Sym(4) \times Sym(4)$ |
| | 9 | $A_5$ | $Sym(3)$ | 1 | $Sym(5)$ |
| | | $Sym(5)$ | $Sym(3) \times \mathbb{Z}_2$ | 1 | $Sym(5)$ |
| | 11 | $A_5$ | $\mathbb{Z}_5$ | $\mathbb{Z}_2$ | $Sym(5) \times \mathbb{Z}_2$ |
| | 19 | $A_5$ | $\mathbb{Z}_3$ | $\mathbb{Z}_2$ | $Sym(5) \times \mathbb{Z}_2$ |
| | 29 | $A_5$ | $\mathbb{Z}_2$ | $\mathbb{Z}_2$ | $Sym(5) \times \mathbb{Z}_2$ |
| | 59 | $A_5$ | 1 | $A_5$ | $(A_5 \times A_5)2$ |
| 4 | 2 | $A_6$ | $Sym(4)$ | 1 | $Sym(6)$ |
| | 2 | $A_7$ | $L_3(2)$ | 1 | $A_7$ |
| | 3 | $A_5 \times \mathbb{Z}_4$ | $Sym(3)$ | $\mathbb{Z}_4$ | $Sym(5) \times \mathbb{Z}_4$ |
| | | $Sym(5) \times \mathbb{Z}_4$ | $Sym(3) \times \mathbb{Z}_2$ | $\mathbb{Z}_4$ | $Sym(5) \times \mathbb{Z}_4$ |
| | | $E_{2^4} : \mathbb{Z}_5$ | $\mathbb{Z}_2$ | $E_{2^3}$ | $((E_{2^4} : \mathbb{Z}_5) \times E_{2^3})\mathbb{Z}_3 : \mathbb{Z}_4$ |
| | | $E_{2^4} : \mathbb{Z}_5 : \mathbb{Z}_2$ | $\mathbb{Z}_2\mathbb{Z}_2$ | $\mathbb{Z}_2$ | $((E_{2^4} : (\mathbb{Z}_{15} : \mathbb{Z}_2)) \times \mathbb{Z}_2)\mathbb{Z}_2$ |
| | | $E_{2^4} : \mathbb{Z}_5 : \mathbb{Z}_4$ | $\mathbb{Z}_2\mathbb{Z}_4$ | 1 | $E_{2^4} : (\mathbb{Z}_5 : \mathbb{Z}_4)$ |
| | | $E_{2^4} : A_5$ | $\mathbb{Z}_2 A_4$ | 1 | $E_{2^4} : Sym(5)$ |
| | | $E_{2^4} : Sym(5)$ | $\mathbb{Z}_2 Sym(4)$ | 1 | $E_{2^4} : Sym(5)$ |
| 6 | 3 | $L_2(13)$ | $\mathbb{Z}_3$ | $E_4$ | $(L_2(13) \times E_4)2$ |

Let $p = 2$. Then $2^{(t-1)/2} + 1 = t$. So $t \leq 5$. But then $q = 2$ or 4, which are exceptional cases in the assertion. Let $p = 3$, then $3^{(t-1)/2} + 1 = 2t$ and we get $t \leq 5$. Then $q = 3$ or $q = 9$ and again we have the exceptional cases. This proves the lemma. □

# 3 On gluings

## 3.1 The construction

Let $(\mathcal{A}_0, ||_0)$ and $(\mathcal{A}_2, ||_2)$ be two affine spaces isomorphic to $AG(n,q)$ and let $\Gamma^\infty = \Gamma_i^\infty$ be the set of parallel classes of $||_i$ $(i = 0, 2)$. Then Pasini introduced in [Pas2] the gluing $AS(n,q) \circ_\alpha AS(n,q)$ of the points and lines of $\mathcal{A}_0$ and of $\mathcal{A}_2$ via the permutation $\alpha$ of $\Gamma^\infty$ as follows:
The points are the points of $\mathcal{A}_0$, the planes the points of $\mathcal{A}_2$ and the lines the pair of lines $(l_0, l_2)$ with $l_i$ a line in $\mathcal{A}_i$ such that $l_0^{\infty\alpha} = l_2^\infty$. Any point is incident with any plane and a point $p$ (resp. plane $u$) is incident with a line $(l_0, l_2)$ if and only if $p$ (resp. $u$) is incident to $l_0$ (resp. $l_2$) in $\mathcal{A}_0$ (resp. $\mathcal{A}_2$).

Let $\Gamma = AS(n,q) \circ_\alpha AS(n,q)$ be the gluing of two desarguesian affine spaces and let $A_i = Aut((\mathcal{A}_i, ||_i))$, $K_i^\infty$ the kernel of the action of $A_i$ on $||_i$ and $A_i^\infty$ the group induced by $A_i$ on $\Gamma_i^\infty$ $(i = 0, 2)$. Then $A_0^\infty = A_2^\infty = A^\infty$. Moreover, let $K_i = O_p(K_i^\infty)$ and $Z_i$ be the stabilizer in $K_i^\infty$ of the zero-point of $\mathcal{A}_i$, for $i = 0, 2$. Then $K_i \cong E_{q^n}$ and $Z_i \cong \mathbb{Z}_{q-1} (i = 0, 2)$.

**Lemma 3.1** *Let $\Gamma = AS(n,q) \circ_\alpha AS(n,q)$. Then $G = Aut(\Gamma) = (K_0^\infty \times K_2^\infty)X$, where $X = A^\infty \cap (A^\infty)^\alpha$, $G_0 = (Z_0 \times K_2^\infty)X$, $G_2 = (Z_2 \times K_0^\infty)X$ and $G_1 = (((\langle x_1 \rangle \times (\langle \alpha(x_1) \rangle))(Z_0 \times Z_2)N_X(\langle x_1 \rangle)$ with $\langle x_1 \rangle \le K_2$, $|\langle x_1 \rangle| = q$. In particular, $G$ acts flag-transitively on $\Gamma$ if and only if $G_0/Z_0$ and $G_2/Z_2$ are doubly transitive permutation groups of degree $q^n$.*

*Proof:* See [BHP, Theorem (3.6)]. □

In the following we will use the notation just introduced. According to [BHP, Theorem (3.9)], we have:

**Lemma 3.2** *Let $\Gamma = AS(n,q) \circ_\alpha AS(n,q)$ and $\Gamma' = AS(n,q) \circ_{\alpha'} AS(n,q)$. Then $\Gamma$ and $\Gamma'$ are isomorphic if and only if $\alpha' \in A^\infty \alpha A^\infty$.*

**Remark.** Lemma 3.2 implies that, if $\Gamma$ is the canonical gluing $AS(n,q) \circ_\alpha AS(n,q)$, then we may assume that $\alpha$ is the identity on $\Gamma^\infty$ and that $\alpha$ induces a $GF(q)$-linear mapping from $K_2$ onto $K_0$.

## 3.2 Examples

Let $H = O_p(H) : Y \le A\Gamma L_n(q)$ such that $O_p(H)$ is the translation group of $A = A\Gamma L_n(q)$ and such that $Y$ acts transitively on $O_p(H)^\#$. Let $\Gamma^\infty$ be the set of 1-spaces of $O_p(H)$, $A^\infty$ the subgroup of $Sym(\Gamma^\infty)$ induced by $A$ on $\Gamma^\infty$ and let $N$ be the normalizer of $Y^\infty$ in $Sym(\Gamma^\infty)$. Suppose $N > N_{A^\infty}(Y^\infty)$. Then $\Gamma = AS(n,q) \circ_\alpha AS(n,q)$ is a non-canonical gluing of the two affine spaces isomorphic to $AS(n,q)$ via the bijection $\alpha \in N \setminus N_{A^\infty}(Y^\infty)$. Moreover,

$$Aut(\Gamma) = (K_0^\infty \times K_2^\infty)X, \text{ where } K_i^\infty \cong E_{q^n}\mathbb{Z}_{q-1} \text{ and } Y^\infty \le X < A^\infty,$$
$$\text{with } X = A^\infty \cap (A^\infty)^\alpha.$$

If $C_{A^\infty}(Y^\infty) = 1$, but $C_{Sym(\Gamma^\infty)}(Y^\infty) \neq 1$, then we may chose in particular $\alpha \in C_{Sym(\Gamma^\infty)}(Y^\infty)^\#$. The following examples are constructed in this way.

(a) Let $n = 2$, $q \in \{5, 7, 11, 23\}$ and let $Y$ be a group listed in [Hu]. By the Lemma of Schur $C_{A^\infty}(Y^\infty) = 1$, but by Lemma 2.1 and Table 1 $C_{Sym(\Gamma^\infty)}(Y^\infty) \neq 1$. Let $\alpha \in C_{Sym(\Gamma^\infty)}(Y^\infty)^\#$. According to Table 1 either $X \cong Sym(4)$ or $q = 11$ and $X \cong A_4$.

(b) Let $n = 2$, $q \in \{11, 19, 29, 59\}$ and let $Y$ be a group listed in [He, Paragraph 5, E1]. Then $Y^\infty \cong A_5$ and $C_{A^\infty}(Y^\infty) = 1$. By Lemma 2.1 and Table 1 $C_{Sym(\Gamma^\infty)}(Y^\infty) \neq 1$ and for $\alpha \in C_{Sym(\Gamma^\infty)}(Y^\infty)^\#$ we have $X \cong A_5$.

(c) Let $n = 4$, $q = 3$ and let $Y \cong 2^{1+4}\mathbb{Z}_5$.
Again $C_{P\Gamma L_4(3)}(Y^\infty) = 1$. By Table 1 $C_{Sym(\Gamma^\infty)}(Y^\infty) \cong 2^3$ and $X/O_2(X) \cong \mathbb{Z}_5$ or $Frob(5:2)$.

(d) Let $n = 6$, $q = 3$ and let $Y \cong SL_2(13)$. Then $C_{A^\infty}(Y^\infty) = 1$ and according to Table 1 $N_{Sym(\Gamma^\infty)}(Y^\infty) \cong (L_2(13) \times 2^2)2$. So we may choose $\alpha$ either such that $\alpha$ commutes with $Y^\infty$ or such that $Y^\infty\langle\alpha\rangle \cong PGL_2(13)$. In both cases $X \cong L_2(13)$.

(e) Let $Y$ be naturally embedded into $\Gamma L_1(q^n)$ and assume that $Y^\infty$ contains a Singer-cycle $T$ of $A^\infty$. Moreover, let $n > 2$ or $n = 2$ and $q \notin \{2, 3, 4, 5, 8, 9\}$. Recall that $T$ is cyclic of order $(q^n-1)/(q-1)$. Let $q = p^r$. By Lemma 2.6 $|Aut(T)| > r \cdot n$. Therefore $N_{A^\infty}(T) \cong T : \mathbb{Z}_{r \cdot n}$ is a proper subgroup of $N_{Sym(\Gamma^\infty)}(T) \cong T : Aut(T)$ (see Lemma 2.1). Let $\alpha \in N_{Sym(\Gamma^\infty)}(T) \setminus A^\infty$. Since $N_{Sym(\Gamma^\infty)}(T)/T$ is abelian, $X$ is the normalizer of $T$ in $A^\infty$.

(f) Let $T$ be a Singer-cycle of $A^\infty$. Let $q = p^r$ and set $m = r \cdot n$. Suppose $gcd(m, (q^n - 1)/(q - 1)) \neq 1$.
Let $s$ be a prime which divides $gcd(m, (q^n - 1)/(q - 1))$ and let $U$ be a subgroup of $T$ of order $|T|/s$. Let $t$ be an element of $T$ of order $|T|_s = s^b$ and let $g$ be an element of $N_{A^\infty}(T)$ of order $s$ which fixes a point in $\Gamma^\infty$. Set $Y^\infty = U\langle tg\rangle$. Then $|Y^\infty| = |T|$.

We claim that $Y^\infty$ acts regularly on $\Gamma^\infty$ (and thereby the preimage $Y$ of $Y^\infty$ in $\Gamma L_n(q)$ acts transitively on $O_p(A\Gamma L_n(q))^\#$ ). If $\langle t, g\rangle$ is an abelian group, then $tg$ is of order $o(t) = s^b$. If $\langle t, g\rangle$ is not an abelian group and if $s$ is odd, then $\langle t, g\rangle$ is isomorphic to $Mod_{s^{b+1}}$ (for the definition of the modular group $Mod_{s^{b+1}}$ see [As, p. 107]) and therefore $tg$ is again of order $s^b$. Hence any Sylow $s$-subgroup of $Y^\infty$ is cyclic. Therefore, any element of order $s$ of $Y^\infty$ is contained in $U$ which implies that $Y^\infty$ acts regularly on $\Gamma^\infty$. If $\langle t, g\rangle$ is not an abelian group and $s = 2$, then $g$ and $tg$ are not conjugate. Thus also in this case $Y^\infty$ acts regularly, as claimed.

By Lemma 2.1 $C = C_{Sym(\Gamma^\infty)}(Y^\infty) \cong Y^\infty$ and $N_{Sym(\Gamma^\infty)}(Y^\infty) \leq CAut(T)$. We claim that $C_{A^\infty}(Y^\infty)$ is a proper subgroup of $C$. If there exists a Zsigmondy prime $t$ for $p^m - 1 = q^n - 1$, then according to [HuI, II. (7.3)] $C_{A^\infty}(U) = T$ which implies $C_{A^\infty}(Y^\infty) = C_T(g) < C$. If there exists no Zsigmondy prime for $p^m - 1$, then $(m,p) = (6,2)$ or $m = 2$ and $p$ is a Mersenne prime [Zsig]. If $(m,p) = (6,2)$, then $s = 3$, $U \cong \mathbb{Z}_{21}$ and $C_{A^\infty}(Y^\infty) = Z(Y^\infty) < C$. Now let $m = 2$ and $p$ be a Mersenne prime. Then $m = n = s = 2$, $U \cong \mathbb{Z}_{p+1/2}$ and $C_{A^\infty}(Y^\infty) = Z(Y^\infty) < C$. Thus our claim turns out to be true in all cases and therefore, there exists a non-canonical gluing via $\alpha \in C \setminus A^\infty$ such that $Y^\infty \leq X$.

Clearly, the above construction can be generalized in such a way that $[Y^\infty : T \cap Y^\infty]$ will not be a prime.

# 4 The main theorem

**Theorem 4.1** *Let $\Gamma$ be a non-canonical gluing $AS(n,q) \circ_\alpha AS(n,q)$, where $q = p^r$. Then $\Gamma$ is one of the examples listed above, i.e. one of the following holds.*

*(1) (i) $n = 2$, $q \in \{5, 7, 11, 23\}$ and $X \cong S_4$;*

*(ii) $n = 2$, $q = 11$ and $X \cong A_4$;*

*(2) $n = 2$, $q \in \{11, 19, 29, 59\}$ and $X \cong A_5$;*

*(3) $n = 4$, $q = 3$, $O_2(X) \cong 2^4$ and $\mathbb{Z}_5 \leq X/O_2(X) \leq Frob(5:2)$;*

*(4) $n = 6$, $q = 3$ and $X \cong L_2(13)$;*

*(5) $n > 2$ or $n = 2$ and $q \notin \{2, 3, 4, 5, 8, 9\}$; $X = N_{\Gamma L_n(q)}(T)$, with $T$ a Singer cycle of $\Gamma L_n(q)$;*

*(6) $gcd(r \cdot n, (q^n - 1)/(q-1)) \neq 1$, $X$ is naturally embedded into $\Gamma L_1(q^n)$, but not of type (5).*

*In cases (1), (2) and (3) we may choose $\alpha$ such that $[X, \alpha] = 1$. In case (4) there exists just two examples: one with $[X, \alpha] = 1$ and the other with $X\langle\alpha\rangle \cong PGL_2(13)$.*

*Proof:* By Lemma 3.1 $G_0/Z_0 \cong K_2^\infty X$ is a doubly transitive permutation group of degree $q^n$. Therefore $X$ and $X^\alpha$ are subgroups of $A := (A^\infty)^\alpha \cong P\Gamma L_n(q)$ which act both transitively on $\Gamma^\infty$ and the preimages $\hat{X}$ and $\hat{X}^\alpha$ of $X$ and $X^\alpha$ in $\Gamma L_n(q)$ act transitively on $O_p(G_2)^\#$.

Notice, since $\Gamma$ is a non-canonical gluing, $n$ is at least 2. If moreover, $\alpha$ normalizes $X$, then $N_{Sym(\Gamma^\infty)}(X)$ is not contained in $N_A(X)$. Suppose that $\hat{X}$ is naturally embedded into $\Gamma L_m(t)$. Then $m$ divides $n$ and $q^n = t^m$.

Suppose first $m \geq 2$. According to Lemma 2.2 $\hat{X}$ and $\hat{X}^\alpha$ are conjugate in $\Gamma L_n(q)$ which implies $X = X^{\alpha g}$ for some $g \in A$. Hence by Lemma 3.2, we may assume $\alpha \in N_{Sym(\Gamma^\infty)}(X)$. Since $\Gamma$ is a non-canonical gluing Lemma 3.2 yields $N_A(X)$ is a proper subgroup of $N_{Sym(\Gamma^\infty)}(X)$. Now Lemma 2.3 and Corollary 2.5 imply (1) – (4) of the theorem.

Now suppose that $m = 1$. We distinguish the cases that $X$ contains a Singer-cycle $T$ or not. First suppose $T \leq X$. Then $X \leq N_A(T)$. As $X^\alpha$ has also a

Singer-cycle $\widetilde{T}$ and is also contained in $N_A(\widetilde{T})$, $X$ and $X^\alpha$ are conjugate in $A$. Thus we may assume $\alpha \in N_{Sym(\Gamma^\infty)}(X)$. Since $N_{Sym(\Gamma^\infty)}(T)$ is not contained in $N_A(T)$, Lemma 2.6 implies $n > 2$ or $n = 2$ and $q \notin \{2,3,4,5,8,9\}$. Hence (5) of the theorem holds. Finally suppose that $X$ does not contain a Singer-cycle. Since $X \leq \Gamma L_1(q^n)$ acts transitively on $\Gamma^\infty$, we obtain that $gcd(rn, (q^n-1)/(q-1)) \neq 1$ and that $\Gamma$ is as in (6).

Let $\Gamma$ be an example of type (1), (2) or (3). In all these examples we have $N_{Sym(\Gamma^\infty)}(X) = C_{Sym(\Gamma^\infty)}(X)N_A(X)$. Therefore the last assertion follows with Lemma 3.2. This proves the theorem. □

**Corollary 4.2** *Let $\Gamma = AS(n,q) \circ_\alpha AS(n,q)$ be a non-canonical gluing not of type (4),(5) or (6) of Theorem 1.1. Then we may choose $\alpha$ such that $[X,\alpha] = 1$.* □

# 5 Exceptional coverings

In this chapter we are going to describe the universal covers of the non-canonical gluings obtained in the previous chapter. Here we just determine those which are not related to the affine operating group. The question for the latter is still open.

The basic idea reads as follows. Let $G$ be the group of the universal covering. Then $G$ is generated by the three parabolic subgroups $G_0$, $G_1$, $G_2$, where $G_0 = (Z_0 \times K_2^\infty)X$ and $G_2 = (Z_2 \times K_0^\infty)X^\alpha$. As has been seen in (4.2) we may choose $\alpha$ as a permutation inside of $\Sigma_n$ which centralizes $X$ besides of the case $X \cong SL_2(13)$. From the structure of $G_1 = \langle\langle x_1\rangle, \alpha(\langle x_1\rangle)\rangle X_1$, where $\langle x_1\rangle$ is a 1-space in $K_2$, we see that $[\langle x_1\rangle, \alpha(\langle x_1\rangle)] = 1$. Now $G = \langle K_0^\infty, K_2^\infty, X\rangle$, where the action of $X$ on $K_0^\infty$ and $K_2^\infty$ is known. The only additional relation we have is that there is some bijective mapping $\alpha$ from the 1-spaces of $K_2$ onto the 1-spaces of $K_0$ which commutes with the action of $X$ and satisfies $[U, \alpha(U)] = 1$ for any 1-space $U$ in $K_2$. Hence we are going to determine precisely those groups.

**Lemma 5.1** *Let $p$ be a prime, $K_0 \cong K_2 \cong E_{q^2}, q = p^r$, be subgroups of a group $G$, $G = \langle K_0, K_2\rangle$. Suppose that $K_0$ and $K_2$ admit $GF(q)$-vector space structures such that there is a $GF(q)$-linear isomorphism $f : K_2 \to K_0$. Moreover, suppose that $[U, f(U)] = 1$ for any 1-dimensional subspace $U$ of $K_2$. Then $G' \leq Z(G)$ and $|G'| \leq q$.*

*Proof:* Let $U$ and $V$ be two different subspaces of $K_2$. Then $K_2 = U \times V$. Observe that $[xy, f(x)f(y)] = 1$ $(x, y \in K_2)$ implies $[x, f(y)] = [y, f(x)]^{-1}$. Let $u_1, u_2 \in U$ and $v_1, v_2 \in V$. As $[v_1, f(v_2)] = 1$, we see $[u_1, f(v_1), v_2] = 1$. As $[u_1, f(u_2)] = 1$, we see $[v_1, f(u_1), u_2] = 1$. So $[u_1, f(v_1), K_2] = 1$. The same argument also shows $[u_1, f(v_1), K_0] = 1$. So $[u_1, f(v_1)] \in Z(G)$ and $G' \leq Z(G)$. Furthermore $[u_1, f(v_2)]^p = [u_1^p, f(v_2)] = 1$.

Let $\lambda \in GF(q)$ such that $\langle\lambda\rangle = GF(q)^\#$ and let $x_1 \in U^\#$ and $y_1 \in V^\#$. Set $x_{i+1} = x_1^{\lambda^i}$ and $y_{i+1} = y_1^{\lambda^i}$ for $i = 1, \ldots, r-1$. Then $\{x_1, \ldots, x_r\}$ and $\{y_1, \ldots, y_r\}$ are a basis of the $GF(p)$-vector spaces $U$ and $V$, respectively.

Set $D = \langle [x_1, f(y_i)] \mid i = 1, \dots, r \rangle$. We claim that $G' = D$. Since $G' \leq Z(G)$, we have $G' = [U, f(V)]$. The fact that $x_i y_i = (x_1 y_1)^{\lambda^{i-1}}$ implies $[x_1 y_1, f(x_i) f(y_i)] = 1$ $(i = 1, \dots, r)$.

Therefore $[y_1, f(x_i)] = [x_1, f(y_i)]^{-1}$ and, as $[x_i, f(y_1)] = [y_1, f(x_i)]^{-1}$, it follows $[x_i, f(y_1)] \in D \quad (i = 1, \dots, r)$. Now suppose that $[x_i, f(y_j)] \in D$ for $i = 1, \dots, r$ and $j \leq k$ and $[x_i, f(y_{k+1})] \in D$ for $i = 1, \dots, l-1$. Since $(x_l y_k)^\lambda = x_l^\lambda y_{k+1}$ we obtain $[x_l y_k, f(x_l^\lambda) f(y_{k+1})] = 1$ and $[x_l, f(y_{k+1})] = [y_k, f(x_l^\lambda)]^{-1}$.

If $l < r$, then $x_l^\lambda = x_{l+1}$ and $[x_l, f(y_{k+1})] = [y_k, f(x_{l+1})]^{-1} = [x_{l+1}, f(y_k)] \in D$. If $l = r$, then $x_r^\lambda = \prod_{i=1}^r x_i^{a_i}$, $a_i \in GF(p)$, and

$$[x_r, f(y_{k+1})] = \textstyle\prod_{i=1}^r([y_k, f(x_i)]^{a_i})^{-1} = \prod_{i=1}^r [x_i, f(y_k)]^{a_i} \in D.$$

This proves $G' = D$. So $|G'| \leq q$. □

Using the same method as in the first paragraph of the previous proof we obtain the following more general result in the case $q = p$.

**Lemma 5.2** *Let $p$ be a prime and $K_0 \cong K_2 \cong E_{p^2}$ be subgroups of a group $G$, $G = \langle K_0, K_2 \rangle$ and $f$ be an isomorphism from $K_2$ onto $K_0$. Let $K_2 = \langle x_1, x_2 \rangle$ and suppose $[x_1, f(x_1)] = [x_2, f(x_2)] = [x_1 x_2, f(x_1) f(x_2)] = 1$, then $|G'| \mid p$.*

**Lemma 5.3** *Let $p$ be a prime and $K_0 \cong K_2 \cong E_{p^2}$ be subgroups of a group $G$, $G = \langle K_0, K_2 \rangle$, where $G$ is extraspecial of order $p^5$. Then there is some isomorphism $f : K_2 \to K_0$ with $[x, f(x)] = 1$ for all $x \in K_2$.*

*Proof:* Let $K_2 = \langle x_1, x_2 \rangle$ and $K_2 = \langle y_1, y_2 \rangle$ with $[x_1, y_1] = [x_2, y_2] = z$ and $[x_1, y_2] = [x_2, y_1] = 1$, $\langle z \rangle = Z(G)$. Now define $f$ as follows $f(x_1^i x_2^j) = y_2^i y_1^{-j}$, $i, j \in \{0, \dots, p-1\}$. Obviously $f$ is an isomorphism. Using the fact $G' \leq Z(G)$ we now have

$$\begin{aligned} [x_1^i x_2^j, f(x_1^i x_2^j)] &= [x_1^i, f(x_1^i x_2^j)][x_2^j, f(x_1^i x_2^j)] \\ &= [x_1, f(x_1^i x_2^j)]^i [x_2, f(x_1^i x_2^j)]^j \\ &= [x_1, f(x_1)]^{i^2} [x_1, f(x_2)]^{ji} [x_2, f(x_1)]^{ij} [x_2, f(x_2)]^{j^2} \\ &= [x_1, y_2]^{i^2} [x_1, y_1]^{-ij} [x_2, y_2]^{ij} [x_2, y_1]^{-j^2} = 1. \end{aligned}$$

□

**Lemma 5.4** *Let $p$ be a prime and $K_0 \cong K_2 \cong E_{p^2}$ be subgroups of a group $G$, $G = \langle K_0, K_2 \rangle$, where $G$ is extraspecial of order $p^5$. Let $\alpha$ be a bijective mapping from the 1-spaces of $K_2$ onto the 1-spaces of $K_0$ such that $[U, \alpha(U)] = 1$ for each 1-space $U$ of $K_2$. Then there is an isomorphism $f : K_2 \to K_0$ with $[x, f(x)] = 1$ for $x \in K_2$, such that $f(U) = \alpha(U)$ for all 1-spaces $U$ of $K_2$.*

*Proof:* Let $K_2 = \langle x_1, x_2 \rangle$ and $K_0 = \langle y_1, y_2 \rangle$ with $[x_1, y_1] = [x_2, y_2] = z$ and $[x_1, y_2] = [x_2, y_1] = 1$, $\langle z \rangle = Z(G)$. Then for each $x \in K_2$ we see that $\alpha(\langle x \rangle)$ is uniquely determined. So $\alpha(\langle x \rangle) = f(\langle x \rangle)$, where $f$ is the mapping from 5.3. □

**Proposition 5.5** *Let $\Gamma$ be a non-canonical gluing $AS(2,p) \circ_\alpha AS(2,p)$, $p$ an odd prime. Then $\Gamma$ is simply connected.*

*Proof:* By (3.1) we have $G_0 = Z_0 K_2^\infty X$, $G_2 = Z_2 K_0^\infty X$ and $G_1 = Y(Z_0 \times Z_2)X_1$, where $X_1 \leq X$ and stabilizes in the elementary abelian group $K_2$ a point $\langle x_1 \rangle$ and in $K_0$ a point $\alpha(\langle x_1 \rangle)$. Hence $Y = \langle \langle x_1 \rangle, \alpha(\langle x_1 \rangle) \rangle$ is elementary abelian of order $p^2$. So we get $[\langle x \rangle, \alpha(\langle x \rangle)] = 1$ for all $x \in K_2$.

Let $K_2 = \langle x_1, x_2 \rangle$ and $y_1, y_2 \in K_2$ with $\alpha(\langle x_i \rangle) = \langle y_i \rangle$, $i = 1, 2$. Let $\alpha(\langle x_1 x_2 \rangle) = \langle y_1^a y_2^b \rangle$. Define the linear mapping $f : K_2 \to K_0$ by $f(x_1) = y_1^a$, $f(x_2) = y_2^b$. Then by 5.2 we have $|\langle K_0, K_2 \rangle'| \mid p$. If $\langle K_0, K_2 \rangle$ is nonabelian, then it is extraspecial of order $p^5$. Now by 5.4 there is some linear mapping $g : K_2 \to K_0$ with $g(\langle x \rangle) = \alpha(\langle x \rangle)$ for all $x \in K_2$. But this means that $g = \alpha$ is in $\Gamma L_2(p) \leq Sym(p+1)$, which contradicts (4.2) and example (3.2)(a),(b). So we have $[K_0, K_2] = 1$ and $\Gamma$ is simply connected. □

**Proposition 5.6** *Let $\Gamma$ be a non-canonical gluing $AS(4,3) \circ_\alpha AS(4,3)$, then $\Gamma$ is simply connected.*

*Proof:* By (3.1) $G_0$ contains a subgroup $H_0 = K_2 Y$, where $K_2$ is elementary abelian of order 81 and $Y$ is an extension of an extraspecial group of order 32 by $Z_5$. Furthermore $G_2$ contains $H_2 = K_0 Y$, where $K_0$ is elementary abelian of order 81 too. The action of $Y$ on $K_2$ is described in [Hu]. Let $\{x_1, x_2, x_3, x_4\}$ be the basis used there and $\{y_1, y_2, y_3, y_4\}$ the corresponding basis of $K_0$. We have $[\langle x_1 \rangle, \alpha(\langle x_1 \rangle)] = 1$. As $\alpha(\langle x_1 \rangle)$ has to be fixed by the one point stabilizer of $Y$ on $K_0$ and $\alpha$ centralizes $Y$ by (4.2) we see that $\alpha(\langle x_1 \rangle) \subseteq \langle y_1, y_3 \rangle$. As $\alpha$ is not the identity we see that $\alpha(\langle x_1 \rangle) \in \{\langle y_3 \rangle, \langle y_1 y_3 \rangle, \langle y_1^{-1} y_3 \rangle\}$. Set $Y_1 = G_1 \cap Y$ and $H_1 = \langle \langle x_1 \rangle, \alpha(\langle x_1 \rangle) \rangle Y_1$. Let $H = \langle H_1, H_2, H_3 \rangle$. Using the relations from $H_1$ and $H_3$ and the additional relation $[x_1, y] = 1$, for some $y \in \{y_3, y_1 y_3, y_1 y_3^{-1}\}$, we get in any case by the Todd-Coxeter algorithm that $H = (K_0 \times K_2)Y$. This shows that $G = K_0^\infty K_2^\infty X$ and so $\Gamma$ is simply connected. □

**Proposition 5.7** *Let $\Gamma$ be a non-canonical gluing $AS(6,3) \circ_\alpha AS(6,3)$. Then one of the following holds*

*(i) If $\alpha$ induces an outer automorphism on $X$, then $\Gamma$ is simply connected.*

*(ii) If $[X, \alpha] = 1$, then the universal cover of $\Gamma$ is a 3-fold covering.*

*Proof:* Let $H_0 = K_2 X$, where $K_2$ is elementary abelian of order 729 and $X \cong SL_2(13)$. We first determine the structure of $H_0$. According to [HPl] the following is a presentation of $X$ :

$$\langle a, x, y, r \mid x^{13} = r^3 = 1, x^r = x^3, y^4 = 1, x^y = x^{-1}, [y, r] = 1, a^2 = y^2, y^a = y^{-1}, r^a = r^{-1}, (ax)^3 = 1, x^6 a x^{-1} a x a x^7 a = (ry)^{-1} \rangle$$

We are going to calculate the action of $X$ on $K_2$. First of all, $\langle x, r\rangle$ decomposes $K_2$ into two 3-dimensional modules. As $GL(3,3)$ does not contain a subgroup of order $13 \cdot 12$, we see that $y$ has to interchange these two modules. So we get the following matrices according to a basis $\{x_1, x_2, x_3, x_4, x_5, x_6\}$ of $K_2$

$$x \to \begin{pmatrix} 0 & 1 & 0 & 0 & 0 & 0 \\ 0 & 0 & 1 & 0 & 0 & 0 \\ 1 & -1 & -1 & 0 & 0 & 0 \\ 0 & 0 & 0 & 1 & 1 & 1 \\ 0 & 0 & 0 & 1 & 0 & 0 \\ 0 & 0 & 0 & 0 & 1 & 0 \end{pmatrix}, \quad r \to \begin{pmatrix} 1 & 0 & 0 & 0 & 0 & 0 \\ 1 & -1 & -1 & 0 & 0 & 0 \\ -1 & 1 & 0 & 0 & 0 & 0 \\ 0 & 0 & 0 & 1 & 0 & 0 \\ 0 & 0 & 0 & 1 & -1 & -1 \\ 0 & 0 & 0 & -1 & 1 & 0 \end{pmatrix}$$

$$y \to \begin{pmatrix} 0 & 0 & 0 & 1 & 0 & 0 \\ 0 & 0 & 0 & 0 & 1 & 0 \\ 0 & 0 & 0 & 0 & 0 & 1 \\ -1 & 0 & 0 & 0 & 0 & 0 \\ 0 & -1 & 0 & 0 & 0 & 0 \\ 0 & 0 & -1 & 0 & 0 & 0 \end{pmatrix}$$

Now we have that $a$ has to act on the fixed points of $r$ i.e. $\langle x_1, x_4\rangle$. So $\langle y, a\rangle$ acts as a quaternion group on $\langle x_1, x_4\rangle$ and so it is uniquely determined. Using the relations above we see $x_1^a = x_1^{-1}x_4$ and $x_4^a = x_1x_4$. This then yields

$$a \to \begin{pmatrix} -1 & 0 & 0 & 1 & 0 & 0 \\ 1 & -1 & 0 & 1 & 1 & 0 \\ 1 & 1 & 1 & 1 & -1 & -1 \\ 1 & 0 & 0 & 1 & 0 & 0 \\ 1 & 1 & 0 & -1 & 1 & 0 \\ 1 & -1 & -1 & -1 & -1 & -1 \end{pmatrix}$$

So let now $K_0 = \langle y_1, y_2, y_3, y_4, y_5, y_6\rangle$. If $[X, \alpha] = 1$ then $X$ acts by the same matrices on $K_2$ as on $K_0$. Let $X_1$ be the stabilizer of $\langle x_1\rangle$ in $X$. Then $\alpha(\langle x_1\rangle)$ is contained in the set of fixed points of $X_1$ on $K_0$ and so is some 1-space in $\langle y_1, y_4\rangle$. Hence we get $\alpha(\langle x_1\rangle) \in \{\langle y_4\rangle, \langle y_1^{-1}y_4\rangle, \langle y_1y_4\rangle\}$. Let $H = \langle K_0X, K_2X\rangle$ with the defining relations of $K_0X$ and $K_2X$ and the additional relation $[x_1, t] = 1$ for some $t \in \{y_4, y_1^{-1}y_4, y_1y_4\}$. Then application of the Todd-Coxeter algorithm in all three cases shows that $H = \langle K_0, K_2\rangle X$ where $\langle K_0, K_2\rangle$ is extraspecial of order $3^{13}$, the assertion.

So assume now that $\alpha$ induces an outer automorphism on $X$. Then $x$ acts as $x^8$ and $a$ as $ay$ on $K_0$, while the action of $r, y$ is unchanged. Hence the fixed points of $X_1$ on $K_0$ are the same as before. So we now choose $t \in \{y_1, y_4, y_1^{-1}y_4, y_1y_4\}$ and add the relation $[x_1, t]$. Then we apply the Todd-Coxeter algorithm to the group $\langle K_0, K_2, X\rangle$ again. This then shows that $\langle K_0, K_2\rangle$ is elementary abelian in all four cases, hence $\Gamma$ is simply connected. □

# 6 Canonical coverings

In this section we are going to determine the universal coverings of the canonical gluings.

**Lemma 6.1** *Let $G = \langle K_0, K_2 \rangle$ with elementary abelian p-groups $K_0, K_2$ of order at most $p^3$, $p$ an odd prime. Let $f : K_2 \to K_0$ be an isomorphism with $[x, f(x)] = 1$ for all $x \in K_2$. Then $G' \leq Z(G)$ and $|G'| \leq |K_2|$.*

*Proof:* The assertion is trivial for $|K_2| = p$. Let $|K_2| = p^2$. Then the assertion follows from 5.1. So we may assume that $|K_2| = p^3$. Set $K_2 = \langle x_1, x_2, x_3 \rangle$ and $K_0 = \langle y_1, y_2, y_3 \rangle$, where $f(x_i) = y_i$, $i = 1, 2, 3$. Let $X$ be the free group in the generators $x_1, x_2, x_3, y_1, y_2, y_3$. We extend $f$ to a mapping $g$ of $\{x_1, x_2, x_3, y_1, y_2, y_3\}$ by setting $g(x_i) = y_i$ and $g(y_i) = x_i$, $i = 1, 2, 3$. As $X$ is free $g$ can be extended to an automorphism of $X$ which we will denote by $g$ too. Let $N = \langle x_i^p, y_i^p, [x_i, x_j], [y_i, y_j], i, j = 1, 2, 3, [x, f(x)], [f(x), x], x \in K_2 \rangle$. Then we see that $N$ is invariant under $g$. So $g$ is an automorphism of $H = X/N$. As $G$ is a factor-group of $H$ it is enough to prove the assertion for $H$. Hence we may assume that $g$ is an automorphism of $G$. By definition we have $g^2 = 1$.

As $[ab, g(a)g(b)] = 1$, we see $[a, g(b)] = [b, g(a)]^{-1}$ for all $a, b \in K_2$. Now $g([b, g(a)]) = [g(b), a] = [b, g(a)]$ and so $[G', g] = 1$. Set

$$L = \langle x_1 g(x_1)^{-1}, x_2 g(x_2)^{-1}, x_3 g(x_3)^{-1} \rangle.$$

Then $L \leq [G, g]$. Let $a \in G$ with $g(a) = a^{-1}$. Let $x \in G'$. Then $[x, a] = g([x, a]) = [x, g(a)] = [x, a^{-1}]$. This implies $x^{-1}a^{-1}xa = x^{-1}axa^{-1}$ or $[a^2, x] = 1$. If $a$ is of odd order we get $[a, x] = 1$. Now choose for $a$ one of the elements $x_i g(x_i)^{-1}$. Then $o(a) = p$ is odd and so we have $[a, G'] = 1$. In particular $[L, G'] = 1$. This shows

(1) $\quad L' \leq Z(L)$.

We have

(2) $[x_1 g(x_1)^{-1}, x_2 g(x_2)^{-1}] = [x_1, x_2 g(x_2)^{-1}]^{g(x_1)^{-1}} [g(x_1)^{-1}, x_2 g(x_2)^{-1}]$
$= [x_1, g(x_2)^{-1}][g(x_1)^{-1}, x_2]$.

By 5.1 we have $[g(x_i), x_j]^{g(x_i)} = [g(x_i), x_j]$ for all $i, j$. So we get

(3) $[g(x_i)^n, x_j] = [g(x_i), x_j]^n$ for all $n \in \overline{Z}$ and all $i, j$.

This now shows

$$[x_1, g(x_2)^{-1}][g(x_1)^{-1}, x_2] = [x_1, g(x_2)]^{-1}[g(x_1), x_2]^{-1} = [x_1, g(x_2)]^{-2}.$$

So (1) - (3) imply

(4) $L' = \langle [x_i, g(x_j)]^{-2} \mid i, j = 1, 2, 3 \rangle$

By 5.1 we have that $o([x_i, g(x_j)]) \mid p$. Hence we get

(5) $L' = \langle [g(x_1), x_2], [g(x_1), x_3], [g(x_2), x_3] \rangle$.

In particular, we have $|L'| \leq p^3$.

It remains to show that $L' = G' \leq Z(G)$. Obviously we have $\langle x_j g(x_j)^{-1}, x_k\rangle \leq \langle x_j, x_k, g(x_j), g(x_k)\rangle = Y$. By 5.1 we see that $Y' = \langle [x_j, g(x_k)]\rangle$. So by (5) $L$ is normal in $G$. We have that $\langle x_1, x_3\rangle$ acts on $\langle [g(x_1), x_2], [g(x_3), x_2]\rangle = C_{L'}(x_2)$. Hence there are $i, j$ with $x_1^i x_3^j \neq 1$ and $[x_1^i x_3^j, C_{L'}(x_2)] = 1$. This shows that $[g(x_1), x_2, x_3^j] = 1 = [g(x_3), x_2, x_1^i]$ by 5.1. We may assume $[g(x_1), x_2, x_3] = 1$. By the Witt identity we have $1 = [g(x_1), x_2, x_3]^{x_2^{-1}}[x_3, g(x_1), x_2]^{g(x_1)^{-1}}$ and so $[g(x_1), x_3, x_2] = 1$. Also we get $1 = [g(x_2), x_3, x_1]^{x_3^{-1}}[x_1, g(x_2), x_3]^{g(x_2)^{-1}}$. As $[x_1, g(x_2)] = [x_2, g(x_1)]^{-1}$ and $[[x_2, g(x_1)]^{-1}, x_3] = 1$, we get $[g(x_2), x_3, x_1] = 1$. Hence we have $[K_2, L'] = 1$. As $[g, L'] = 1$, we now get $[G, L'] = 1$. This implies $G' = L' \leq Z(G)$. □

If $q = 2$, then the universal cover of the canonical gluing $AS(n,2) \circ_\alpha AS(n,2)$ is well known, see for instance [Ba], [BaPa] or [BPM]. It is a $2^{2^n-n-1}$-fold cover with automorphism group $V : Sym(2^n)$, $V$ an elementary abelian group of order $2^{2^n-1}$, see for a geometric description the remark after Theorem 6.5.

**Lemma 6.2** *Let $\overline{\Gamma}$ be the universal cover of the canonical gluing $AS(3,q) \circ_\alpha AS(3,q)$, $q = 2^r$, $r > 1$, with automorphism group $\overline{A} = Aut(\overline{\Gamma})$ . Then*

$$\langle O_2(\overline{A_0}), O_2(\overline{A_2})\rangle' \leq Z(\langle O_2(\overline{A_0}), O_2(\overline{A_2})\rangle).$$

*Proof:* Set $K_0 = O_2(\overline{A_2})$, $K_2 = O_2(\overline{A_0})$ and $K = \langle K_2, K_0\rangle$. Let $A = Aut(\overline{\Gamma})$. Then $O_2(\overline{A_i})$ is mapped isomorphically onto $O_2(A_i)$ by the covering ($i = 0, 2$). Therefore we will identify $O_2(\overline{A_i})$ and $O_2(A_i)$ for $i = 0, 1$.

Since $\Gamma$ is a canonical gluing, the subgroup $G = (K_2 \times K_0)X$ of $Aut(\Gamma)$ acts flag-transitively on $\Gamma$, where $X \cong GL_3(q)$, $K_0 \cong K_2 \cong E_{q^3}$ are isomorphic $GF(q)X$-modules with isomorphism $\alpha : K_2 \to K_0$. We may choose notation such that $G_0 = K_2X$, $G_2 = K_0X$ and $G_1 = (\langle u\rangle_{GF(q)} \times \langle \alpha(u)\rangle_{GF(q)})N_X(\langle u\rangle_{GF(q)})$, where $u \in K_2$ and where $\langle u\rangle_{GF(q)}$ is the $GF(q)$-span of $u$ in $K_2$. Then $G_0 \cap G_1 = \langle u\rangle_{GF(q)}N_X(\langle u\rangle_{GF(q)})$, $G_0 \cap G_2 = X$, $G_1 \cap G_2 = \langle \alpha(u)\rangle_{GF(q)}N_X(\langle u\rangle_{GF(q)})$ and $B = N_X(\langle u\rangle_{GF(q)})$.

The universal cover of $\Gamma$ is the group geometry $\overline{\Gamma} = \Gamma(\overline{G}, (G_0, G_1, G_2))$, with $\overline{G}$ the universal completion of the amalgam $(G_i, G_i \cap G_j, B \mid i, j = 0, 1, 2)$.

Since $K_0$ and $K_2$ are $GF(q)X$-modules, we may also read them as $GF(2)X$-modules.

Consider the following amalgam:

$$H_0 = G_0,\ H_2 = G_2,\ H_1 = (\langle u\rangle_{GF(2)} \times \langle \alpha(u)\rangle_{GF(2)})N_X(\langle u\rangle_{GF(2)})$$

with

$$H_0 \cap H_1 = \langle u\rangle_{GF(2)}N_X(\langle u\rangle_{GF(2)}), \quad H_0 \cap H_2 = X,$$
$$H_1 \cap H_2 = \langle \alpha(u)\rangle_{GF(2)}N_X(\langle u\rangle_{GF(2)}) \quad \text{and} \quad B = N_X(\langle u\rangle_{GF(2)}).$$

For $H = \langle H_0, H_1, H_2\rangle = G$ the group geometry $\widetilde{\Gamma} = \Gamma(H, (H_0, H_1, H_2))$ is a canonical gluing $AS(3r, 2) \circ_\alpha AS(3r, 2)$ with flag-transitive group $H = G$.

We may lift $H$ to a flag-transitive group of automorphism $\widetilde{H}$ of the universal cover of $\widetilde{\Gamma}$. Then $\overline{G}$ is a quotient of $\widetilde{H}$. Let $\widetilde{K_0}$ and $\widetilde{K_2}$ be the preimages of $K_0$ and $K_2$ in $\widetilde{H}$, respectively. Then according to [Ba] or [BPM] $\widetilde{K} = \langle \widetilde{K_0}, \widetilde{K_2} \rangle$ is a 2-group and $\widetilde{K}' = [\widetilde{K_0}, \widetilde{K_2}]$ is elementary abelian of order $2^{2^{3r}-3r-1}$.

We will prove that $[K,K,K] = K^{(4)}$ holds in $\overline{G}$. Since $K$ is a 2 group, it will imply $K' \leq Z(K)$, as asserted. Thus we have to show $[K,K,K] = 1$ under the assumption that $K^{(4)} = 1$. By [Ba] the isomorphism $\alpha$ is induced by an involution $i$ of $G$, so $K_2^{\alpha} = K_0$ and $K_0^{\alpha} = K_2$. Therefore, it just remains to show $[K_0, K_2, K_2] = 1$.

Let $u, v, w \in K_2$. Let $\lambda \in GF(q)$. Then

$$1 = [\alpha(uv), (uv)^{\lambda}] = [\alpha(u)\alpha(v), u^{\lambda}v^{\lambda}].$$

Applying Lemma 5.1 we obtain

$$1 = [\alpha(u)\alpha(v), u^{\lambda}v^{\lambda}] = [\alpha(u), v^{\lambda}][\alpha(v), u^{\lambda}].$$

Thus $[\alpha(u), v^{\lambda}] = [\alpha(v), u^{\lambda}]$. By symmetry, $[\alpha(v), u^{\lambda}] = [\alpha(u^{\lambda}), v]$. This yields $[\alpha(u), v^{\lambda}] = [\alpha(u^{\lambda}), v]$ (see the proof of Lemma 5.1) .

Next we claim that $[\alpha(u), v, w^{\lambda}] = [\alpha(u)^{\lambda}, v, w]$. Since $[\alpha(uw), (uw)^{\lambda}] = 1$, we have $1 = [\alpha(uw), v, (uw)^{\lambda}]$. Using the fact that $[ab, c] = [a, c][b, c][a, c, b]$ for any $a, b, c \in K$ and using Lemma 5.1, we obtain

$$[\alpha(uw), v, (uw)^{\lambda}] = [\alpha(u), v, w^{\lambda}][\alpha(w), v, u^{\lambda}].$$

Setting $\lambda = 1$, we see that $[\alpha(u), v, w] = [\alpha(w), v, u]$. Thus we obtain $[\alpha(u), v, w^{\lambda}] = [\alpha(w), v, u^{\lambda}] = [\alpha(u)^{\lambda}, v, w]$, as asserted.

Clearly, $[\alpha(u), v^{\lambda}] = [\alpha(u^{\lambda}), v]$ implies $[\alpha(u)^{\lambda^{-1}}, (vw)^{\lambda}] = [\alpha(u), vw]$. On the other hand

$$[\alpha(u)^{\lambda^{-1}}, (vw)^{\lambda}] = [\alpha(u)^{\lambda^{-1}}, v^{\lambda}][\alpha(u)^{\lambda^{-1}}, w^{\lambda}][\alpha(u)^{\lambda^{-1}}, v^{\lambda}, w^{\lambda}]$$

$$= [\alpha(u), v][\alpha(u), w][\alpha(u), v, w^{\lambda}].$$

Therefore $[\alpha(u), v, w^{\lambda}] = [\alpha(u), v, w]$ for all $\lambda \in GF(q)$. If $\lambda \neq 1$, then

$$[\alpha(u), v, w] = [\alpha(u), v, w^{\lambda + (1-\lambda)}]$$

$$= [\alpha(u), v, w^{\lambda}][\alpha(u), v, w^{(1-\lambda)}] = [\alpha(u), v, w^{\lambda}]^2 = 1.$$

Hence $[K_0, K_2, K_2] = 1$ and the assertion holds. □

**Lemma 6.3** *Let $\Gamma$ be a canonical gluing $AS(n,q) \circ_{\alpha} AS(n,q)$, $q = p^r$, $q > 3$, with automorphism group $G$. Let $\tilde{G}$ be the group belonging to the universal cover. Then $G = \tilde{G}/N$, where $N$ is an elementary abelian $p$-group of order at most $q^{\binom{n}{2}}$.*

*Proof:* We have $\tilde{G}_0 = (Z_0 \times K_2^\infty)X$, $\tilde{G}_2 = (Z_2 \times K_0^\infty)X$ and $\tilde{G}_1 = Y(Z_0 \times Z_2)X_1$, where $K_0 = O_p(K_0^\infty)$ and $K_2 = O_p(K_2^\infty)$ are elementary abelian of order $q^n$. Further $|Y \cap K_0| = |K_2 \cap Y| = q$, and $Y = (K_0 \cap Y)(K_2 \cap Y)$ is elementary abelian of order $q^2$. Finally there is a $GF(q)$-linear isomorphism $\alpha : K_2 \to K_0$ which commutes with the action of $X$ and maps $K_0 \cap Y$ onto $K_2 \cap Y$. Hence we have $[U, \alpha(U)] = 1$ for every 1-dimensional $GF(q)$-subspace $U$ of $K_2$.

Set $H = \langle K_0, K_2 \rangle$. Let $u, v, w \in K_2$. Then

$$[u, \alpha(v), w] \in \langle u, v, w, \alpha(u), \alpha(v), \alpha(w) \rangle.$$

Therefore Lemmas 6.1 and 6.2 imply $[u, \alpha(v), w] = 1$, so $[u, \alpha(v), K_2] = 1$ for all $u, v \in K_2$. Using the same argument we also obtain $[u, \alpha(v), K_0] = 1$ for all $u, v \in K_2$. This shows $H' \leq Z(H)$. In particular, $H'$ is elementary abelian and $H' = \langle [u, \alpha(v)] \mid u, v \in K_2 \rangle$.

Let $U_1, \ldots, U_n$ be 1-dimensional $GF(q)$-subspaces of $K_2$ with $K_2 = U_1 \times \ldots \times U_n$. Since $[U_i, \alpha(U_j)] \leq \langle U_i, U_j, \alpha(U_i), \alpha(U_j) \rangle$ $(1 \leq i, j \leq n)$, Lemma 5.1 implies $[U_i, \alpha(U_j)] = [U_j, \alpha(U_i)]$ is a group of order at most $q$ $(1 \leq i, j \leq n)$. Hence $H' = \langle [U_i, \alpha(U_j)] \mid i < j \leq n \rangle$ is of order at most $q^{\binom{n}{2}}$. As $\tilde{G} = H(Z_0 \times Z_2)X$, the assertion follows. □

**Proposition 6.4** *Let $\Gamma$ be a canonical gluing $AS(n,q) \circ_\alpha AS(n,q)$, $q = p^r$, $q > 2$, with group $G$. Let $\tilde{G}$ be the group belonging to the universal covering. Then $G = \tilde{G}/N$, where $N$ is an elementary abelian $p$-group of order $q^{\binom{n}{2}}$.*

*Proof:* Let $W$ be an $(n+1)$-dimensional vectorspace over $GF(q)$ with basis $\{e_1, \ldots, e_{n+1}\}$ and let $V = \langle e_1, \ldots, e_n \rangle$. Let $S = SL(W) \cong SL_{n+1}(q)$. Then $N_S(V) = C_S(V)N_S(\langle e_{n+1} \rangle)$ with $K_2 = C_S(V)$ an elementary abelian group of order $q^n$ and $X = N_S(\langle e_{n+1} \rangle) \cong GL_n(q)$. Set $U = W \wedge W$ and $N = V \wedge V$. Then $N$ is elementary abelian of order $q^{\binom{n}{2}}$ and $U/N \cong \langle e_i \wedge e_{n+1} \mid i = 1, \ldots, n \rangle$ is elementary abelian of order $q^n$.

Set $K_0 = \langle e_i \wedge e_{n+1} \mid i = 1, \ldots, n \rangle$. Let $H = U : K_2X$. Then $H$ is generated by $H_0 = K_2X$ and $H_2 = K_0X$.

Let $\alpha$ be the isomorphism from $K_2$ onto $K_0$ with $\alpha(a) = [a, e_{n+1}] \wedge e_{n+1}$ for all $a \in K_2$. Then $\alpha$ is an $GF(q)X$-isomorphism. Observe that $[R, \alpha(R)] = 1$ for all 1-dimensional subspaces $R$ of $K_2$. Set $H_1 = (T \times \alpha(T))N_X(T)$, where $T$ is a $GF(q)$-subspace of $K_2$. Then $\{H_0, H_1, H_2\}$ generates a cover of $\Gamma$. By 6.3 this is the universal cover of $\Gamma$. □

**Theorem 6.5** *Let $\Gamma$ be a canonical gluing $AS(n,q) \circ_\alpha AS(n,q)$, $q = p^r$ with group $G$ and $\tilde{\Gamma}$ the universal covering with group $\tilde{G}$. Then $\tilde{G}/N = G$, where $N$ is an elementary abelian $p$-group and*

*(1) If $q = 2$, then $|N| = 2^{2^n - n - 1}$*

*(2) If $q > 2$, $|N| = q^{\binom{n}{2}}$*

*Proof:* The assertion (1) follows from [BMP] and (2) follows from 6.4. □

**Remark.** The covers of $AS(n,q)\circ_\alpha AS(n,q)$, $q$ odd or $q=2$, which turned out to be universal (Theorem 6.5), were already constructed in [Pas2]. For the convenience of the reader we include the constructions of the universal covers.

First let $q=2$. Let $\Gamma_m$ be the Coxeter complex of type $D_m$, $m=2^n$ and let $Tr(\Gamma_m)$ be the $\{+,0,-\}$-truncation of $\Gamma_m$ formed by the elements of type $+,0$ and $-$, where $+,0$ and $-$ are as follows:

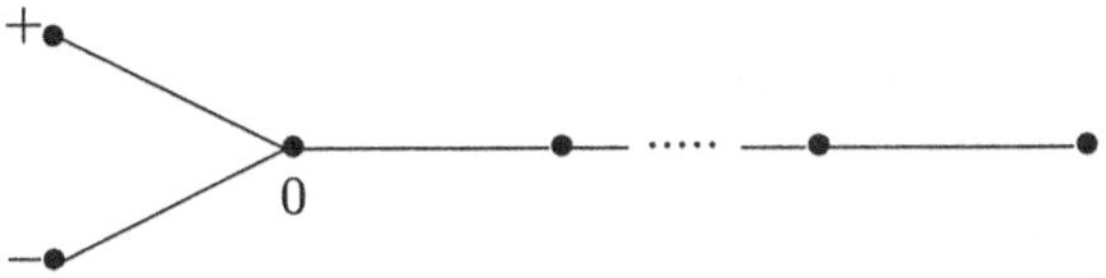


Then $Tr(\Gamma_m)$ is the universal cover of the canonical gluing $AS(n,2)\circ_\alpha AS(n,2)$.
This universal cover is isomorphic to the $GF(2)$-expansion $D(A)$, where $A$ is the affine space $AG(n,2)$, see [BPM, Corollary 9].

Now let $q>2$. Let $\Delta$ be the building of type $D_{n+1}$, $n\geq 2$. As above take the types as follows:

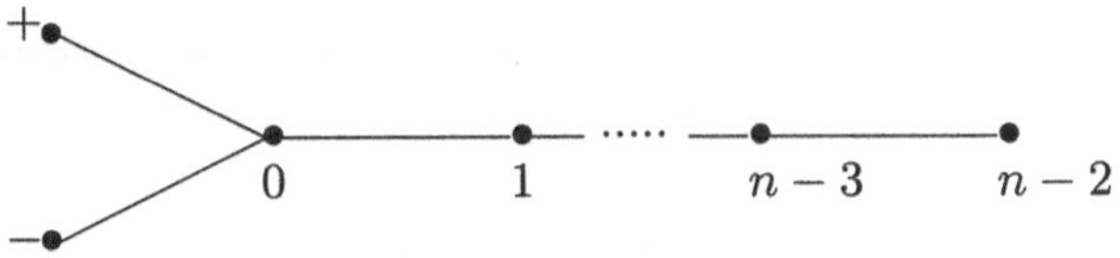


Let us write $\varepsilon$ to denote any of the two types $+$ or $-$. For $\varepsilon\in\{+,-\}$ let $\Delta^\varepsilon$ be the half-spin geometry relative to the type $\varepsilon$, that is $\Delta^\varepsilon$ is the geometry of rank 2 having the elements of $\Delta$ of type $\varepsilon$ as points and those of type 0 as lines, with the incidence inherited from $\Delta$. The diameter of $\Delta^\varepsilon$ is $n/2$ (resp. $(n+1)/2$) for $n$ even (resp. $n$ odd). Let $a^+$ and $a^-$ be elements of type $+$ and $-$ respectively which are incident. Let $H^\varepsilon$ be the set of those points of $\Delta^\varepsilon$ which are not at maximal distance from $a^\varepsilon$ for $\varepsilon\in\{+,-\}$. Then $H^\varepsilon$ is a hyperplane in $\Delta^\varepsilon$.

Let $\overline{\Delta}$ be the following geometry: the elements of $\overline{\Delta}$ are the elements of type $\varepsilon$ which are not in $H^\varepsilon$, $\varepsilon\in\{+,-\}$, and the elements of type 0 which meet $H^\varepsilon$ in exactly one element for $\varepsilon\in\{+,-\}$; the incidence is inherited from $\Delta$.
Then $\overline{\Delta}$ is the universal cover of the canonical gluing $AS(n,q)\circ_\alpha AS(n,q), q>2$.

## References

[As] M. Aschbacher, *Finite Group Theory*, Cambridge University Press, Cambridge, London 1986.

[Ba] B. Baumeister, On flag-transitive $c.c^*$-geometries, in *Proceeedings of the Ohio State Conference* (May 1993), ed. by K.T. Arasu et all, 3–23.

[BaPa] B. Baumeister, A. Pasini, On flat flag-transitive $c.c^*$-geometries, *J. Alg. Comb.*, **6**, No 1, (1997), 5–27.

[BPM] B. Baumeister, Th. Meixner, A. Pasini, $GF(2)$-expansions, to appear in *Geometriae Dedicata.*

[BDPM] B. Baumeister, A. Del Fra, Th. Meixner, A. Pasini, Flag-transitive $c.Af^*$-geometries, *Beiträge zur Algebra und Geometrie* **37**, No 2, (1996), 231–258.

[BHP] F. Buekenhout, C. Huybrechts, A. Pasini, Parallelism in diagram geometry, *Bull. Belg. Math. Soc.* **1** (1994), 355–397.

[CCNPW] J.H. Conway, R.T. Curtis, S.P. Norton, R.A. Parker and R.A. Wilson, *Atlas of Finite Groups*, Clarendon Press, Oxford 1985.

[DP] A. Del Fra, A. Pasini, On $Af.Af^*$-geometries, *J.of Geom.*, **54** (1995), 15-29.

[He] C. Hering, Transitive linear groups and linear groups which contain irreducible subgroups of prime order, II, *J. Algebra* **93** (1985), 151–164.

[HPl] D. Holt, W. Plesken, Perfect groups, Oxford University Press, 1989

[Hu] B. Huppert, Zweifach transitive, auflösbare Permutationsgruppen, *Math. Z.* 68 (1957), 126 - 150

[HuI] B. Huppert, *Endliche Gruppen I*, Springer–Verlag, Berlin, Heidelberg 1979.

[Kl] P.B. Kleidman, The maximal subgroups of the finite 8–dimensional orthogonal groups $P\Omega_8^+(q)$ and of their automorphism groups, *J. Algebra* **117** (1988), 30–71.

[Li] M.W. Liebeck, The affine permutation groups of rank three, *Proc. London Math. Soc.* (3) **54** (1987), 477–516.

[Pas1] A. Pasini, *Diagram Geometries*, Oxford University Press, 1994.

[Pas2] A. Pasini, Gluing two affine spaces, *Bull. Belg. Math. Soc.* **3** (1996), 25–40.

Barbara Baumeister
Fachbereich Mathematik und Informatik
Martin-Luther Universität Halle Wittenberg
06099 Halle, Germany
e-mail: baumeis@coxeter.mathematik.uni-halle.de

Gernot Stroth
Fachbereich Mathematik und Informatik
Martin-Luther Universität Halle Wittenberg
06099 Halle, Germany
e-mail: stroth@coxeter.mathematik.uni-halle.de

Trends in Mathematics, © 1998 Birkhäuser Verlag Basel/Switzerland

# The Geometry Far from a Residue

Rieuwert J. Blok and Andries E. Brouwer

**Abstract**

We show that in general the subgeometry of a building of spherical type induced by all objects in general position with respect to a given residue is a residually connected geometry with a Buekenhout-Tits diagram resembling the original diagram, but with certain strokes replaced by the corresponding 'affine' strokes. The exceptions (where connectedness fails) are discussed in some detail.

## 1 Introduction

Given a geometry $\Gamma$, and some substructure $S$, and a suitable notion of 'far from', we can study the subgeometry $\Delta$ of $\Gamma$ consisting of objects far from $S$. We show that $\Delta$ has a Buekenhout-Tits diagram related to that of $\Gamma$, and study the question whether $\Delta$ will be connected.

For the geometry $\Gamma$ we shall take a building of spherical type. We focus on the following three cases.

(i) Fix a chamber or residue $R$ of $\Gamma$, and look at Far($R$), the collection of all chambers opposite to $R$ (or, equivalently, at maximal distance from $R$), with the induced adjacency.

(ii) Fix a type $i$ in the diagram of $\Gamma$, and look at the points at maximal distance from a given point in the corresponding point-line geometry.

(iii) Fix a type $i$ in the diagram of $\Gamma$, and look at the complement of a geometric hyperplane in the corresponding point-line geometry.

The special case of buildings of rank 2 was treated in Brouwer [3] using eigenvalue arguments. Roughly speaking, the result is that the subgeometries considered are always connected, except in the cases $B_2(2)$, $G_2(2)$, $G_2(3)$, ${}^2F_4(2)$. For a precise statement, see Section 2.

In the present note we consider the higher rank case, find that it can be reduced to the rank 2 case, and hence are left only with the geometries that have $B_2(2)$ residues, namely $Sp(2n, 2)$ and $F_4(2)$.

We find that far from a chamber the $Sp(2n, 2)$ geometry has $2^{n-1}$ connected components, while in $F_4(2)$ there are 4 connected components.

In the interests of brevity, the arguments will be somewhat sketchy in places. We hope to write down a fuller treatment elsewhere.

# 2 The rank 2 case

The results in the finite rank 2 case are as follows:

**Theorem 2.1** (Brouwer [3]) *Let $(X, \mathcal{L})$ be a thick finite generalized $n$-gon of order $(s,t)$. Then*

(i) *The subgeometry induced by the points and lines in general position w.r.t. a given point $x$ (or, indeed, the complement of an arbitrary geometric hyperplane in $(X, \mathcal{L})$) is connected, except possibly in the cases $(n, s, t) = (6, 2, 2)$, $(8, 2, 4)$.*

(ii) *The subgeometry induced by the points and lines in general position w.r.t. a given flag $(x, L)$ is connected, except possibly in the cases $(n, s, t) = (4, 2, 2)$, $(6, 2, 2)$, $(6, 3, 3)$, $(8, 2, 4)$, $(8, 4, 2)$.*

(iii) *For the stated possibly exceptional parameter sets actual exceptions do occur: in case (i) for $G_2(2)$ (short root geometry) and ${}^2F_4(2)$ there are two connected components, and in case (ii) for $B_2(2)$, $G_2(2)$, $G_2(3)$ and ${}^2F_4(2)$ there are 2, 4, 3, 2 connected components, respectively.*

**Problem** Show a direct connection with the fact that $G_2(2)$, ${}^2F_4(2)$ and ${}^2B_2(2)$, ${}^2G_2(3)$ are not simple.

For infinite generalized $n$-gons with $n \leq 4$ the subgeometries considered are all connected ([3]). On the other hand, for $n \geq 5$ one may use free constructions to obtain generalized $n$-gons such that the subgeometry on the points far from a given point is disconnected (Abramenko [1], Proposition 9). The thick Moufang $n$-gons distinct from those belonging to $B_2(2)$, $G_2(2)$, $G_2(3)$ and ${}^2F_4(2)$, are connected far from a chamber (Abramenko [1], Proposition 7).

# 3 The higher rank case

If $A$ and $B$ are two sets of chambers, then let $d(A, B) := \min(d(a, b) \mid a \in A, b \in B)$.

**Theorem 3.1** (Mühlherr & Ronan [6], Theorem 1.5) *Let $\Gamma$ be a spherical building of rank at least 3. If $R$ is a residue such that in every rank 2 residue $X$ the subgeometry $\mathrm{Far}_X(\mathrm{proj}_X(R))$ is connected, then in $\Gamma$ the subgeometry $\mathrm{Far}_\Gamma(R)$ is connected.*

(One may formulate a version that is also valid in the non-spherical case: then the statement is that if in $X$ any two chambers can be joined by a gallery, such that no chamber in this gallery is closer to $\mathrm{proj}_X(R)$ than its extremities, then in $\Gamma$ any two chambers can be joined by a gallery, such that no chamber in this gallery is closer to $R$ than its extremities.)

We shall see below that the converse holds as well: if some rank 2 residue $X$ has a disconnected $\mathrm{Far}_X(\mathrm{proj}_X(R))$, then also $\mathrm{Far}_\Gamma(R)$ is disconnected.

# 4 The symplectic case

**Theorem 4.1** *Fix a chamber $c$ in the $Sp(2n, 2)$ building, where $n > 0$, and consider the geometry $\Delta$ of objects far from $c$. Then $\Delta$ has $2^{n-1}$ connected components.*

There are various ways to understand the occurrence of components here. Maybe the best way to describe these components is as corresponding to $O^+(2n, 2)$ subgeometries of the $Sp(2n, 2)$ geometry.

Fix an $Sp(2n, 2)$ geometry and an $O^+(2n, 2)$ subgeometry. In the $Sp(2n, 2)$ chamber system $C$, an $O^+(2n, 2)$ chamber corresponds to a pair of $n$-adjacent chambers $(c', c'')$ (for the usual labeling of the $C_n$ diagram), two chambers $(c', c'')$ and $(d', d'')$ being $(n-1)$- (or $n$-) adjacent, when $c' \neq d'$ but $c'' = d''$ (or $c'' \neq d''$ but $c' = d'$). Let $D$ be the collection of chambers in $C$ occurring in such pairs. Let $\{c, c', c''\}$ be an $n$-panel in $C$, with $c \in C \setminus D$, $c', c'' \in D$. Then a chamber $d' \in D$ is far from $c$ if and only if it occurs in an $O^+(2n, 2)$ chamber $(d', d'')$ opposite to $(c', c'')$. Since the projection of $c$ into the $n$-panel $\{d, d', d''\}$ is not far from $c$, we see that $n$-edges in $\mathrm{Far}(c)$ remain in $D$. Since it is clear that $D$ is closed for $j$-edges, $j < n$, we see that any connected component of $\mathrm{Far}(c)$ meeting $D$ is contained in $D$. But $D \cap \mathrm{Far}(c)$ is connected (for example because the $D_n$ diagram has no $B_2$ subdiagrams), so is a connected component.

Let us redo the above in geometric terms. Let '$i$-object' stand for 'totally isotropic (t.i.) $i$-space'. Two chambers $c$ and $c'$ in an $Sp(2n, q)$ geometry are far (opposite) if and only if whenever $A$ and $A'$ are $i$-objects in $c$ and $c'$, respectively, we have $A^\perp \cap A' = 0$.

**Proposition 4.2** *Fix an $n$-object $U$ with hyperplane ($(n-1)$-object) $H$ in the $Sp(2n, 2)$ geometry. Let $F$ be the flag $(H, U)$. The connected component of the geometry far from $F$ containing a given $n$-object $V$ is the geometry far from $H$ in the $O^+(2n, 2)$ geometry defined by the quadric $Q$ that defines the given symplectic form, for which $H$ and $V$ are totally singular, but $U$ is not. (That is, $H = Q \cap U$.)*

The conditions that $Q$ defines the given symplectic form, and vanishes identically on $H$ and $V$ but not on $U$, and that $Q$ defines a hyperbolic quadric, indeed determine $Q$ uniquely. If we choose a basis, then $Q$ is determined by the symplectic form, except for its diagonal elements. Requiring that $Q$ vanishes on $H$ determines $n-1$ diagonal elements; one more is fixed since $Q$ must not vanish on $U$; one more is fixed since $Q$ must be hyperbolic, so that $H^\perp/H$ is a hyperbolic line; finally there are $n-1$ diagonal elements to choose freely, giving $2^{n-1}$ connected components.

# 5 The stabilizer of a component

In this section, we assume that $\Gamma$ is a spherical building with a connected diagram of rank at least 3. We employ the result by Tits, that $\Gamma$ is of Moufang type. Thus, fix an apartment $\Sigma$ with root system $\Phi$ and a chamber $c$ of $\Sigma$ and let $G$

be the group of automorphisms of $\Gamma$ generated by the root groups $U_\alpha$ ($\alpha \in \Phi$). Then, the pair $(B = \mathrm{Stab}_G(c), N = \mathrm{Stab}_G(\Sigma))$ is a $(B, N)$-pair for $G$ yielding the building $\Gamma$. For a residue $R$, let $\mathrm{op}_\Sigma(R)$ denote the residue (in $\Gamma$) of the same type as, and containing $\mathrm{op}_\Sigma(R \cap \Sigma)$. Let us denote the type of this latter residue by $\mathrm{op}_I(\mathrm{type}\,(R))$.
For a residue $R$ on $c$, define the following sets of roots:

$$\begin{aligned} \Phi(R,i) &= \{\alpha \in \Phi \mid c \in \alpha,\ d(R,-\alpha) = i\} \quad (i \geq 0), \\ \text{and} \quad \Phi(R) &= \{\alpha, -\alpha \mid \alpha \in \Phi(R,0)\}. \end{aligned}$$

Put $H = B \cap N$ and consider the following subgroups of $G$:

$$\begin{aligned} L(R) &= \langle H, U_\alpha | \alpha \in \Phi(R) \rangle \\ U(R) &= \langle U_\alpha | \alpha \in \Phi(R,i),\ i > 0 \rangle \\ P(R) &= U(R).L(R) \\ U(R,1) &= \langle U_\alpha | \alpha \in \Phi(R,1) \rangle \\ P(R,1) &= U(R,1).L(R) \end{aligned}$$

The subgroup $P(R)$ is just the ordinary parabolic subgroup of $G$ stabilizing $R$ with Levi component $L(R)$ and unipotent radical $U(R)$. The definition of $P(R,1)$ is justified by the following theorem:

**Theorem 5.1** *The subgroup $P(R)$ of $G$ is the full stabilizer of* $\mathrm{Far}_\Gamma(R)$ *in $G$, and it acts transitively on the chambers of* $\mathrm{Far}_\Gamma(R)$. *The subgroup $P(R,1)$ of $P(R)$ is the full stabilizer in $G$ of the connected component of* $\mathrm{Far}_\Gamma(R)$ *containing* $\mathrm{op}_\Sigma(R)$, *and it acts transitively on the chambers of this component. In particular,* $\mathrm{Far}_\Gamma(R)$ *is connected if and only if $P(R,1) = P(R)$.*

*Proof.* The parabolic subgroup $P(R)$ is the full stabilizer of $R$ in $G$. It is a semidirect product $P(R) = U(R)\,.\,L(R)$ (see [7], Ch. 6). The subgroup $L(R)$ is the full stabilizer of the pair $(R, \mathrm{op}_\Sigma(R))$ and it acts transitively on the chambers of each of these residues. The subgroup $U(R)$ acts regularly on the set of residues of type $\mathrm{op}_I(\mathrm{type}\,(R))$ opposite $R$. This implies that $P(R)$ acts transitively on the chambers of $\mathrm{Far}_\Gamma(R)$ and that the stabilizer of $c' = \mathrm{op}_\Sigma(c)$ in $P(R)$ is contained in $L(R)$.

Let $C$ be the connected component of $\mathrm{Far}_\Gamma(R)$ containing $c'$. Let $u \in U_\alpha$, $u \neq 1$. If $\alpha \in \Phi(R)$, then $uc' \in \mathrm{op}_\Sigma(R)$ and if $\alpha \in \Phi(R,1)$, then there is a panel $\pi$ that determines $\alpha$ and that has precisely one chamber $r$ in $R$. The opposite panel $\pi' = \mathrm{op}_\Sigma(\pi)$ is the disjoint union of $\mathrm{Far}_\Gamma(r) \cap \pi'$ and $\alpha \cap \pi' = \{\mathrm{proj}_{\pi'}(r)\}$ and hence $u\,\mathrm{op}_\Sigma(R) \subseteq C$. Thus, $P(R,1)$ preserves $C$.

Let $c' = x_0, ..., x_l = x$ be a gallery in $C$. We show by induction on $l$ that there exists an element $u \in P(R,1)$ which sends $c'$ to $x$. For $l = 0$ this is clear. Suppose $v \in P(R,1)$ sends $c'$ to $x_{l-1}$ and let $\pi$ be the panel on $c$ such that $\pi' = \mathrm{op}_\Sigma(\pi)$ contains both $c'$ and $x' = v^{-1}x$. If $\pi' \subset \mathrm{Far}(R)$, then there exists an element $u' \in L(R)$ with $u'c' = x'$. If $\pi' \not\subset \mathrm{Far}(R)$, then $\pi$ only has the chamber $c$ in common with $R$, and so it determines a root $\alpha$ in $\Phi(R,1)$. Like above, $\pi'$ is the disjoint union of $\mathrm{Far}(R) \cap \pi'$ and $\alpha \cap \pi'$, and so there again exists an element $u' \in U_\alpha$ with $u'c' = x'$. In both cases the element $u = u'v \in P(R,1)$ sends $c'$ to $x$. Since the stabilizer of $c'$ in $P(R)$ is contained in $L(R) \subset P(R,1)$, the subgroup $P(R,1)$ is the full stabilizer of the component of $\mathrm{Far}_\Gamma(R)$ containing $\mathrm{op}_\Sigma(R)$. □

Let $S$ be another residue on $c$ and let $\Sigma_S = \Sigma \cap S$. This is a building of Moufang type and the group $G_S = \langle U_\alpha \mid \alpha \in \Phi(S) \rangle$ is the automorphism group generated by the root groups of $S$. Thus, $(B_S = \mathrm{Stab}_{G_S}(c), N_S = \mathrm{Stab}_{G_S}(\Sigma_S))$ is a $(B,N)$-pair for $G_S$ and we put $H_S = B_S \cap N_S$. Define

$$\begin{aligned} \Phi_S(R,i) &= \{\alpha \in \Phi(S) \mid c \in \alpha,\ d(S \cap R, -\alpha) = i\} \quad (i \geq 0), \\ \text{and} \quad \Phi_S(R) &= \{\alpha, -\alpha \mid \alpha \in \Phi_S(R,0)\}. \end{aligned}$$

Furthermore, define the subgroups $L_S(R)$, $U_S(R)$, $P_S(R)$, $U_S(R,1)$ and $P_S(R,1)$ of $G_S$ just as we did for $G$.

Clearly, Theorem 5.1 is applicable to the building $S$ and its automorphism group $G_S$.

The connection between the stabilizer of (a component of) $\mathrm{Far}_\Gamma(R)$ and the stabilizer of (a component of) $\mathrm{Far}_S(S \cap R)$ is given in the following proposition.

**Proposition 5.2** *For any two residues $R$ and $S$ on $c$, we have*
*(i)* $L_S(R) = G_S \cap L(R)$
*(ii)* $U_S(R) = G_S \cap U(R)$
*(iii)* $P_S(R) = G_S \cap P(R)$
*(iv)* $U_S(R,1) = G_S \cap U(R,1)$
*(v)* $P_S(R,1) = G_S \cap P(R,1)$

In order to prove this, we need similar properties for the corresponding root systems.

**Lemma 5.3** *For any two residues $R$ and $S$ on $c$, we have*
*(i)* $\Phi_S(R) = \Phi(S \cap R) = \Phi(S) \cap \Phi(R)$ *and*
*(ii)* $\Phi_S(R,i) = \Phi(S) \cap \Phi(R,i)$.

The following corollary, which follows easily from Theorem 5.1 and Proposition 5.2, is almost the converse to Theorem 3.1.

**Corollary 5.4** *Let $R$ and $S$ be residues on a common chamber $c$. If* $\mathrm{Far}_S(S \cap R)$ *is disconnected, then so is* $\mathrm{Far}_\Gamma(R)$.

**Corollary 5.5** *Let $\Gamma$ be a spherical building of rank at least 3, with a naturally labeled diagram $M$, and let $R$ be a $J$-residue. Then* $\mathrm{Far}_\Gamma(R)$ *is disconnected when either*
*(i) $M = C_n$, $\Gamma$ is defined over the field $\mathbf{F}_2$ and $J \cap \{n-1, n\} = \emptyset$, or*
*(ii) $M = F_4$, $\Gamma$ is defined over the field $\mathbf{F}_2$ and $J \cap \{2,3\} = \emptyset$.*

From the theory above, it is very easy to reprove that in the symplectic case there are $2^{n-1}$ connected components, and after some fiddling it also follows that $F_4(2)$ has 4 connected components far from a chamber.

# 6 The Buekenhout-Tits diagram of the geometry far from a residue

In this section we show that the subgeometry of a building of spherical type obtained by fixing a flag $F$ and taking all objects 'far away from' (or 'in general position w.r.t.') $F$ has a Buekenhout-Tits diagram (cf. [5]).

**Theorem 6.1** *Let $\Gamma$ be (the geometry of) a building of spherical type, and fix a flag $F$. The subgeometry $\mathrm{Far}_\Gamma(F)$ of $\Gamma$ consisting of all elements of $\Gamma$ far away from (that is, opposite to) $F$ (with inherited type and incidence) is a geometry with a Buekenhout-Tits diagram obtained from that of $\Gamma$ by adding arrows pointing towards the nodes in* $\mathrm{op}(\mathrm{typ}\, F)$.

As always in Buekenhout geometry, the theorem is a claim for geometries of rank at least 3, and is a definition of the strokes involved for geometries of rank 2. Let us repeat this definition explicitly.

The three diagrams

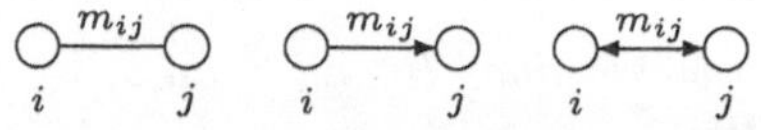


denote, respectively, (i) the class of all generalized $m_{ij}$-gons, (ii) the class of all subgeometries of a generalized $m_{ij}$-gon found by taking all objects in the incidence graph at distance $m_{ij} - 1$ or $m_{ij}$ from a given $i$-object when $m_{ij}$ is odd, or from a given $j$-object when $m_{ij}$ is even, and (iii) the class of all subgeometries of a generalized $m_{ij}$-gon found by taking all objects in flags at distance $m_{ij}$ from a given flag in the flag graph (the line graph of the incidence graph). As customary, we delete all (arrowed) edges labeled 2 (with or without arrows these all represent generalized digons), we delete labels 3, and use double edges instead of edges labeled 4.

Thus, the diagrams

represent the classes of projective planes, affine planes, dual affine planes and biaffine planes, respectively. Instead of using arrows, one usually uses labels Af and Af* for affine and dual affine planes.

The above theorem was inspired by a question of S. Shpectorov, who asked for the classification of geometries with diagram

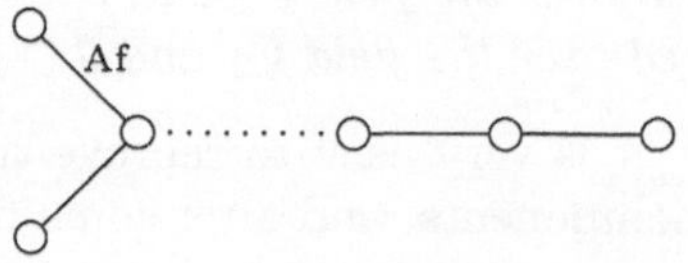

Before proving the theorem, let us define the terms more precisely. Two chambers $c, c'$ of a spherical building are called *opposite* when $\delta(c, c') = w_0$, where, as usual, $w_0$ is the longest element of the Coxeter group $W$, and $\delta$ is the $W$-valued metric on the spherical building. Since $w_0 = w_0^{-1}$, this is a symmetric relation. Two residues are called opposite when they contain opposite chambers. *Far* is synonymous with opposite.

A Coxeter chamber system $W$ becomes a building with the distance function given by $\delta(x, y) = x^{-1}y$, and we see that each object has a unique opposite. Let $\mathrm{op}_W$ be the map defined on the Coxeter building $W$ by $x \mapsto xw_0$. This is an involution sending residues of type $J$ to residues of type $\mathrm{op}_I(J)$.

**Lemma 6.2** *Let $W$ be a Coxeter building of spherical type. Then for any two residues $R$ and $S$ we have* $\mathrm{op}_R(\mathrm{proj}_R(S)) = \mathrm{proj}_R(\mathrm{op}_W(S))$.

*Proof.* Since we defined $\mathrm{op}_R$ and $\mathrm{proj}_R$ elementwise, it suffices to show that $\mathrm{op}_R(\mathrm{proj}_R(x)) = \mathrm{proj}_R(\mathrm{op}_W(x))$ for a chamber $x \in W$. But both projection and opposite are invariant under left multiplication, so we may assume that $R = W_J$ is a subgroup of $W$. Now $\mathrm{proj}_R(x) = p$ is the element of $R$ such that $p^{-1}x$ is the unique shortest coset representative of the right coset $Rx$. So we have to show that if $a$ is the shortest element of $Ra$, i.e., is left $R$-reduced, then $w_0(R)aw_0$ is shortest in $Raw_0$. Or again, that $w_0(R)a$ is longest in $Ra$. But that holds, since $l(ra) = l(r) + l(a)$ for all $r \in R$. □

Our first concern is to show that $\Delta := \mathrm{Far}_\Gamma(F)$ is a geometry. Let $C$ be the chamber system of $\Gamma$, and consider the collection $\mathcal{F}$ of all residues in $C$ (flags in $\Gamma$) far from $F$. We want to show that $\mathcal{F}$ is the flag complex of $\Delta$. Now $\mathcal{F}$ will be a flag complex if any three pairwise incident residues in $\mathcal{F}$ have a common chamber in $\mathcal{F}$. But this follows immediately from the lemma below.

**Lemma 6.3** *Let $R$ and $S$ be incident residues belonging to $\mathcal{F}$. Then $R \cap S$ also belongs to $\mathcal{F}$.*

*Proof.* Let $A$ be an apartment incident with $F$ and $R \cap S$. Then in $A$ the three residues $\mathrm{op}F$ and $R$ and $S$ are pairwise incident, and hence have a common element. □

**Lemma 6.4** *Let $R$ and $S$ be opposite. Then the set of residues meeting* $\mathrm{Far}_\Gamma(S)$ *and contained in $R$ equals* $\mathrm{Far}_R(\mathrm{proj}_R(S))$. *In particular, every object in $R$ belongs to* $\mathrm{Far}_\Gamma(S)$ *if and only if* $\mathrm{proj}_R(S) = R$, *that is, if and only if* $\mathrm{op}(\mathrm{typ}\, S) \subseteq \mathrm{typ}\, R$.

*Proof.* It suffices to prove this for a Coxeter complex $\Sigma$ instead of $\Gamma$ – then the statements about $\Gamma$ follow by taking the union over all $\Sigma$ containing both $R$ and $S$. By Lemma 6.2 we are done. □

*Proof.* (of the theorem). Follows directly from the lemma above. □

# 7 Geometric hyperplanes and far subgeometries

Often, geometric hyperplanes generalize subgeometries far from a point. In particular, when one classifies the geometries having one of the diagrams found in the previous section, then usually the result one finds is that they are complements of a geometric hyperplane in a building.

And indeed, the complement of a hyperplane in the point-line geometry for node $i$ has diagram obtained from the original diagram by inserting arrows pointing into $i$.

Suppose we have a building of spherical type, select a type $i$ to be the point type, and consider the corresponding point-line geometry (where the lines are the flags of cotype $i$). If $P$ is a point, then does 'not far from $P$' define a geometric hyperplane?

**Proposition 7.1** *(i) The set of points of a point-line geometry as above, not far from a given point $P$ is a geometric hyperplane if and only if conjugation by $w_0$ does not move node $i$. (ii) The set of points of a point-line geometry as above, not far from a given residue $R$ is a geometric hyperplane when $R$ has type* op$(i)$.

*Proof.* □

Note that $w_0$ acts like the identity on all diagrams, but flips $A_n$, $D_{2m+1}$, $E_6$. So, we always find a geometric hyperplane, except for non-middle nodes of $A_n$, for the dual polar geometries for $D_{2m+1}$, for $E_{6,1}$ and $E_{6,2}$.

Now in a point-line geometry we have at least two natural concepts of 'far'. Is it true that being opposite is equivalent to being at maximal distance in the point-line geometry?

**Proposition 7.2** *Yes.*

*Proof.* First of all, we may restrict ourselves to the thin case. Look at the double coset diagram of the point-line geometry. The distance from 1 of the double coset represented by $w$ is the minimum number of factors $r$ (where $r$ is the reflection belonging to node $i$) in any expression of $w$. Since $w < w_0$ in Bruhat order, for each reduced expression of $w_0$ there is a reduced expression for $w$ obtained from that of $w_0$ by cancelling factors. Thus, no double coset is farther from 1 than $w_0$. If $w_0$ does leave type $i$ invariant, then the double coset diagram of the point-line geometry is reflected by left multiplication by $w_0$ (see [4], 10.2.11). Since the double coset of 1 has a unique neighbour in the diagram, this means that also the double coset of $w_0$ has a unique neighbour in the diagram. Consider a double coset represented by $w$ different from that represented by $w_0$. There is an expression $w_0 = wu$ with $l(w_0) = l(w) + l(u)$, and $u$ describes a walk from $w$ to $w_0$ through the double coset diagram in which the distance to 1 never decreases. But it must pass through the unique neighbour of $w_0$, which is closer to 1, so also $w$ is closer to 1.

On the other hand, if $w_0$ moves type $i$, then we either have a Grassmann graph, or a dual polar graph, or $E_{6,1}$ (all distance regular), or $E_{6,2}$. Only the latter requires further investigation, and we find that also there there is a unique double coset at maximal distance, see the diagram below. □

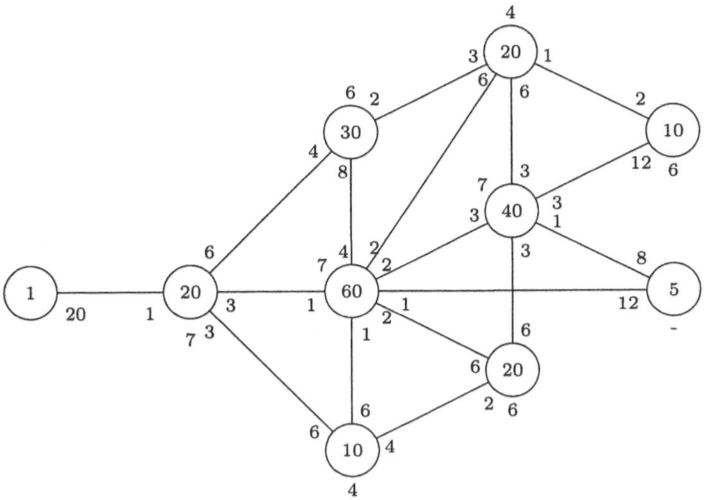


$v = 216$

**Problem** Is there a nicer proof, without having to look at $E_{6,2}$?

**Theorem 7.3** *In a thick near polygon with quads, the complement of a hyperplane is connected.*

*Proof.* Let $H$ be a hyperplane, and $x$, $y$ two points outside. Let $i = d(x, y)$. We show that the distance can be diminished while walking in the complement of $H$. Pick a line $L$ on $x$ that contains a point $x'$ with $d(x', y) = i - 1$. Pick a line $M$ on $y$ parallel to $L$. Then $M$ contains a point $y'$ with $d(y', x) = i - 1$. If either $x'$ or $y'$ is outside $H$, then we are done. Otherwise, we can pick a third point $x''$ on $L$; it has distance $i - 1$ to some point $y''$ of $M$, and $y'' \neq y'$ so that $x'', y'' \notin H$. Thus, walking from $x$ to $x''$ and from $y$ to $y''$ we have diminished the distance. □

**Theorem 7.4** *In a point-line geometry $\Gamma$ defined by selecting type $i$ in a spherical building, the complement of a hyperplane is connected (and has diameter at most $2d - 1$ when $\Gamma$ has diameter $d$) when $\mu$ (the number of points collinear to each of two points at mutual distance 2) is larger than 1.*

Thus, the conclusion of the theorem holds in the cases $A_{n,i}$ ($n \geq i \geq 1$), $B_{n,1}$ ($n \geq 3$), $B_{n,n}$ ($n \geq 2$), $D_{n,1}$ ($n \geq 4$), $D_{n,n}$ ($n \geq 4$), $E_{6,1}$, $E_{7,7}$.

*Proof.* Similar to the above, using the existence of 'parallel' lines. In Weyl group terms this means that

$$i^W = i^{W_{I \setminus \{i\}}}.$$

(The same condition is necessary and, in cases with single bonds only, sufficient to have an apartment generate the geometry.) □

## References

[1] P. Abramenko, *Twin buildings and applications to S-arithmetic groups*, Lecture Notes in Math. 1641, Springer, Berlin, 1996.

[2] N. Bourbaki, *Groupes et algèbres de Lie, Chap. IV, V, VI*, Hermann, Paris, 1968.

[3] A.E. Brouwer, *The complement of a geometric hyperplane in a generalized polygon is usually connected*, pp. 53–57 in: Finite geometry and combinatorics – Proc. Deinze 1992, F. De Clerck et al., London Math. Soc. Lect. Note Ser. 191, Cambridge Univ. Press, 1993.

[4] A.E. Brouwer, A.M. Cohen & A. Neumaier, *Distance-regular graphs*, Ergebnisse der Mathematik 3.18, Springer, Heidelberg, 1989.

[5] F. Buekenhout, *The basic diagram of a geometry*, pp. 1–29 in: Geometries and Groups, M. Aigner & D. Jungnickel (eds.), Lecture Notes in Math. 893, Springer, Berlin, 1981.

[6] B. Mühlherr & M. Ronan, *Local to global structure in twin buildings*, Invent. Math. **122** (1995) 71–81.

[7] M. Ronan, *Lectures on buildings*, Academic Press, Boston etc., 1989.

[8] J. Tits, *Buildings of spherical type and finite BN-pairs*, Lecture Notes in Math. 386, Springer, Berlin, 1974.

[9] J. Tits, *A local approach to buildings*, pp. 519–547 in: The Geometric Vein (the Coxeter Festschrift), C. Davis, B. Grünbaum, F.A. Sherk (eds.), Springer Verlag, Berlin, 1981.

Rieuwert J. Blok
Department of Mathematics,
Delft University of Technology,
P.O. Box 5031, 2600 GA Delft, The Netherlands.
e-mail: `R.J.Blok@twi.tudelft.nl`

Andries E. Brouwer
Department of Mathematics,
Eindhoven University of Technology,
P.O. Box 513, 5600 MB Eindhoven, The Netherlands.
e-mail: `aeb@cwi.nl`

Trends in Mathematics, © 1998 Birkhäuser Verlag Basel/Switzerland

# On Flag–transitive Incidence Geometries of Rank 6 for the Mathieu Group $M_{12}$

Francis Buekenhout*, Michel Dehon and Dimitri Leemans

**Abstract**

We show that the Mathieu group $M_{12}$ does not have geometries of rank greater or equal to 6, satisfying the RWPRI and $(IP)_2$ conditions. Our proof of this result is based on classifications of geometries of some subgroups of $M_{12}$ which have been obtained using MAGMA programs.

# 1 Introduction

## 1.1 General context of the work

This context is the geometric interpretation of finite simple groups (and their automorphism groups), especially as to their exceptional and sporadic nature going back to the ideas and work of Tits since 1954 (see Tits 1954,1981,1962,1974, Buekenhout 1986, 1995a, 1995b). The most developed and systematic approach in order to classify geometries in this context has been to choose some diagram and to look for all geometries in it satisfying some additional conditions, like for instance the presence of a flag-transitive automorphism group. This has given plenty of remarkable results. For a survey see Buekenhout 1995b, Chapter 22. In the meantime, many more of such results have been obtained. Here, we want to deal with another approach discussed briefly at the end of the latter reference and especially in Buekenhout 1986, 1995a. It starts with a group G and asks for all geometries on which G acts flag-transitively (property (FT)) and which satisfy some given conditions as being firm (property (F)) and residually connected (property (RC)). Definitions are recalled or given in Section 2.

## 1.2 Investigations

On the basis of Buekenhout 1995a, two slightly different conditions called PRI and RWPRI have been intensively and systematically investigated by various persons in an algorithmic spirit with the help of computer systems, mainly CAY-

*We gratefully acknowledge financial support of the "Fonds National de la Recherche Scientifique de Belgique" with this project.

LEY (see Cannon 1982, Cannon-Bosma 1991) and MAGMA (see Cannon-Bosma 1994, Cannon-Playoust 1993). Experience has led us to add property $(IP)_2$ in all cases which means that every residue of rank 2 satisfies the intersection property. Let us refer to Buekenhout-Cara-Dehon 1995, Dehon 1994 and Buekenhout-Dehon-Leemans 1996b for results on the condition PRI+$(IP)_2$ providing exhaustive lists of geometries for 24 almost simple groups including the groups $G$ with $PSL(2,q) \leq G \leq Aut(PSL(2,q))$ and $5 \leq q \leq 19$, the Mathieu group $M_{11}$ and $PSU(4,2)$. This has produced 562 group-geometry pairs of rank 2 at least, up to isomorphism.

## 1.3 RWPRI + $(IP)_2$

In this paper we are interested only in the case RWPRI+$(IP)_2$. Here the main references are Buekenhout-Dehon-Leemans 1996a, Gottschalk 1995,1996a,1996b, Gottschalk-Leemans 1997, Miller 1996 (joint work with Dehon leading to Dehon-Miller 1997a,b), Leemans 1996 and further work to be written up. The references cover 15 almost simple groups namely the $Alt(n)$, $Sym(n)$ with $5 \leq n \leq 7$, the groups $G$ with $PSL(2,q) \leq G \leq Aut(PSL(2,q)), q = 7,8$, the Mathieu group $M_{11}$, $PSU(4,2)$, $PSL(3,4)$, $Sz(8)$ and the Janko group $J_1$. This has produced 555 group-geometry pairs of rank 2 at least, up to isomorphism. During the last year, building in part on our improved theoretical understanding of the problem, Dehon and Miller wrote new programs in 3 stages, the first building the subgroup lattice of the group $G$ in CAYLEY, the second classifying boolean configurations (amalgams) written in the programming language C++ and the third checking which of these configurations lead to actual geometries with (FT) in MAGMA (see Miller 1996). Rather spectacular progress has quickly been made thanks to this work. In Miller 1996, we find the results on $M_{11}$ and on $PSU(4,2)$. These results will appear in Dehon-Miller 1997a,b.

## 1.4 Maximal ranks

Let $G$ be a finite nonabelian almost simple group. Let $r(G)$ be the maximal rank of a geometry for $G$ with RWPRI+$(IP)_2$. If $G$ is a group of Lie-Chevalley type of Lie rank $k$ we know that $k \leq r(G)$ thanks to the building derived from $G$. Actually, on the examples treated exhaustively we get slightly larger bounds. For $G \cong Sym(n)$, we know that $r(G) \geq n-1$ thanks to Buekenhout-Cara-Dehon 1996 and Buekenhout-Cara 1997. A complete classification of the corresponding geometries of rank $n-1$ with a connected diagram is available. They are thin geometries. The equality $r(G) = n-1$ holds almost certainly. For $G \cong Alt(n)$ it is almost certain that $r(G) = n-2$.
As to sporadic groups, $r(M_{11}) = 5$ by Dehon-Miller 1997b and $r(J_1) = 4$ by Gottschalk-Leemans 1997. Using the catalogue of a hundred of geometries in Buekenhout 1985 for the other sporadic groups and some work to check wether

they are RWPRI the answer is positive in most (but not all) cases and so lower bounds are obtained for $r(G)$.
At this time,

- 3 is the highest rank we can assign to $J_2, J_3, He, Ru, HN, LyS, Th$;
- 4 is the highest for $M_{12}, M_{22}, M_{23}, HS, M^cL, Sz, O'N, Co_2, Fi_{22}, Co_1, J_4$;
- 5 is the highest for $M_{24}, Co_3, Fi_{23}, BM, M$;
- 6 is the highest for $Fi'_{24}$.

Using those results it is natural to wonder about $r(M_{12})$ and perhaps to expect that it may be equal to 6, in view of $r(M_{11}) = 5$ and the usual relationships between those groups. In principle, but more unlikely $r(M_{12})$ could even be larger. Our main result here can now be stated and it is a little surprising.

**Theorem 1.1** *The sporadic group $M_{12}$ has no geometries of rank greater or equal to 6 that are RWPRI + $(IP)_2$.*

Our proof is strongly computer dependent. It could be less dependent at the cost of many pages of lengthy arguments.

## 1.5 Remarks

Our approach applies to some extent to the possible rank 5 geometries but we did not push this case to its end yet.
Our result might be an encouraging example in order to attack some larger sporadic groups.
We wonder whether we could get rid of $(IP)_2$.
Our proof does not require to introduce the group $G \cong M_{12}$ in a computer. But it heavily relies on the computer for a number of subgroups of $G$ and for the subgroup lattice of $G$.

## 1.6 Idea of the proof

From now on we put $G \cong M_{12}$.
The term geometry stands for (FT) under $G$, with (F), (RC) and RWPRI + $(IP)_2$. We assume that $\Gamma$ is such a geometry, with $\Gamma = \Gamma(G; G_0, G_1, G_2, G_3, G_4, G_5)$. We must pay attention also to ranks greater or equal to seven. We may assume without loss of generality that $G_0$ is maximal in $G$ and that $G_{01}$ is maximal in $G_0$, etc.

1. Among the maximal subgroups of $G$ we eliminate as many as we can as $G_0$-candidates for the fact that they do not have a geometry of rank greater or equal to 4. The remaining maximal subgroups are "$G_0$-candidates".
2. Given a "$G_0$-candidate" we show that at least one $G_i$ is one of $Sym(6)$ or $M_{11}$.

3. Assuming that some $G_i$ is $Sym(6)$ we show that some other must be $M_{11}$. At this stage the problem has been reduced to the case $G_0 \cong M_{11}$.

4. Assuming that $G_0 \cong M_{11}$ derive a contradiction.

### 1.7 Acknowledgement

We thank Ms. Peggy Leroy for her help with the conversion of our paper in a LaTeX file.

## 2 Definitions and notation

### 2.1 Geometries and their basic properties

We assume some knowledge of this and refer to Buekenhout 1995b. Here, a geometry $\Gamma$ requires a finite set of types $I$, a set of elements $X$, a relation of incidence $*$ on $X$, a type function $t$ from $X$ to $I$ with $t(x)=t(y)$ and $x * y$ implies $x = y$. We freely use the concepts of rank, flag, residue, chamber (maximal flag) and we assume (require) that all maximal flags are of type $I$. We assume that $\Gamma$ is residually connected (the incidence graph of any flag-residue of rank 2 at least is connected) and firm (every nonmaximal flag is contained in at least two chambers). These are the properties (RC) and (F).
We assume furthermore that $\Gamma$ has a group of type-preserving automorphisms $G$ acting transitively on the set of chambers of $\Gamma$. For shortness we say abusively that $\Gamma$ is "flag-transitive" or that it has property (FT). As a consequence, each flag-residue together with the group induced on it by the flag-stabiliser in $G$ is itself (FT), (RC) and (F).
Property $(IP)_2$ means that every rank 2 residue of $G$ either is a generalized digon or that any 2 distinct elements of the same type in it are both incident to at most one element of the other type.

### 2.2 Group-geometry pairs in group-theoretical terms

We freely use standard concepts and notation for groups and follow most often the Atlas (see Conway e.a. 1985). The cyclic group of order $n$ is sometimes denoted by $\mathbb{Z}_n$ and sometimes, simply by $n$.
One of the deep observations due to Tits (see Tits 1962 ) is that a group-geometry pair can be translated entirely as a structure in the group.
Let $(G,\Gamma)$ be a group-geometry pair over the set of types $I$ and fix a chamber $(C_i)$, $i \in I$ where $C_i$ is an element of type $i$. Let $G_i$ be the stabilizer of $C_i$ in $G$. Then each element of type $i$ of $G$ can be identified with a left coset $gG_i$ of $G_i$ in $G$. Moreover, it can be shown that elements $gG_i$ and $hG_j$ are incident if and only if the cosets $gG_i$ and $hG_j$ have a non-empty intersection in $G$. The action of $g \in G$ on the elements is simply translated by the fact that $g$ maps $xG_i$ on

$(gx)G_i$. There are further necessary conditions on the subgroups $G_i$ on which we come back somewhat later.

Quite remarkably, this process can be reversed. We start with $I$, a group $G$ and a subgroup $G_i$ of $G$ for each $i \in I$. We define the *pre-geometry* $\Gamma = \Gamma(G, (G_i)_{i \in I})$ as follows.

The set of elements of $G$ of type $i$ consists of all cosets $gG_i$, for all $i$. Elements $gG_i$ and $hG_j$ are called incident if and only if $gG_i$ and $hGj$ have a non-empty intersection and the action of $g \in G$ on $G$ is defined by the fact that $g$ maps $xG_i$ onto $(gx)G_i$.

All properties that we need further, that $G$ be a geometry, that it be flag-transitive (FT), firm (F) and residually connected (RC) can now be translated further as conditions on the subgroups $G_i$.

These results make it possible to develop algorithms of a group-theoretical nature in order to study and classify group-geometry pairs $(G, \Gamma)$ in which $G$ is a given group and $G$ may be a variable.

It is clear now that an automorphism of $G$ (in particular conjugation) provides isomorphic geometries namely : if each $G_i$ is sent on $H_i$ by some automorphism of $G$ then the corresponding (group-geometry) pairs are isomorphic. A similar statement can be made in a straightforward way for correlations.

We were mentioning further necessary and sufficient conditions on the subgroups $G_i$ in order that $\Gamma = \Gamma(G, (G_i)_{i \in I})$ be a (FT) geometry. In the rank 2 cases this set of conditions is empty. In the rank three case there is a unique condition (with equivalent variations) namely $G_0G_1 \cap G_0G_2 = G_0(G_1 \cap G_2)$. In the rank 4 case the last condition has to be required from every triple of subgroups among the four $G_i$ and one additional property is required namely $(G_0 \cap G_1 \cap G_2)(G_0 \cap G_1 \cap G_3) = ((G_1 \cap G_2)(G_2 \cap G_3)) \cap G_0$ (in a version due to Buekenhout-Hermand 1991, see also Dehon 1994). We send to those references for the higher rank situations that are quite similar with one additional condition involving all $G_i$.

## 2.3 The boolean lattice of a group-geometry pair

Consider a geometry $\Gamma(G, (G_i)_{i \in I})$ with (F) and (RC) of rank $n$. Then the $2^n$ subgroups $\bigcap(G_j), j \in J$ where $J$ is a subset of $I$, are distinct, constitute a boolean lattice and the subgroup generated by any two of its members is actually their lowest upper bound in the lattice. In Gottschalk-Leemans 1997, this is called a *strongly boolean lattice.*

From now on we always assume that this condition is satisfied. It implies the properties (F) and (RC) of 2.1.

We are especially interested in inclusions inside that lattice that are of maximal type namely pairs consisting of a subgroup and one of its maximal subgroups.

### 2.4 The conditions PRI and RWPRI

A geometry $\Gamma(G,(G_i)_{i\in I})$ is called *primitive* or PRI if each of the $G_i$ is a maximal subgroup of $G$.
A geometry $\Gamma(G,(G_i)_{i\in I})$ is called *weakly primitive* or WPRI if at least one of the subgroups $G_i$ is maximal in $G$ and it is called *residually weakly primitive* or RWPRI if it is WPRI and each of the residue group-geometry pairs of rank at least equal to one is WPRI as well.

### 2.5 The maximal rank $r(G)$

This is the largest value taken by a geometry of $G$ which is (F.T), RWPRI+$(IP)_2$.

## 3 Preliminary results

### 3.1 The subgroup pattern of $M_{12}$

1. The subgroup pattern of $G$ is used (Buekenhout-Rees 1988) together with its notation. We can of course not repeat that here. We recall that $G$ has 147 conjugacy classes of subgroups, that the subgroup pattern consists of the partially ordered set of these 147 classes, of an ID for each class like $K1$, $K2$, etc., that each class has a name reflecting the structure of its members like $2^3.Sym(3)$, that the number of members (the length) of the class is provided, as well as the numbers of maximal subgroups and minimal overgroups of each member in the different possible other classes. Two classes fused under $Aut(M_{12})$ are called $Ki$ and $K\bar{i}$ respectively. For each class there is some additional information that we leave out here.
A typical example goes as follows.
$K72$ $2\times Sym(4)$ 1980 $4K35$, $3K42$, $1K55$, $1K\overline{55}$, $1K58$ $1K94$, $1K97$
It shows that a member of $K72$ is contained in exactly 2 subgroups as a maximal subgroup and a look at $K94$, $K97$ shows that these are maximal subgroups of $G$.
A mistake about the length of classes $K17$ and $K\overline{17}$ (cyclic of order 8) has been detected by M. Bianchi and A. Gillio from the University of Milano (1989). The subgroup pattern is given also in GAP (see Schönert 1994). Using this, we have used a corrected version of the Buekenhout-Rees list which becomes reliable in view of the fact that its approach and that used in GAP are completely different. An interesting feature of the Buekenhout-Rees approach is that it does not require to introduce $G$ itself in the computer but it relies on computer work for its maximal subgroups and assumes that the maximal subgroups are known.
Finally, as we shall see, not so much information from the subgroup pattern is required here. It is quite the contrary in a computer search of all geometries of a given group.

**Corrections** to Buekenhout-Rees.
Page 597. A member of class $K6$ (respectively $K\overline{6}$) admits 2 rather than 1 overgroups in class $K17$ (respectively $K\overline{17}$).

Page 598. Classes $K17$ and $K\overline{17}$ are of length 2970 rather than 1485 and their normalizers are in $K65$ and $K\overline{65}$ respectively rather than in $K81$.
Page 603. A member of class $K\overline{80}$ contains 10 rather than 1 members of class $K\overline{16}$. A member of class $K84$ (respectively $K\overline{84}$) contains 3 rather than 1 members of class $K69$ (respectively $K\overline{69}$). A member of class $K90$ contains 6 rather than 65 members of class $K\overline{52}$.

**Comment:**
We like to insist on the interest of subgroup patterns for various purposes.
Pioneering work of that kind was done for the non-abelian groups of order at most 100 by Neubüser in the 1960's. We had similar work at our disposal in tables established in Delwiche 1992 for all of such groups except those of orders 64 and 96. This became an easy though lengthy exercise with CAYLEY. At this time GAP (see Schönert 1994) gives a library of all non-abelian groups of order 100 at most. In the theoretical algorithmic work implemented in CAYLEY, a key role was played by V.Felsch who, as an application, got the subgroup pattern of a group as large as $Alt(8)$ already in 1982, with 137 classes of subgroups.
In Buekenhout 1985 the subgroup patterns were given for $M_{11}$ and $J_1$ then corrected by H. Pahlings (private communication). There are 39 classes in $M_{11}$ and 40 in $J_1$. Later on Pahlings 1985 produced the subgroup pattern of the Janko group $J_2$ getting 147 classes. More recently, Pfeiffer 1991 dealing with the much larger Janko group $J_3$ got only 137 classes. All of this has been extended in GAP as a table of marks presumably by Pfeiffer and to mention but sporadic groups it contains all Mathieu groups and the group $M^cL$. We notice that $M_{24}$ has 1500 classes.

## 3.2 The rank 5 geometries of $M_{11}$

This is the highest rank of geometries for $M_{11}$. At the rank 5 there are four such. Each of them admits at least 2 subgroups $G_i$ isomorphic to $PSL(2,11)$. In the final step (4.9) of the proof we also use the fact that the two first of these four geometries admit a $G_i$ isomorphic to $Sym(3) \times Sym(3)$.
These results are due to Dehon-Miller 1996b (also Miller 1996). It would be worth trying a computer-free proof.

## 3.3 The rank 5 geometries of $Sym(6)$

We only need to know that $r(Sym(6)) = 5$ and that each rank 5 geometry has at least one $G_i$ isomorphic to $Sym(5)$. This follows from Buekenhout-Dehon-Leemans 1996a. Here too the search for a computer-free proof would be welcome.

## 3.4 There are no geometries of rank $> 5$ for $H = 2 \times Sym(5)$

In the subgroup pattern of $G = M_{12}$ we can read the maximal subgroups of $H$ and treat them according to the strategy applied here to $G$. Here they are.

1. $r(2 \times Sym(4)) = 4$ according to work done by Dehon some years ago and not written up.

2. $r(2^2 \times Sym(3)) = 4$ as we explain now.

   Indeed, the maximal subgroups give $r(D_{12}) = r(6 \times 2) = r(2^3) \leq 3$ which is straightforward.

3. $r(2 \times (5:4)) \leq 3$ as we explain now.

   Indeed, the maximal subgroups give $r(2\times 4) = 2$, $r(5:4) = 2$ by Buekenhout-Dehon-Leemans 1996a and $r(D_{20}) \leq 3$ because 20 is a product of 3 primes. Therefore, assuming that $M \cong 2 \times (5:4)$ has a rank 4 geometry, it requires $M_0 \cong D_{20}$. Then among the maximal subgroups of $M_0$, 10 lifts uniquely, a contradiction, each $D_{10}$ lifts uniquely to $M_1 \cong 5:4$ contradicting $r(5:4) = 2$ and finally $M_{01} \cong 2^2$ lifting necessarily to $M_1 \cong 2\times 4$ another contradiction.

4. $r(Sym(5)) = 4$ by Buekenhout-Dehon-Leemans 1996a.

5. $r(2 \times Alt(5)) \leq 4$ because for each of its maximal subgroups we get $r(2 \times Alt(4)) = r(D_{12}) = r(D_{20}) = r(Alt(5)) = 3$ the latter by Buekenhout-Dehon-Leemans 1996a and the other being met earlier in 2 and 3, except $2 \times Alt(4)$ . Assuming that the latter could have a rank 4 geometry, we see that its maximals satisfy $r(6) = 2 = r(Alt(4))$ the latter by Buekenhout-Dehon-Leemans 1996a and $r(2^3) = 3$. Therefore, a rank 4 geometry for $N \cong 2 \times Alt(4)$ requires $N_0 \cong 2^3$ and there a unique $2^2$ lifts in a second way, namely to $Alt(4)$ a contradiction with $r(Alt(4)) = 2$.

**Remark.** An exhaustive computer search shows that $H$ has respectively 5, 12, 16, 15, 10, 0 geometries of respective ranks $1, 2, 3, 4, 5, \geq 6$. See Cara-Leemans 1997.

### 3.5 There are no geometries of rank > 4 for $PSL(2, 11)$

This is a consequence of an exhaustive search by Dehon to be published. We need no more from it and again a computer-free proof may be suitable.

### 3.6 There are no geometries of rank > 4 for $K = P\Gamma L(2, 9)$

In the subgroup pattern of $G = M_{12}$ we can read the maximal subgroups of $K$ and treat them according to the strategy applied here to $G$. Here they are.

1. $r(3^2:8:2) = 3$ by Buekenhout-Dehon-Leemans 1996a.

2. $r(2 \times (5:4)) = 3$ by case 3 in 3.4.

3. $r(D_{16}:2) \leq 3$ since this subgroup has all but two maximal subgroups distributed in $K45, K47$ and $K48$. The latter have maximal rank 2 in view of the structure of their own maximal subgroups.

4. $r(Sym(6)) = 5$ by Buekenhout-Dehon-Leemans 1996a.

5. $r(M_{10}) = 3$ by an exhaustive search (see 3.12).

6. $r(PGL(2,9)) = 3$ as we show in 3.14.

If there is a geometry of rank 5 or more for $K$ the preceding list shows that $K_0 \cong Sym(6)$. Then Buekenhout-Dehon-Leemans 1996a applied to $Sym(6)$ imply that $K_{01}$ is one of $Sym(5)$ or $Alt(6)$. Lifting these one derives a contradiction, using the bounds on the maximal ranks for $M_{10}$ and $PGL(2,9)$ given in 3.12 and 3.14.

**Remark.** An exhaustive computer search gives respectively 6, 15, 14, 3, 0 geometries of respective ranks 1, 2, 3, 4, $\geq 5$.

## 3.7 There are no geometries of rank $> 4$ for $AGL(2,3)$

This results from Buekenhout-Dehon-Leemans 1996a.

## 3.8 No geometries of rank $> 4$ for $L = Alt(4) \times Sym(3)$

In the subgroup pattern of $G \cong M_{12}$ we can read the maximal subgroups of $L$ and treat them according to the strategy applied here to $G$. Here they are.

1. $r(3 \times Sym(3)) \leq 3$, since its maximal subgroups satisfy $r(3^2) = r(Sym(3)) = r(6) = 2$.

2. $r(2 \times Alt(4)) \leq 3$ as we show in 3.5.

3. $r(2^2 \times Sym(3)) \leq 4$ as we show in 3.4, 2.

4. $r(3 \times Alt(4)) \leq 3$ because its maximal subgroups satisfy $r(3^2) = r(Alt(4)) = 2$ as we saw earlier and $r(6 \times 2) \leq 3$ by 3.4, 2. If $Q \cong 3 \times Alt(4)$ has a rank 4 geometry it requires $Q_0 \cong 6 \times 2$ in which every 6 lifts uniquely so that $Q_{01}$ can be a unique $2^2$ and that there is no candidate left as $Q_{02}$.

The statement for $L$ follows.

**Remark.** An exhaustive computer search shows that $L$ has respectively 4, 7, 6, 2, 0 geometries of respective ranks $1, 2, 3, 4, \geq 5$.

## 3.9 The group $R = Sym(4) \times 2$

In this group, every rank 4 geometry requires $R_0 \cong Sym(4)$. This follows from an exhaustive computer search providing 4 geometries.

## 3.10 The subgroup $M_8.Sym(4)$

It has 78 conjugacy classes and respectively 4, 8, 11, 2, 0 geometries of respective ranks $1, 2, 3, 4, \geq 5$. This is the result of an exhaustive computer search.

## 3.11 The subgroup $2^{2+3}.Sym(3)$

It has 83 conjugacy classes and respectively 4, 8, 9, 4, 0 geometries of respective ranks $1, 2, 3, 4, \geq 5$. This is the result of an exhaustive computer search.

## 3.12 The group $M_{10}$ satisfies $r(M_{10}) = 3$

This is the result of an exhaustive computer search.

## 3.13 The group $M_9.2$ satisfies $r(M_9.2) = 3$

This is the result of an exhaustive computer search and is in Buekenhout-Dehon-Leemans 1996a.

## 3.14 No geometries of rank $> 3$ for $P = PGL(2, 9)$

Applying once more our method, we get the following.

1. $r(Alt(6)) = 4$ by Buekenhout-Dehon-Leemans 1996a.
2. $r(3^2 : 8) = 2$ by Buekenhout-Dehon-Leemans 1996a.
3. $r(D_{20}) \leq 3$ by the fact that 20 is a product of 3 primes.
4. $r(D_{16}) \leq 2$ because there are only 3 maximal subgroups one of which is cyclic and has maximal rank 1.
5. If $P$ has a rank 5 geometry, $P_0 \cong Alt(6)$. Then Buekenhout-Dehon-Leemans 1996a shows that we may assume $P_{01} \cong Alt(5)$ however this group lifts in a unique way in $P$, a contradiction.

   Thus we suppose there is a geometry of rank 4 for $P$.
6. As a first case, we consider $P_0 \cong D_{20}$. Among the maximals of the latter $\mathbb{Z}_{10}$ and one $D_{10}$ are of no use as possible $P_{01}$ because they lift uniquely in the pattern of $P$. This leaves 2 cases.

   (a) $P_{01} \cong D_{10}$ which lifts in $P_1$ as $Alt(5)$ or $Alt(6)$ but by Buekenhout-Dehon-Leemans 1996a the latter has no rank 3 geometry with a $G_i$ isomorphic to $D_{10}$. Thus $P_1 \cong Alt(5)$. Now necessarily $P_{02} \cong P_{03} \cong 2^2$. Also, $P_{012} \cong P_{013} \cong P_{023} \cong 2$ and $P_{0123} = 1$. Since $P_2$ and $P_1$, also $P_3$ and $P_1$ generate $P$ rather than $Alt(6)$ we see that $P_2 \cong P_3 \cong D_{20}$. Then the strongly boolean lattice completes itself in a unique way with the subgroup pattern of $P$. Thus, $P_{23} \cong 2^2$, $P_{12} \cong P_{13} \cong D_{10}$, the diagram comes out with residues of a 1-element having only pentagons as rank 2 residues. Now Buekenhout-Dehon-Leemans 1996a shows that $P_1$ has no such geometry.

(b) $P_{01} \cong 2^2$. The latter can only lift in $P_1 \cong D_{20}$ because $D_8$ and $D_{16}$ are ruled out by their maximal rank 2. There easily follows that all $P_{ij}$ are isomorphic to $2^2$ and that $P_2 \cong P_3 \cong D_{20}$. Then the diagram is totally disconnected with orders equal to 1 while there are 36 elements of each type, a contradiction with the "direct sum theorem" of geometries (see Buekenhout 1995a).

7. As a final case we take $P_0 \cong Alt(6)$. Then $P_{01}$ cannot be $Alt(5)$ or $Sym(4)$ that lift uniquely nor $3^2 : 4$ because it lifts uniquely to $3^2 : 8$ whose maximal rank is 2.

# 4 Proof of the theorem

**4.1.** From the list of maximal subgroups of $G$ and 3.5 to 3.7 we see that the only $G_0$-candidates are $2^{2+3}.Sym(3)$, $M_8.Sym(4)$, $2 \times Sym(5)$ and $M_{11}$.

**4.2.** From the results in 3.2 to 3.11 we can see that $G$ has no geometry of rank 7 or more.

**4.3.** Hence we are reduced to the search of rank 6 geometries for $G$.

**4.4.** Assume that $G_0 \cong 2^{2+3}.Sym(3)$. We could rely on 3.11 but we prefer to give an argument that depends less on the computer.
We see in the subgroup pattern of $G$ that each maximal of $G_0$ lifts uniquely in $G$ which is contradictory, except for $2 \times Sym(4)$ and $Q_8D_8$. First we assume that $G_{01} \cong 2 \times Sym(4)$. By 3.8, we may assume further that $G_{012} \cong Sym(4)$ a member of class $K55$. In the subgroup pattern (3.1) we see that this subgroup lifts to $G$ along exactly 2 paths of types $K87$, $K94$ and $K91$, $K97$ respectively. This is a contradiction with respect to the boolean lattice of the geometry. Hence $G_{01} \cong Q_8D_8$. Also, there are three 1-elements in $G_0$. The pattern of $G$ shows that $G_1$ must be a unique $M_8.Sym(4)$ in $K95$.
How about $G_{02}$ (and every $G_{0i}$ with $i = 2, 3, 4, 5$) ?
It acts transitively on the three 1-elements in $G_0$. Hence, it is not in $Q_8D_8$.
Now $G_{012}$ is of index 2 in $K81$ (nilpotent) hence of order 32, so $G_{02}$ is of index 2 or 3 in $G_0$ and it is maximal in it. It cannot be in $K86$, $K87$ and $K\overline{87}$ whose members lift uniquely. Therefore it is in $K72$ of structure $2 \times Sym(4)$. However, the latter does not have a subgroup of order 32. A contradiction.

**4.5.** Assume that $G_0 \cong M_8.Sym(4)$. Then $G_{0i} (i = 2, 3, 4, 5)$ cannot be $Q_8D_8$ nor $2^3.Alt(4)$. The remaining possibilities are $2 \times Sym(4)$ in $K73$, $GL(2, 3)$ in $K74$ and $GL(2, 3)$ in $K\overline{74}$. These lift either to $Sym(6)$, $AGL(2, 3)$ that cannot occur or $M_{11}$. Hence one of $Sym(6)$ or $M_{11}$ occurs as a $G_i$.

**Remark.** We could have eliminated this case just by 3.10.

**4.6.** Assume that $G_0 \cong 2 \times Sym(5)$. Considering the maximals of it, the maximal rank of their geometries remains below 5 or they lift in a unique way.

**4.7.** As a conclusion so far we may assume that $G_0$ is either $Sym(6)$ or $M_{11}$.

**4.8.** Assume that some $G_i$ is $Sym(6)$.
Going to the BDL-Atlas we see that $Sym(5)$ occurs in all possible $G_0$. Hence we may assume that $G_{01} \cong Sym(5)$ and so that $G_1 \cong M_{11}$.

**4.9.** Assume that $G_0 \cong M_{11}$.

By 3.2 we may assume $G_{01} \cong PSL(2,11)$ and similarly for $G_{02}$. These lift uniquely to $G_1 \cong M_{11} \cong G_2$ (of the second class). Now, the $G_{ij}$ with $i = 0,1,2$ cannot be $M_{10}$ or $M_9.2$ by 3.12 and 3.13. This leaves $GL(2,3)$, $Sym(5)$, $PSL(2,11)$ and possibly some of their subgroups. The Dehon-Miller list of rank 5 geometries for $M_{11}$ shows that the last condition eliminates their geometries except possibly numbers 1 and 2 that require $Sym(3) \times Sym(3)$. This is not in $GL(2,3)$, $Sym(5)$, $PSL(2,11)$. It is the final contradiction.

# 5 Comments

At this stage, we have no rank 5 geometry for G and we can state a number of facts obtained along arguments as the above, with more labor. The case where $G_0 \cong PSL(2,11)$ was reduced to a single configuration in which all $G_i$ were conjugate to $G_0$ and this was killed by a computer search. While looking for all geometries all of whose $G_i$ are conjugate to $PSL(2,11)$ we got an interesting rank 4 geometry described below.

There are two rank 4 geometries for $G$ known to us and given here.

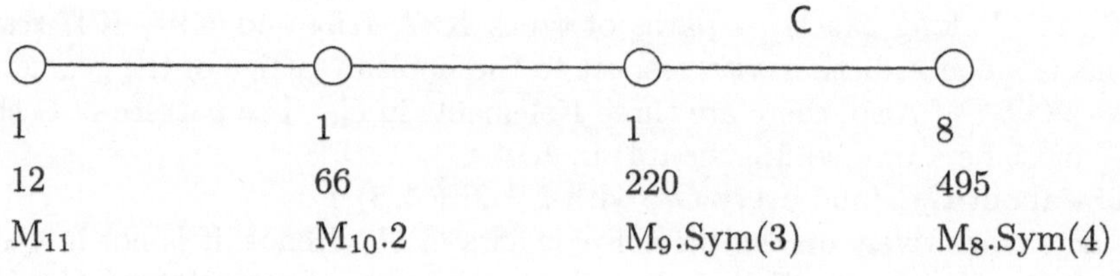


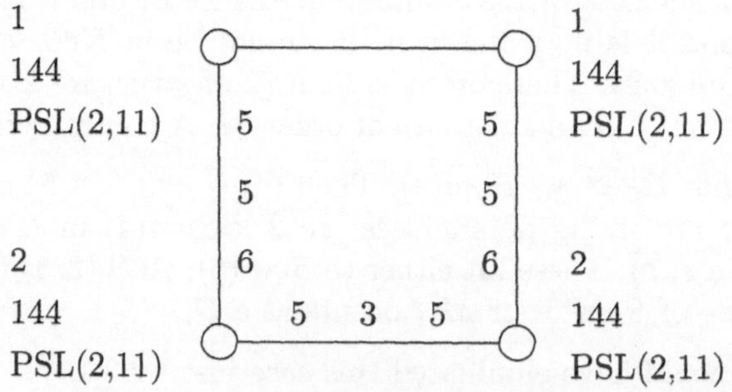


A rank 5 geometry which is not RWPRI but PRI was produced by Pasechnik (private communication). It is described here.

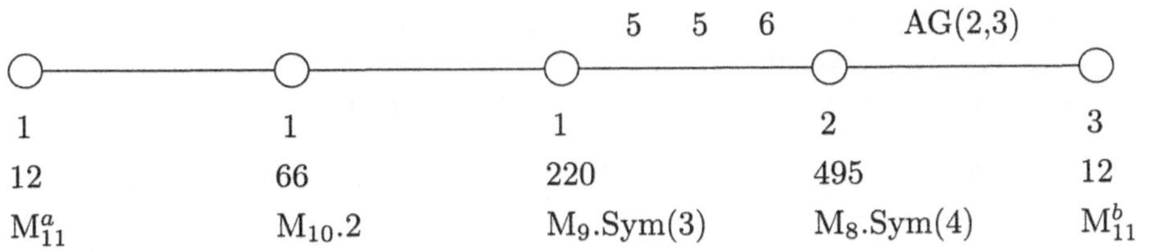


It may be worth to recall two other rank 5 geometries that are neither RWPRI nor PRI and whose diagrams appear in Buekenhout 1979, 1982. These geometries are related to the 5-transitive action of $G$ (this is the "Mathieu geometry") and to the Steiner system extending the affine plane of order 3 (this is the "Carmichael-Witt geometry").

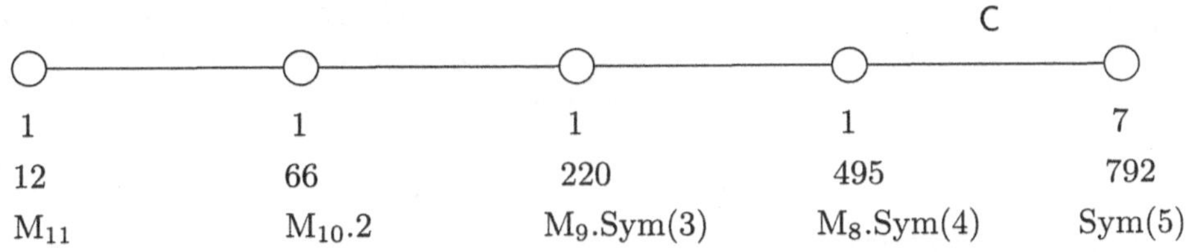


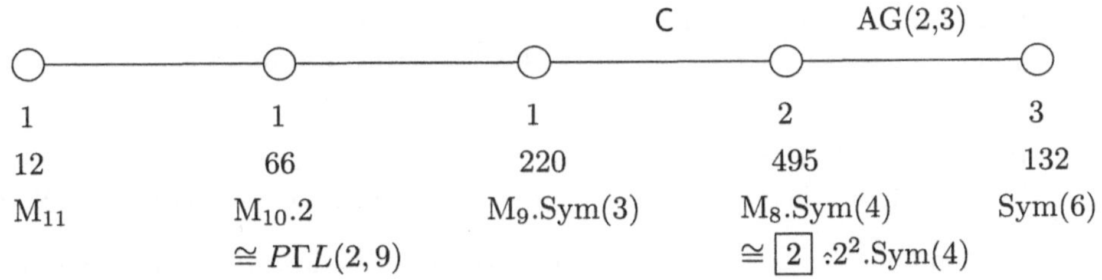


## References


[1] Buekenhout F. 1979. *Diagrams for geometries and groups.* Jour. Comb. Theory (A). **27**. 121–151.

[2] Buekenhout F. 1982. *Geometries for the Mathieu group $M_{12}$.* Proceedings Conference Combinatorics. Lecture Notes in Mathematics **1969**. Springer-Verlag. Berlin. 74–85.

[3] Buekenhout F. 1985. *Diagram geometries for sporadic groups.* Contemp. Math. **45**. 1–32.

[4] Buekenhout F. 1986. *The geometry of the finite simple groups.* In "Buildings and the geometry of diagrams, Ed. Rosati L.A.". Lect. Notes in Math. **1181**. 1–78. Springer-Verlag. Berlin.

[5] Buekenhout F. 1995a. *Finite groups and geometry: A view on the present state and the future.* "In Groups of Lie type and their geometries. Ed. Kantor W.M. and Di Martino L.". Cambridge University Press. 35–42.

[6] Buekenhout F. (Editor). 1995b. *Handbook of Incidence Geometry. Buildings and Foundations.* Elsevier, Amsterdam.

[7] Buekenhout F., Cara Ph. 1997. *Some properties of inductively minimal geometries.* Preprint.

[8] Buekenhout F., Cara Ph., Dehon M. 1995. *Geometries of small almost simple groups based on maximal subgroups.* Preprint. 157 pages.

[9] Buekenhout F., Cara Ph., Dehon M. 1996a. *Inductively minimal flag-transitive geometries.* In N.L.Jonhson (Editor), Mostly Finite Geometries. Marcel Dekker. New York, 185–190. To appear.

[10] Buekenhout F., Dehon M., Leemans D. 1996a. *An Atlas of residually weakly primitive geometries for small groups.* Mém. Acad. R. Belg. Classe des Sciences. To appear.

[11] Buekenhout F., Dehon M., Leemans D. 1996b. *All geometries of the Mathieu group $M_{11}$ based on maximal subgroups.* Experimental Mathematics. **5** (1996) 101–110.

[12] Buekenhout F., Hermand M. 1991. *On flag-transitive geometries and groups.* Travaux de Mathématiques. Université Libre de Bruxelles. Fascicule I, 45–78.

[13] Buekenhout F., Rees S. 1988. *The subgroup structure of the Mathieu group $M_{12}$.* Math. Comp. **50**. 595–605.

[14] Cannon J. 1982. *A language for group theory.* Department of pure mathematics. University of Sydney.

[15] Cannon J., Bosma W. 1991. CAYLEY*: quick reference guide.* University of Sydney.

[16] Cannon J., Bosma W. 1994. *Handbook of* MAGMA *functions.* University of Sydney.

[17] Cannon J., Playoust C. 1993. *An introduction to* MAGMA. University of Sydney.

[18] Cara Ph., Leemans D. 1997. *The residually weakly primitive geometries of $S_5 \times 2$.* Preprint.

[19] Conway J.H., Curtis R.T., Norton S.P., Parker R.A., and Wilson R.A. 1985. *Atlas of Finite Groups.* Clarendon Press. Oxford.

[20] Dehon M. 1994a. *Classifying geometries with* CAYLEY. Jour. Symbolic Computation. **17**. 259–276.

[21] Dehon M., Miller X. 1997a. *The RWPRI and (IP)$_2$ geometries of U(4,2).* In preparation.

[22] Dehon M., Miller X. 1997b. *The RWPRI and (IP)$_2$ geometries of $M_{11}$.* In preparation.

[23] Delwiche V. 1992. *The subgroup structure of the non-abelian groups of order up to 100.* Mémoire de licence. Université Libre de Bruxelles.

[24] Gottschalk H. 1995. *A classification of geometries associated with PSL(3,4).* Diplomarbeit. Mathematisches Institut der Justus-Liebig Universität. Giessen.

[25] Gottschalk H. 1996a. *Rank three geometries associated with PSL(3,4).* Bull. Soc. Math. Belg. Simon Stevin. **3**. 147–160.

[26] Gottschalk H. 1996b. *A connection between six distinguished diagrams.* Atti Sem. Mat. Fis. Modena. To appear.

[27] Gottschalk H. and Leemans D. 1997. *The residually weakly primitive geometries of the Janko group $J_1$.* In, Editor A.Pasini e.a., Groups and Geometries. Birkhäuser. Basel.

[28] Leemans D. 1996. *The residually weakly primitive geometries of the Suzuki simple group $S_z(8)$.* Preprint.

[29] Miller X. 1996. *Construction algorithmique des géométries associées à un groupe.* Mémoire de licence. Université Libre de Bruxelles.

[30] Pahlings H. 1985. *The subgroup structure of the Hall-Janko group $J_2$.* Bayreuth Math. Schr. **23**. 135–165.

[31] Pfeiffer G. 1991. *Von Permutationscharakteren und Markentafeln.* Diplomarbeit. RWTH Aachen.

[32] Schönert M. e.a. 1994. GAP*: Groups, Algorithms and Programming.* Lehrstuhl D für Mathematik. Aachen.

[33] Tits J. 1954. *Espaces homogènes et groupes de Lie exceptionnels.* Proc. Internat. Congr. Math., Amsterdam, sept. vol. I, 495–496.

[34] Tits J. 1962. *Géométries polyédriques et groupes simples.* Deuxième réunion du Groupement de mathématiciens d'expression latine, Florence, 66–88.

[35] Tits J. 1974. *Buildings of spherical type and finite BN-pairs.* Lecture Notes in Mathematics. **386**. Springer-Verlag, Berlin.

[36] Tits J. 1981. *Buildings and Buekenhout geometries.* In Ed. M. Collins, Finite Simple Groups II, Academic Press, New-York, 309–320.

Francis Buekenhout
Département de Mathématique - CP 216
Université Libre de Bruxelles
Bd. du Triomphe, B-1050 Bruxelles, Belgium
e-mail: `fbueken@ulb.ac.be`

Michel Dehon
Département de Mathématique - CP 216
Université Libre de Bruxelles
Bd. du Triomphe, B-1050 Bruxelles, Belgium
e-mail: `mdehon@ulb.ac.be`

Dimitri Leemans
Département de Mathématique - CP 216
Université Libre de Bruxelles
Bd. du Triomphe, B-1050 Bruxelles, Belgium
e-mail: `dleemans@cso.ulb.ac.be`

Trends in Mathematics, © 1998 Birkhäuser Verlag Basel/Switzerland

# On a Connection between Ovoids on the Hyperbolic Quadric $\mathcal{Q}^+(10, q)$ and the Lie Incidence Geometry $E_{6,1}(q)$

Bruce N. Cooperstein

**Abstract**

It is demonstrated that the existence of an ovoid on the hyperbolic quadric $\mathcal{Q}^+(10, q)$ implies the existence of a co-clique of maximal possible cardinality in the collinear graph of the Lie Incidence Geometry $E_{6,1}(q)$.

## 1 Introduction

### 1.1 Basic Concepts and General Definitions

By a *graph* we mean a set $P$ whose elements are called *vertices* together with a symmetric, antireflexive relation $\sim$ referred to as adjacency. The pairs $\{x, y\}$ from $P$ with $x \sim y$ are called *edges*.

An *incidence system* is a triple $(P, L, I)$ consisting of a set $P$ whose elements are called *points*, a set $L$ whose members are called *lines*, and a symmetric relation $I \subset (P \times L) \cup (L \times P)$. If $x \in P$, $l \in L$ and $(x, l) \in I$ then we say $x$ is *incident* with or on $l$. $(P, L, I)$ is said to be a *linear incidence system* or a *point-line geometry* if two points are incident with at most one line. In this case we may identify each line with the set of points with which it is incident and replace $I$ with the symmetrization of the relation $\in$ and then we will write $(P, L)$ in place of $(P, L, I)$. The *collinearity graph* of a linear incidence system $(P, L)$ is the graph $(P, \sim)$ where $x \sim y$ for $x, y \in P$ if and only if $x$ and $y$ are collinear. For a pair of points $x, y$ a path from $x$ to $y$ is a sequence $x = x_0, x_1, \ldots, x_t = y$. The distance between $x$ and $y$, denoted $d(x, y)$ is the length of a shortest path, if one exists, otherwise it is $+\infty$. When $d(x, y) = d$ is finite we refer to a path of length $d$ from $x$ to $y$ as a geodesic.

A subspace of $(P, L)$ is a subset $X$ of $P$ with the property that if $l \in L$ and $|l \cap X| \geq 2$ then $l \subseteq X$. Clearly an intersection of subspaces is a subspace. For a subset $Y$ of $P$ the subspace generated by $Y$ is the intersection of all subspaces which contain $X$. A subspace $X$ is *convex* if, for all $x, y \in X$ every geodesic joining

$x$ to $y$ is contained in $X$. A subspace, all of whose points are collinear will be called *singular*.

## 1.2 Hyperbolic Quadrics and Ovoids

Let $V$ be a vector space of dimension $m$ over a field $\mathbb{K}$ with basis $v_1, v_2, \ldots, v_m$. We refer to the 1-spaces of $V$ as *points* and the set of all 1-spaces as the projective space $PG(V) = PG(m-1, \mathbb{K})$. The 2-dimensional subspaces of $V$ will be referred to as *lines* and as above we will adopt the convention that a line is identified with the points to which it is incident.

Now assume that $m = 2n$ and that $V$ is equipped with the quadratic form $Q : V \to \mathbb{K}$ given by

$$Q(\sum_{1}^{2n} \alpha_i v_i) = \alpha_1\alpha_2 + \alpha_3\alpha_4 + \ldots \alpha_{2n-1}\alpha_{2n}.$$

The bilinear form associated to $Q$ is the form, $B : V \times V \to \mathbb{K}$ given by

$$B(v, w) = Q(v + w) - Q(v) - Q(w).$$

A point $\langle v \rangle$ of $PG(V)$ will be said to be *singular with respect to* $Q$ if $Q(v) = 0$. The set of all singular points is known as a *hyperbolic quadric* in $PG(V)$ and is denoted by $\mathcal{Q}^+(2n, \mathbb{K})$ (we use the linear dimension here, rather than the projective dimension). When $\mathbb{K} = \mathbb{F}_q, q = p^e, p$ a prime, we will write $\mathcal{Q}^+(2n, q)$.

Two points $\langle v \rangle, \langle w \rangle$ are *orthogonal* if $B(v, w) = 0$. For a vector $v$ let $v^\perp = \{u \in V | B(v, u) = 0\}$. This is a hyperplane of $V$. When the points $\langle v \rangle, \langle w \rangle$ are singular and orthogonal to one another then all the points on the line $\langle v, w \rangle$ are singular. We refer to such lines as *singular lines.* More generally, we will refer to a subspace $U$ of $V$ which satisfies $Q(U) = \{0\}$ as *singular.* The maximal singular subspaces have dimension $n$ ($\langle x_1, x_3, \ldots, x_{2n-1} \rangle$ is an example of such a subspace) and are referred to as *generators* of the quadric.

Now if we set $P = \mathcal{Q}^+(2n, q)$ and $L = \{\langle v, w \rangle | \langle v \rangle, \langle w \rangle \in P, v \perp w\}$ we get an incidence geometry which we denote by $D_{n,1}(q)$.

By a *cap* in $P$ we mean a subset $C$ of $P$ (singular points of $PG(V)$ with the property that no two are orthogonal. The bound on the cardinality of a cap is $q^{n-1} + 1$ (cf [10], [16]). When this bound is realized, then every maximal singular subspace of $V$ contains a unique point of $C$ and $C$ is said to be an *ovoid.* When $n = 3$ (dim $V = 6$), via the Klein correspondence, an ovoid is nothing more than an affine translation plane of dimension two over its kernal (see [11]). Ovoids are much rarer when $n = 4$ but a number of families have been constructed (see [6], [10], [12],[13]). It has been conjectured that ovoids do not exist for $n \geq 5$. It is known that if an ovoid exists on $\mathcal{Q}^+(2n, q)$ then an ovoid exists on $\mathcal{Q}^+(2n-2, q)$

(cf [16]) and therefore the conjecture is equivalent to the non-existence of ovoids on $\mathcal{Q}^+(10,q)$. Heretofore some partial results have been obtained:

Kantor [10] demonstrated that there is no ovoid on $\mathcal{Q}^+(10,2)$.

Shult [15] proved that there is no ovoid on $\mathcal{Q}^+(10,3)$.

Using $p$- ranks of the incidence matrix for the incidence system $(\mathcal{P},\mathcal{H})$ where $\mathcal{H}=\{v^\perp|\langle v\rangle\in\mathcal{P}\}$, Blokhuis and Moorhouse [2] demonstrated that for a cap $C$ on a quadric $\mathcal{Q}^+(2n,q)$

$$|C|\leq [\binom{p+2n-2}{2n-1}-\binom{p+2n-4}{2n-1}]^e+1$$

and as a consequence proved the non-existence of ovoids in $\mathcal{Q}^+(10,p^e), p=2,3$ and $\mathcal{Q}^+(12,p^e), p=5,7$.

Independently, Cooperstein [8] proved the non-existence of ovoids in $\mathcal{Q}^+(10,2^e)$ making use of tensor products and a result of Dye's [9] giving a bound on a maximal cap of $\mathcal{Q}^+(2m,2)$.

The purpose of this note is to suggest a new line of attack towards establishing the nonexistence of ovoids in $\mathcal{Q}^+(10,q)$, one which switches the point of view from the orthogonal space to another geometry, specifically, the Lie incidence geometry $E_{6,1}(q)$. The idea is to show that the existence of an ovoid on the quadric $\mathcal{Q}^+(10,q)$ implies the existence of a set of pairwise non-adjacent points in the collinearity graph of the $E_{6,1}(q)$ geometry which realizes the theoretic upper limit for the cardinality of a co-clique. It is hoped that one will be able to use the intrinsic geometry of $E_{6,1}(q)$, its standard embedding, or perhaps even incidence matrix techniques to get further constraints on the ovoid and thereby establish a contradiction.

The organization of this note is as follows: In section two we very briefly define the geometry $D_{5,5}(q)$ and record it cardinality. In section three we record some essential facts about the $E_{6,1}(q)$ geometry and establish a bound on the size of a co-clique in the collinearity graph of this geometry. Finally, in section four we investigate co-cliques in the collinearity graph of $E_{6,1}(q)$; in particular, we establish a bound on the size of such a co-clique. We then go on to indicate how existence of an ovoid in $\mathcal{Q}^+(10,q)$ implies the existence of a co-clique in $E_{6,1}(q)$ which meets this bound.

## 2 The Lie Incidence Geometry $D_{5,5}(q)$.

In this section $V,Q,B$ are as above. Let $\Sigma$ denote the set of all maximal singular subspaces with respect to $Q$. Define a graph on $\Sigma$ as follows: for $M_1,M_2\in\Sigma$, $M_1\sim M_2$ if and only if $dim(M_1/(M_1\cap M_2))=dim(M_1/(M_2\cap M_2))=2$. This graph has two connected components, which we label as $\Sigma^+,\Sigma^-$. Now fix one of the classes

$\Sigma^{\pm}$, say $\Sigma^{+}$, and for convenience of notation let us denote it for the present by $\mathcal{P}$. For an adjacent pair $M, N \in \mathcal{P}$ there are precisely $q+1$ members of $\mathcal{P}$ which contain $M \cap N$. Set $l(M,N) = \{R \in P | R \supset M \cap N\}$ and set $\mathcal{L} = \{l(M,N) | M, N \in P, M \sim N\}$. This gives us an incidence system $(\mathcal{P}, \mathcal{L})$ which we will refer to as a $D_{n,n}(q)$ geometry. These geometries have been studied extensively and characterized. For our purposes all we require is the point-cardinality of $D_{5,5}(q)$ which is given by

$$(2.1) \quad |\mathcal{P}| = (q^4+1)(q^3+1)(q^2+1)(q+1).$$

## 3 The $E_{6,1}(q)$ Geometry

The $E_{6,1}(q)$ geometry does not have as simple a description as does the previous geometries $D_{n,1}(q), D_{n,n}(q)$. It can be realized as an incidence geometry whose points and lines are conjugacy classes of parabolic subgroups of a Chevalley group $E_6(q)$ or alternatively as a shadow geometry of a building of type $E_6(q)$. Perhaps the most concrete model is as follows (see [4]): Let $M_3$ denote the space of three by three matrices over $\mathbb{K}$ and set $M = M_3 \oplus M_3 \oplus M_3$. Let $T : M \to \mathbb{K}$ be defined by

$$T(m_1, m_2, m_3) = det(m_1) + det(m_2) + det(m_3) - trace(m_1 m_2 m_3).$$

This is a cubic form in the twenty seven matrix coordinates which contains 45 monomials, each variable is in 5 such monomials. By taking the partials of $T$ with respect to each of the 27 matrix coordinates we get 27 quadratic forms. The points of the $E_{6,1}(q)$ geometry can be identified with the points of $PG(M)$ which are singular points of every one of these quadratic forms, the lines with the lines which are singular lines with respect each of the forms.

**(3.1) Remarks.** (a) The isometry group, $E = I(M,T)$ of linear transformation of $M$ which preserve $T$, is the universal Chevalley group of type $E_6$ over the field $\mathbb{K}$. This group has three orbits on the points of $PG(M)$ with representatives $(m, 0, 0)$ where $m$ is, respectively, a matrix with rank one, two, and three. The latter is the only class on which $T$ is non-zero. Because $E$ has only three orbits on points of $PG(M)$ it likewise has only three orbits on hyperplanes or, equivalently, on the points of $PG(M^*)$, where $M^* = Hom_{\mathbb{K}}(M, \mathbb{K})$.

(b) When $\mathbb{K} = \mathbb{F}_q$, a finite field, by the Chevalley-Warning theorem (cf page 5 in [14]) if $U$ is a subspace of $M$ such that for all $u \in U \setminus \{0\}, T(u) \neq 0$ then dim $U \leq \deg T = 3$.

Because of the multiplicity of geometries introduced and referred to we shall adopt the following notation for the remainder: the point set of a geometry will be denoted by $\mathcal{P}$ and the line set by $\mathcal{L}$ with an appropriate subscript taken from the set $\{D_{n,1}, D_{5,5}, E_{6,1}\}$. For $I$ such a subscript and $x$ a point in $\mathcal{P}_I$, $\Delta_I(x)$ will denote the points collinear with $x$ and for a pair of points $x, z \in \mathcal{P}_I, \Delta_I(x,z) = \Delta_I(x) \cap \Delta_I(z)$.

For a pair of $x, z \in \mathcal{P}_I$ we will denote the subspace generated by $\{x, y\} \cup \Delta_I(x, z)$ by $S_I(x, z)$. Also, for a point $x \in \mathcal{P}_I, \mathcal{L}_I(x) = \{l \in \mathcal{L}_I | x \in l\}$.

The facts we need for the $E_{6,1}(q)$ geometry are the following. These can be deduced from [1], [5], [7].

(3.2) $|\mathcal{P}_{E_{6,1}}| = \frac{(q^{12}-1)(q^9-1)}{q^4-1)(q-1)}$.

(3.3) For $x \in \mathcal{P}_{E_{6,1}}$,
$|\Delta_{E_{6,1}}(x)| = q|\mathcal{P}_{D_{5,5}}|, |\mathcal{L}_{E_{6,1}}(x)| = |\mathcal{P}_{D_{5,5}}| = (q^4+1)(q^3+1)(q^2+1)(q+1)$.

(3.4) The singular subspaces of $(\mathcal{P}_{E_{6,1}}, \mathcal{L}_{E_{6,1}})$ are isomorphic to projective spaces over $\mathbb{K}$ of dimension less than or equal to five.

(3.5) For $x, z \in \mathcal{P}_{E_{6,1}}$ non-collinear points, $|\Delta_{E_{6,1}}(x, z)| = |\mathcal{P}_{D_{4,1}}| = (q^3 + 1)(q^2+1)(q+1)$. The geometry induced on $\Delta_{E_{6,1}}(x, z)$ is isomorphic to the geometry $D_{4,1}(q)$.

(3.6) For $x, z$ a pair of non-collinear points, $|S_{E_{6,1}}(x,z)| = |\mathcal{P}_{D_{5,1}}| = \frac{(q^5-1)(q^4+1)}{q-1}$. The geometry induced on $S_{E_{6,1}}(x, z)$ is isomorphic to $D_{5,1}(q)$.

The subspaces $S_{E_{6,1}}(x, z), x, z$ a non-collinear pair of points, are convex and are called *symps* of the geometry. Let $\mathcal{S}_{E_{6,1}}$ denote the symps of the $E_{6,1}(q)$ geometry.

(3.7) If $S_1, S_2$ are distinct symps of the $E_{6,1}(q)$ geometry then either $S_1 \cap S_2$ is a point or is a singular subspace isomorphic to $PG(4, q)$ (moreover, in this latter case the singular subspace is a maximal singular subspace, that is, not contained properly in any other singular subspace). For a singular subspace $N$ isomorphic to $PG(4, q)$ which is the intersection of two symps there are exactly $q+1$ symps containing $N$. If we treat the maximal $PG(4, q)$'s as a set of "lines", $L$, we get an incidence geometry $(\mathcal{S}_{\mathcal{E}_{6,1}}, L)$ whose point set is the collection of symps of $(\mathcal{P}_{E_{6,1}}, \mathcal{L}_{E_{6,1}})$. This geometry is isomorphic to the $E_{6,1}(q)$ geometry $(\mathcal{P}_{E_{6,1}}, \mathcal{L}_{E_{6,1}})$. In this incidence geometry the points of $\mathcal{P}_{E_{6,1}}$ play the role of "symps", that is, by taking all the symps on a point $x \in \mathcal{P}_{E_{6,1}}$ (a collection which has cardinality $|\mathcal{P}_{D_{5,1}(q)}|$) we obtain a subspace of this geometry isomorphic to the incidence geometry $D_{5,1}(q)$.

(3.8) If $S$ is a symp, $x \in \mathcal{P}_{E_{6,1}} \setminus S$ then either $S \cap \Delta_{E_{6,1}}(x)$ is empty or a singular subspace isomorphic to $PG(4, q)$ (and clearly a non-maximal singular subspace since it is properly contained in the singular subspace $\langle x, S \cap \Delta_{E_{6,1}}(x)\rangle$).

(3.9) If $S$ is a symp, then the subspace of $M$ spanned by all points $x \in \mathcal{P}_{E_{6,1}}$ such that $S \cap \Delta_{E_{6,1}}(x) \neq \emptyset$ is a hyperplane, which we denote by $H_S$. Such a hyperplane is fixed by a parabolic subgroup of $E$. When $\mathbb{K} = \mathbb{F}_q, |\mathcal{P}_{E_{6,1}} \cap H_S| = |\mathcal{P}_{E_{6,1}}| - q^{16}$.

(3.10) If $S_1, S_2$ are adjacent symps, that is, $S_1 \cap S_2 = PG(4, q)$, then every hyperplane containing $H_{S_1} \cap H_{S_2}$ is of the form $H_S$ for a symp containing $S_1 \cap S_2$. On the other hand, if $S_1, S_2$ are symps which meet in a point $x$ then the only hyperplanes of the form $H_S$ which contain $H_{S_1} \cap H_{S_2}$ are $H_{S_i}, i = 1, 2$. The other hyperplanes which contain this intersection are members of a second class and for such a hyperplane $H$

$$|PG(H) \cap \mathcal{P}_{E_{6,1}}| = |\mathcal{P}_{E_{6,1}}| - q^{16} - q^{12}.$$

(3.11) Finally, if we take a point $\langle x \rangle$ with $T(x) \neq 0$ then $E_{\langle x \rangle}$ is a Chevalley group of type $F_4(\mathbb{K})$ and this group fixes a complementary hyperplane $H_{\langle x \rangle}$. Such a hyperplane is representative of the third class and

$$|PG(H_{\langle x \rangle}) \cap \mathcal{P}_{E_{6,1}}| = \frac{(q^{12}-1)(q^8-1)}{(q^4-1)(q-1)}.$$

We refer to these as $F_{4,4}$ hyperplanes.

**(3.12) Remark.** There exists an $E$ invariant cubic form $T^*$ on $M^*$ such that the set of points of $PG(M^*)$ corresponding to the $F_{4,4}$ hyperplanes is the support of $T^*$.

**(3.13) Lemma.** *Assume that $S_1, S_2$ are two symps of $(\mathcal{P}_{E_{6,1}}, \mathcal{L}_{E_{6,1}})$ whose intersection is a point $x$. Let $y_i \in S_i, i = 1, 2$ be points distinct from $x$. If $y_1$ and $y_2$ are collinear then they are both collinear with $x$.*

*Proof.* Since $y_2 \in S_2 \cap \Delta_{E_{6,1}}(y_1)$ it follows that $S_2 \cap \Delta_{E_{6,1}}(y_1) = Z$ is a singular subspace $PG(4, q)$. Suppose $y_1$ is not collinear with $x$. Since $x \in S_2, x$ is collinear with a hyperplane of $Z$ and the $Z \cap \Delta_{E_{6,1}}(x) \subset \Delta_{E_{6,1}}(x) \cap \Delta_{E_{6,1}}(y_1) \subset S_1$. But then $Z \cap \Delta_{E_{6,1}}(x) \subset S_1 \cap S_2 = \{x\}$, a contradiction. Thus, $x \in \Delta_{E_{6,1}}(y_1)$ and as $x \in S_2$ it follows that $x \in Z$ so that $x$ and $y_2$ are collinear.

**(3.14) Corollary.** *Assume $S_1, S_2, x$ are as above. If $z_i \in S_i \setminus [\{x\} \cup \Delta_{E_{6,1}}(x)], i = 1, 2$ then $z_1, z_2$ are non-collinear.*

# 4 Co-Cliques in the $E_{6,1}(q)$ Collinearity Graph

In this section we continue the notation of section three. We first get a bound on the size of a co-clique in the collinearity graph of $E_{6,1}(q)$ :

**(4.1) Theorem.** *Let $C$ be a co-clique in the collinearity graph of $E_{6,1}(q)$. Then $|C| \leq q^8 + q^4 + 1$.*

*Proof.* From the previous section we can deduce that the collinearity graph of $E_{6,1}(q)$ is a strongly regular graph with parameters

$$k = q(q^4+1)(q^3+1)(q^2+1)(q+1), l = q^8\frac{(q^5-1)}{(q-1)}$$
$$\lambda = q^2(q^4+q^3+q^2+q+1)(q^2+1), \mu = (q^3+q^2+q+1)(q^3+1).$$

The eigenvalues of this graph are $k, r, s$ with $r, s$ the roots of the quadratic equation

$$\theta^2 + (\mu - \lambda)\theta + (\mu - k).$$

From this we obtain

$$r = q^8 + q^7 + q^6 + q^5 + q^4 - 1, s = -(q^3 + 1).$$

Then for the complementary graph the eigenvalues are

$$\bar{r} = q^3, \bar{s} = -(q^8 + q^7 + q^6 + q^5 + q^4) = -q^4\frac{(q^5 - 1)}{(q-1)}.$$

Now a co-clique $C$ of the collinearity graph of $E_{6,1}(q)$ geometry is a clique of the complementary graph and by the Hoffman bound (cf page 10 in [3]) we have

$$|C| \leq 1 + \frac{l}{-\bar{s}} = 1 + q^4(q^4 + 1) = q^8 + q^4 + 1.$$

**(4.2) Remark.** If $C$ is a such a co-clique ($|C| = q^8 + q^4 + 1$) then for any point $x \in \mathcal{P}_{E_{6,1}} \setminus C, |\Delta_{E_{6,1}}(x) \cap C| = -\bar{s} = q^3 + 1$.

In section three we provided a model of the $E_{6,1}(q)$ geometry with its points and lines having a projective embedding in $PG(26, q)$. By an *external flat* to this embedding we mean a subspace $PG(Z), Z \leq M$ such that $PG(Z) \cap \mathcal{P}_{E_{6,1}} = \emptyset$.

**(4.3) Lemma.** *If $PG(Z)$ is an external flat to $\mathcal{P}_{E_{6,1}}$ then dim $Z \leq 18$.*

*Proof.* Assume $Z$ has maximal dimension such that $PG(Z)$ is an external flat and assume that dim $M/Z = t$. Let $U$ be a subspace of $M$ which contains $Z$ as a hyperplane. Since we are assuming that $Z$ is a maximal external flat, $\mathcal{P}_{E_{6,1}}(U) = PG(U) \cap \mathcal{P}_{E_{6,1}} \neq \emptyset$. Now $\mathcal{P}_{E_{6,1}}(U)$ must be a co-clique for otherwise there is a two space $X \leq U$ such that $PG(X) \subset \mathcal{P}_{E_{6,1}}$. But then as $Z$ is a hyperplane of $U$ it follows that $0 \neq X \cap Z \in \mathcal{P}_{E_{6,1}}$ contradicting the fact that $PG(Z)$ is an external flat. By (4.1) $|\mathcal{P}_{E_{6,1}}(U)| \leq q^8+q^4+1$. However, if $x \in \mathcal{P}_{E_{6,1}}$ then $x \in \mathcal{P}_{E_{6,1}}(\langle x, Z\rangle)$. It therefore follows that as $U$ ranges over the $\frac{q^t-1}{q-1}$ subspaces of $M$ which contain $Z$ as a hyperplane then $\mathcal{P}_{E_{6,1}}(U)$ partitions $\mathcal{P}_{E_{6,1}}$. But then we get

$$(q^8 + q^4 + 1)\frac{(q^9 - 1)}{(q-1)} \leq (q^8 + q^4 + 1)\frac{(q^t - 1)}{(q-1)}$$

from which it follows that $t \geq 9$ so that dim $Z \leq 18$.

**(4.4) Lemma.** *Assume that $Z \leq M$, dim $Z = 18$ and $PG(Z)$ is an external flat. Then for any subspace $U$ of $M$ containing $Z$ as a hyperplane $\mathcal{P}_{E_{6.1}}(U)$ is a co-clique with $q^8 + q^4 + 1$ points.*

*Proof.* Since there are $\frac{q^9-1}{(q-1)}$ such subspaces and

$$|\mathcal{P}_{E_{6,1}}| = (q^8 + q^4 + 1)\frac{(q^9 - 1)}{(q - 1)}$$

the average number of points in $\mathcal{P}_{E_{6,1}}(U)$ is $q^8 + q^4 + 1$ which is also the maximum possible number. Hence for any such $U$ we must have $|\mathcal{P}_{E_{6,1}}(U)| = q^8 + q^4 + 1$.

It would be very nice if such subspaces existed, whence examples of maximal co-cliques. However, as we show in our next result this is not the case.

**(4.5) Proposition.** *There are no subspaces $Z$, dim $Z = 18$, such that $PG(Z)$ is an external flat to $\mathcal{P}_{E_{6,1}}$.*

*Proof.* Assume $Z$ exists and let $H$ be any hyperplane of $M$ containing $Z$. Then $\mathcal{P}_{E_{6,1}}(H) = PG(H) \cap \mathcal{P}_{E_{6,1}}$ is partitioned by the points in $\mathcal{P}_{E_{6,1}}(U)$ as $U$ varies over the subspace of $H$ which contain $Z$ as a hyperplane. Since there are $\frac{q^8-1}{q-1}$ such subspaces and for each one $|\mathcal{P}_{E_{6,1}}(U)| = q^8 + q^4 + 1$ it follows that

$$|\mathcal{P}_{E_{6,1}}(H)| = (q^8 + q^4 + 1)\frac{(q^8 - 1)}{(q - 1)}$$

from which it follows that $H$ is an $F_{4,4}(q)$ hyperplane. However, $H$ was arbitrary. By (3.12) $F_{4,4}$ hyperplanes correspond to points which are nonzeros of a cubic form $T^*$ on the dual space $M^*$. This implies that every point of $PG(Z^*)$, $Z^* = \{f \in M^* | f(Z) = 0\}$, is a nonzero of $T^*$ which is impossible by remark (3.1).

An alternative construction of a maximal co-clique in the collinearity graph of $E_{6,1}(q)$ of cardinality $q^8 + q^4 + 1$ arises from the existence of an ovoid on the hyperbolic quadric $\mathcal{Q}^+(10, q)$. Thus, for the remainder of this section assume that there exists an ovoid in the quadric $\mathcal{Q}^+(10, q)$.

**(4.6) Lemma.** *Let $x \in \mathcal{P}_{E_{6,1}}$. There exists a collection $\mathcal{C}$ of $q^4+1$ symps containing $x$ such that for $S, S' \in \mathcal{C}, S \cap S' = \{x\}$.*

*Proof.* Recall from (3.7) that the incidence geometry whose points are the symps of $(\mathcal{P}_{E_{6,1}}, \mathcal{L}_{E_{6,1}})$ and whose lines are the maximal $PG(4, q)$ is also isomorphic to $(\mathcal{P}_{E_{6,1}}, \mathcal{L}_{E_{6,1}})$. Moreover, a "symp" in this incidence geometry consist of all the symps on a point. Thus, if we take the collection of symps on $x$ and define two to be adjacent if they meet in a $PG(4, q)$ then this is the orthogonality graph of the quadric $\mathcal{Q}^+(10, q)$. By our assumption regarding the existence of an ovoid in this quadric, we can find a collection of $q^4 + 1$ non-adjacent symps on $x$, that is, symps which, pairwise, meet in just $x$.

**(4.7) Remark.** The symps in $\mathcal{C}$ cover $\Delta_{E_{6,1}}(x)$. Also, if $T$ is a symp on $x, T \notin \mathcal{C}$ then $T$ is adjacent to $q^3+1$ members of $\mathcal{C}$ (that is, $T \cap S$ properly contains $\{x\}$). This follows from the fact that $\mathcal{C}$ is an "ovoid" in the "quadric" consisting of the symps on $x$.

**(4.8) Theorem.** *Assume an ovoid exists on the quadric $\mathcal{Q}^+(10,q)$. Then there is a co-clique in the collinearity graph of $E_{6,1}(q)$ with $q^8+q^4+1$ points.*

*Proof.* Let $x \in \mathcal{P}_{E_{6,1}}$. By (4.6) we can find a collection $\mathcal{C}$ of symps containing $x$ which pairwise meet in just $x$. Now in each $S \in \mathcal{C}$ we can find an ovoid $O_S$ containing $x$. Suppose $S_i \in \mathcal{C}, i = 1,2$ and $y_i \in O_{S_i}, y_i \neq x, i = 1,2$. Then by (3.14) $y_1, y_2$ are non-collinear. It then follows that

$$C = \bigcup_{S \in \mathcal{C}} O_S$$

is a co-clique in $\mathcal{P}_{E_{6,1}}$. Moreover,

$$|C| = |\mathcal{C}| \times |O_S \setminus \{x\}| + 1 = (q^4+1) \times q^4 + 1 = q^8 + q^4 + 1.$$

**(4.9) Remark.** It is easy to see that for a co-clique $C$ constructed as in (4.8) and a point $y \in \mathcal{P}_{E_{6,1}} \setminus C$ that $|\Delta_{E_{6,1}}(y) \cap C| = q^3+1$ as remarked in (4.2). There are two cases to consider: (i) $y \in \Delta_{E_{6,1}}(x)$ and (ii) $y \notin \Delta_{E_{6,1}}(x)$. In (i) $y$ belongs to an $S \in \mathcal{C}$ and in this case $\Delta_{E_{6,1}}(y) \cap C = \Delta_{E_{6,1}}(y) \cap S$ as cardinality $q^3+1$. Assume (ii) and suppose that $S \in \mathcal{C}$ and $\Delta_{E_{6,1}}(y) \cap S \neq \emptyset$. Then $\Delta_{E_{6,1}}(y) \cap S$ is a maximal singular subspace $S$ and hence contains a unique point of $C \cap S$ different from $x$. Set $T = S_{E_{6,1}}(x,y)$. The symps $S \in \mathcal{C}$ for which $\Delta_{E_{6,1}}(y) \cap S \neq \emptyset$ are precisely those which are adjacent to $T$. By (4.7) there are $q^3+1$ such symps from which it follows that $|\Delta_{E_{6,1}}(y) \cap C| = q^3+1$ as claimed.

# References


[1] M. Aschbacher, *The 27-dimensional module for $E_6$, I,* Inventiones Mathematicae **89** (1987), 159–184.

[2] A. Blokhuis and G.E. Moorhouse, *Some p–ranks related to Orthogonal Spaces,* J. of Algebraic Combinatorics **4** (1995), 529–551.

[3] A. E. Brouwer, A. M. Cohen and A. Neumaier, "Distance Regular Graphs", Springer-Verlag, Berlin, 1989.

[4] A. Cohen, *Point-Line Spaces Related to Buildings* in "Handbook of Incidence Geometry" (F. Buckenhout ed.), Elsevier, New York 1995.

[5] A. Cohen and B. Cooperstein, *A characterization of some geometries of exceptional Lie type,* Geometriae Dedicata **15** (1983), 73–105.

[6] J.H. Conway, P.B.Kleidman and R.A. Wilson, *New Families of ovoids in* $O_8^+$, Geometriae Dedicata **26** (1988), 157–170.

[7] B. Cooperstein, *A characterization of some Lie incidence geometries*, Geometriae Dedicata **6** (1977), 205–258.

[8] B. Cooperstein, *A note on tensor products of polar spaces over finite fields*, Bull. Belg. Math. Soc. **2** (1995), 253-257.

[9] R. H. Dye, *Maximal sets of non-polar points of quadrics and symplectic polarities over* $GF(2)$, Geometriae Dedicata **44** (1992), 281–294.

[10] W. Kantor, *Ovoids and translation planes*, Can. J. of Mathematics **34** (1982), 1195–1207.

[11] G. Mason and E. Shult, *The Klein correspondence and the ubiquity of certain translation planes*, Geometriae Dedicata **21** (1986), 29–59.

[12] G.E. Moorhouse, *Root Lattice Constructions of Ovoids*, in "Finite Geometry and Combinatorics", Cambridge University Press, Cambridge, 1992.

[13] G. E. Moorhouse, *Ovoids from the* $E_8$ *root lattice*, Geometriae Dedicata **46** (1993), 287–297.

[14] J.-P. Serre, "A Course in Arithmetic", Springer-Verlag, New York, 1973.

[15] E. Shult, *The Non-existence of ovoids in* $O^+(10,3)$, J. of Combinatorial Theory Ser. A **51** (1989), 250–257.

[16] J.A. Thas, *Ovoids and spreads of finite classical polar spaces*, Geometriae Dedicata **10** (1981), 135–144.

Bruce N. Cooperstein
University of California, Santa Cruz
Dept. of Mathematics,
Santa Cruz, CA 95064, U.S.A
e-mail: coop@cats.ucsc.edu

Trends in Mathematics, © 1998 Birkhäuser Verlag Basel/Switzerland

# The Residually Weakly Primitive Geometries of the Janko Group $J_1$

Harald Gottschalk and Dimitri Leemans*

**Abstract**

We classify all firm, residually connected coset geometries, on which the group $J_1$ acts as a flag-transitive automorphism group fulfilling the primitivity condition RWPRI: For each flag $\mathcal{F}$, its stabilizer acts primitively on the elements of some type in the residue $\Gamma_{\mathcal{F}}$. We demand also that every residue of rank two satisfies the intersection property.

## 1 Introduction

Inspired by JACQUES TITS' early geometric interpretation of the exceptional complex Lie groups beginning in 1954 (see [23] and [25]), FRANCIS BUEKENHOUT generalized in [2] certain aspects of this theory in order to achieve a combinatorial understanding of all finite simple groups. Since then, two main traces have been developed in diagram geometry. One is to try to classify geometries over a given diagram, mainly over diagrams extending buildings (see e. g. [12] for a survey and [24] for the theory of buildings). Another trace is to classify coset geometries for a given group under certain conditions. Rules for such classifications have been stated by BUEKENHOUT in [5] and [6]. These guidelines recently led to collections of group-geometry-pairs by BUEKENHOUT, CARA and DEHON [9], BUEKENHOUT, DEHON and LEEMANS [10], [11], and GOTTSCHALK [16]. The present paper continues these collections for the sporadic group $J_1$. The study of this particular group has been proposed to the authors by FRANCIS BUEKENHOUT.

Several papers have been written on geometries for $J_1$. These papers mostly deal with the geometry of $J_1$ related to its permutation representation on 266 points, first described in detail by LIVINGSTONE [20], coming from JANKO's original construction of $J_1$ as a subgroup of $GL_7(11)$ and $G_2(11)$ [19]. We refer also to [27] for a study of $J_1$ as a subgroup of $PGL_7(11)$. Geometries related to this representation can be found, e. g., in [1], [2], [3], [4], [12], [17], [18] and [21].

A classification of a somewhat more general concept of amalgam for $J_1$ has been achieved by TSARANOV [26]. He also shows that amalgams of rank at least

*We gratefully acknowledge financial support from the "Fonds National de la Recherche Scientifique de Belgique".

five for $J_1$ cannot occur. We will rely on his classification in the case of geometries of rank at least four.

The paper is organized as follows. In the next section we recall the basic concepts. The third section gives an outline of the classification of geometries since this proof is lengthy and partially done with help of the computer. The results of this (constructive) process are contained in the last section, that is, we describe the geometries by giving their diagrams and the corresponding amalgams of subgroups of $J_1$.

**Acknowledgement.**
We would like to thank FRANCIS BUEKENHOUT and JOHANNES UEBERBERG for fruitful discussions and helpful comments on the original manuscript.

## 2 Basic definitions and preliminaries

In this section, we recall the basic notion of a coset geometry and state the conditions under which we classify such geometries in this paper. A general reference for diagram geometries and their properties is [7].

Let $I = \{0, 1, \ldots, n-1\}$ be a finite set, called the *type set.* Its elements are called *types.* Let $G$ be a group and $(G_i)_{i\in I}$ be a collection of distinct subgroups of $G$, and let $X := \{gG_i : g \in G, G_i \in (G_i)_{i\in I}\}$ be the set of their cosets. We define a *pregeometry* $\Gamma = \Gamma(G; (G_i)_{i\in I}) = (X, t, *)$ provided with a *type function* $t : gG_i \mapsto i$ and an *incidence relation* $* \subset X \times X$, such that

$$gG_i * hG_j :\Leftrightarrow gG_i \cap hG_j \neq \emptyset.$$

The number $n = |I|$ is called the *rank* of $\Gamma$. A *flag* $\mathcal{F}$ of $\Gamma$ is a set of pairwise incident elements, and $t(\mathcal{F}) := \{t(x) : x \in \mathcal{F}\}$ is called its *type.* A flag $\mathcal{C}$ with $t(\mathcal{C}) = I$ is called a *chamber.* Then $\Gamma$ is called a (*coset*) *geometry* provided that any flag is contained in a chamber. We call a geometry *firm* (resp. *thin, thick*) if any flag is contained in at least two (resp. exactly two, at least three) chambers.

The *residue* of a flag $\mathcal{F}$ of $\Gamma$ is the geometry $\Gamma_{\mathcal{F}}$ consisting of the elements of $\Gamma$ incident with all elements of $\mathcal{F}$, together with the restricted type-function and induced incidence relation. Let $\mathcal{F}$ be a flag of type $J \subset I$. Then $\Gamma_{\mathcal{F}}$ is a geometry over the typeset $I - J$. We set $\Gamma_{\emptyset} := \Gamma$.

The *J-truncation* of $\Gamma$ is the geometry consisting of the elements of type $j \in J$, together with the restricted type-function and induced incidence relation. In group-geometry terms, the $J$-truncation of $\Gamma(G; (G_i)_{i\in I})$ is the geometry $\Gamma(G; (G_j)_{j\in J})$.

A coset geometry $\Gamma$ is called *residually connected* if the incidence graph of every residue of rank at least two is connected.

For any $\emptyset \neq J \subset I$ we set $G_J := \bigcap_{j\in J} G_j$, $B := G_I$ and $G_{\emptyset} := G$. Then we call $\mathcal{L}(\Gamma) := \{G_J : J \subset I\}$ the *sublattice* (of the subgroup lattice of $G$) *spanned* by the collection $(G_i)_{i\in I}$. The group $B$ is said to be the *Borel subgroup* of $\mathcal{L}(\Gamma)$. We say that $\mathcal{L}(\Gamma)$ is *strongly boolean* if, for any two elements of $\mathcal{L}(\Gamma)$, their lowest upper bound is the subgroup that they generate in $G$.

A lattice $\mathcal{A} = \{G_J : \emptyset \neq J \subset I\}$ is also called an *amalgam* (see e. g. [26]). Therefore $\mathcal{L}(\Gamma)$ is also an amalgam.

Then we have the following condition to check whether a pregeometry $\Gamma$ is a residually connected geometry.

**Lemma 2.1** *[24] Let $\Gamma = \Gamma(G; (G_i)_{i\in I})$ be a pregeometry. Then $\Gamma$ is a residually connected geometry if and only if $\mathcal{L}(\Gamma)$ is strongly boolean.*

Clearly, if $\Gamma(G; (G_i)_{i\in I})$ is a (pre-)geometry, $G$ acts as an automorphism group on $\Gamma$ by left multiplication. The action involves a *kernel* $K$ which is the largest normal subgroup of $G$ contained in every $G_i$, $i \in I$. If the kernel is the identity, we say that $G$ *acts faithfully* of $\Gamma$. If the subgroup $G_i$ acts with a non-trivial kernel $K_i$ of the residue of the element $G_i$ of $\Gamma$, we describe $G_i$ as $\boxed{K_i}.G_i/K_i$. We call $G$ a *flag-transitive* automorphism group if $G$ acts transitively on the set of flags of type $J$ for any subset $J$ of $I$. However the lemma stated above imposes restrictions to the choice of the family $(G_i)_{i\in I}$, it does not guarantee $G$ to act flag-transitively[1]. A criterion for both properties, namely being a geometry and being flag-transitive is given by the following lemma.

**Lemma 2.2** *[8] Let $\Gamma = \Gamma(G; (G_i)_{i\in I})$ be a pregeometry, and let $\alpha : \mathcal{P}(I) - \{\emptyset\} \to I$ be a map, such that $J\alpha \subset J$ for every non-empty subset of $I$. Then $\Gamma$ is a flag-transitive geometry if and only if, for every $J \subset I$ with $|J| \geq 3$, we have*

$$\bigcap_{j\in J-J\alpha} G_j G_{J\alpha} = \left( \bigcap_{j\in J-J\alpha} G_j \right) G_{J\alpha}.$$

It has been seen in the past, that even flag-transitivity does not impose enough restrictions to reduce the total number of coset geometries for a given group to a reasonable number (see [6] for a survey on the history). We introduce some primitivity conditions in the following definition.

**Definition 2.3** Let $\Gamma = \Gamma(G; (G_i)_{i\in I})$ be a flag-transitive coset geometry.

1. We call $\Gamma$ (or the pair $(\Gamma, G)$) *primitive* (PRI) if $G_i$ is a maximal subgroup of $G$ for every $i \in I$. The geometry is called *residually primitive* (RPRI) if $\Gamma_{\mathcal{F}}$ is PRI for any flag $\mathcal{F}$.

2. We say that $\Gamma$ is *weakly primitive* (WPRI) if $G_i$ is maximal in $G$ for at least one $i \in I$. $\Gamma$ is said to be *residually weakly primitive* (RWPRI) provided that $\Gamma_{\mathcal{F}}$ is WPRI for every flag $\mathcal{F}$.

3. We call $\Gamma$ *quasi-primitive* (QPRI) if every $G_i$ is a quasi-maximal subgroup of $G$ namely, there is a unique maximal chain of subgroups from $G_i$ to $G$. We say that $\Gamma$ is *residually quasi-primitive* (RQPRI) provided that $Gamma_{\mathcal{F}}$ is QPRI for every flag $\mathcal{F}$.

[1] A counter-example is given, e. g., in [16]

We say that a coset geometry satisfies the *intersection property* $(IP)_2$ if every residue of rank two is either a partial linear space or a generalized digon. Note that this condition excludes all $2-(v,k,\lambda)$ designs, $\lambda \geq 2$, except the generalized digons.

We call $\Gamma$ *locally 2-transitive* and we write $(2T)_1$ for this, provided that the stabilizer $G_{\mathcal{F}}$ of any flag $\mathcal{F}$ of rank $n-1$, acts 2-transitively on the residue $\Gamma_{\mathcal{F}}$.

In this paper, we classify the geometries of $G = J_1$ under the following conditions. Let $\Gamma = \Gamma(G; (G_i)_{i \in I})$ be a coset geometry. The geometry must be firm, residually connected, it must satisfy the intersection property $(IP)_2$ and the group $G$ must act flag-transitively and residually weakly primitively on $\Gamma$.

Assume that $I$ is a set of types and that $(G, \Gamma)$ is a group-geometry pair. A *correlation* of $(G, \Gamma)$ is an automorphism of the incidence graph of $\Gamma$ mapping any two elements of equal type onto elements of equal type and normalizing the permutation group $G$. The group of all correlations of $(G, \Gamma)$ is call ed *Cor(G, Γ)*. The automorphism group or group of type-preserving correlations is called *Aut(G, Γ)*.

Throughout this paper, we use the notation of the ATLAS [14] for groups up to slight variations.

# 3 Outline of the proof

The classification of coset geometries for $J_1$ splits into two main parts. The first part is to determine all possible strongly boolean sublattices fulfilling the conditions RWPRI and (in the case of geometries of rank three and four) $(IP)_2$. For this part, we use the *subgroup pattern* of $J_1$, i. e., its subgroup lattice up to conjugacy. This subgroup pattern was first given by FRANCIS BUEKENHOUT in [5]. In 1985, in a private communication to BUEKENHOUT, PAHLINGS corrected that pattern. It is now implemented in the computer package GAP [22] as a table of marks. We give a picture of it at the end of the paper.

In a second step, we use the computer algebra package CAYLEY [13] to check whether our 'candidates' of rank at least three are flag-transitive. In the case of rank two geometries, the computer is used to test if a geometry is $(IP)_2$ and to compute its diagram when it is possible.

### Rank two geometries

Let $G_0$ and $G_1$ be two different subgroups of $J_1$. Then, by 2.2, $J_1$ acts as a flag-transitive automorphism group on $\Gamma = \Gamma(J_1; G_0, G_1)$. The condition RWPRI holds in $\Gamma$ if and only if at least one $G_i$ is a maximal subgroup of $J_1$ and $B = G_0 \cap G_1$ is maximal in $G_0$ and in $G_1$.

To find the 'candidates' for our classification, we start with a maximal subgroup $G_0$ of $J_1$. If a maximal subgroup $B$ of $G_0$ is maximal in a subgroup $G_1 \neq G_0$, then $\Gamma = \Gamma(J_1; G_0, G_1)$ is a firm and connected rank two geometry, such that $J_1$ acts flag-transitively on $\Gamma$ and the condition RWPRI holds. Thus, it remains to check our condition $(IP)_2$. Therefore, we try to construct a circuit of length four in the incidence graph of $\Gamma$ using the computer. For the remaining geometries, we have to determine their diagrams. We give references for every geometry that was previously known. The parameters of the other ones are determined with the help of the computer.

**Rank three geometries**

In the case of rank three geometries we have to proceed in another way to obtain all strongly boolean sublattices. A strong restriction is again imposed by RWPRI: The Borel subgroup $B$ has to be maximal in $G_{01}$, $G_{02}$ and $G_{12}$. Also, each $G_i$ must be a minimal supergroup of at least one $G_{ij}$, and $G_i$ is maximal in $J_1$ for at least one $i \in \{0,1,2\}$. This guarantees a path of maximal inclusions $\{J_1, G_0, G_{01}, B\}$ from $J_1$ to $B$.

We analyse all these paths and try to extend them to a strongly boolean lattice possibly corresponding to a rank three geometry such that RWPRI holds. Along with this process, we use the tables of marks (the subgroup patterns) of $J_1$, $L_2(11)$ and the Frobenius groups that are maximal subgroups of $J_1$, implemented in GAP [22]. We also use GAP to determine possible intersections between subgroups of $J_1$.

We check the condition $(IP)_2$ at this stage since the diagrams for rank two residues can be computed 'by hand' (for small groups) or can be found in [9] (for $L_2(11)$) and in [11] (for $2 \times A_5$, resp. $A_5$).

After this process, flag-transitivity has to be checked by the computer.

**Rank four geometries**

In the case of rank four geometries we rely on the classification by TSARANOV [26]. We only have to check the conditions $(IP)_2$ and RWPRI for his geometries. It turns out that only the examples one and four of his list (the only examples with non-trivial Borel subgroup $B$) satisfy these conditions. But these two examples are also known to be flag-transitive (see next section for references).

# 4 The results

In this section, we give, up to isomorphism, all geometries for $J_1$ that are firm, residually connected, that satisfy the condition $(IP)_2$ and on which $J_1$ acts flag-transitively and residually weakly primitively. We also give the corresponding sublattices. A thick stroke in such a lattice denotes a maximal inclusion. We state also whether our geometries satisfy the more special conditions PRI, RPRI, QPRI and RQPRI or not by writing 'PRI', 'RPRI', 'QPRI' or 'RQPRI' next to the diagram. We mention when a geometry satisfies $(2T)_1$. For a given geometry $\Gamma$, we give the correlation groups *Cor(G, $\Gamma$)* provided it is different from *Aut(G, $\Gamma$)*. The automorphism group *Aut(G, $\Gamma$)* is isomorphic to $J_1$ for every geometry so we do not mention it with the diagrams. We also mention when a geometry is a truncation of one of the geometries we obtained. For all known geometries, we state references. In the case where no reference is stated the geometry was found by the authors during the work on this paper.

**Rank two geometries**

**1.1**
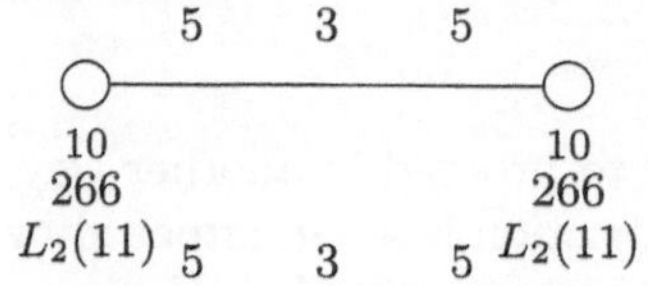


$Cor(\Gamma, G) \cong J_1 \times 2$, $B = A_5^a$, RPRI, $(2T)_1$, truncation of 2.2 and 3.2

**1.2**
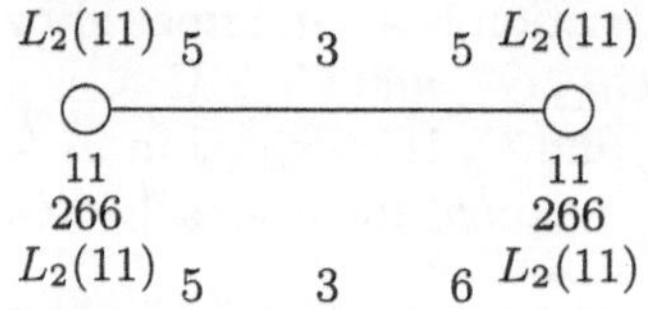


$Cor(G, \Gamma) = J_1 \times 2$, $B = 11 : 5$, RPRI, $(2T)_1$, references: [3], [4]

**1.3**
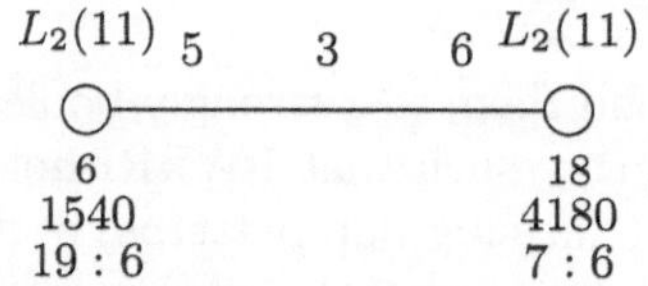


$B = 6$, RPRI

**1.4**
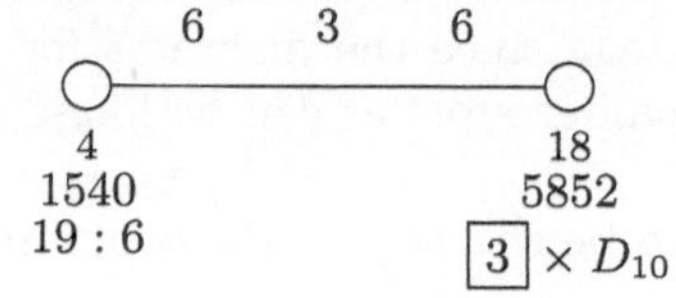


$B = 6$, QPRI

**1.5**
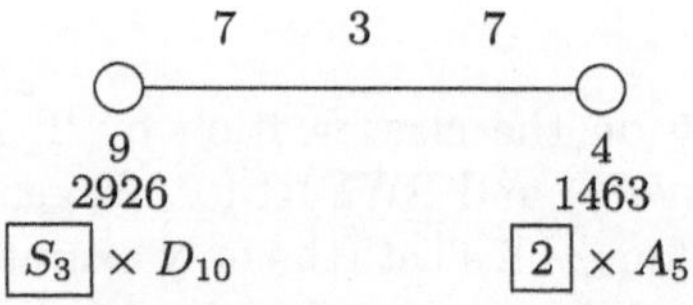


$B = D_{12}$, RPRI

Truncation of 2.1, 2.3, 3.1 and 3.2

**1.6**
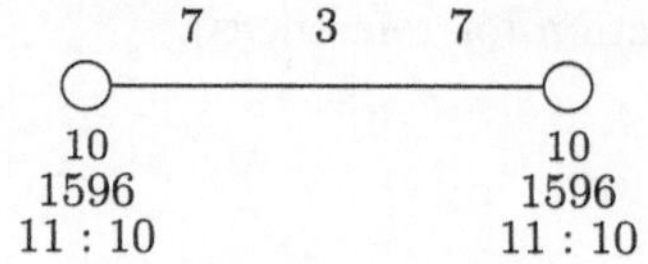


$Cor(G, \Gamma) = J_1 \times 2$, $B = 10$, RPRI, $(2T)_1$

**1.7**
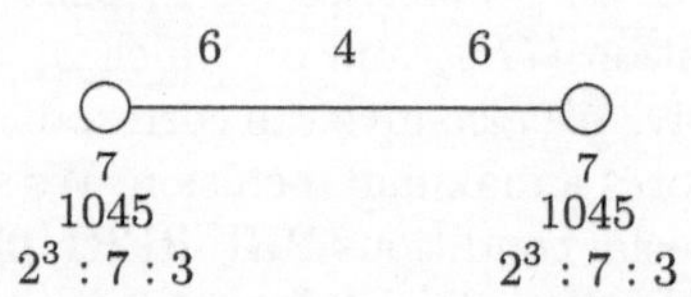

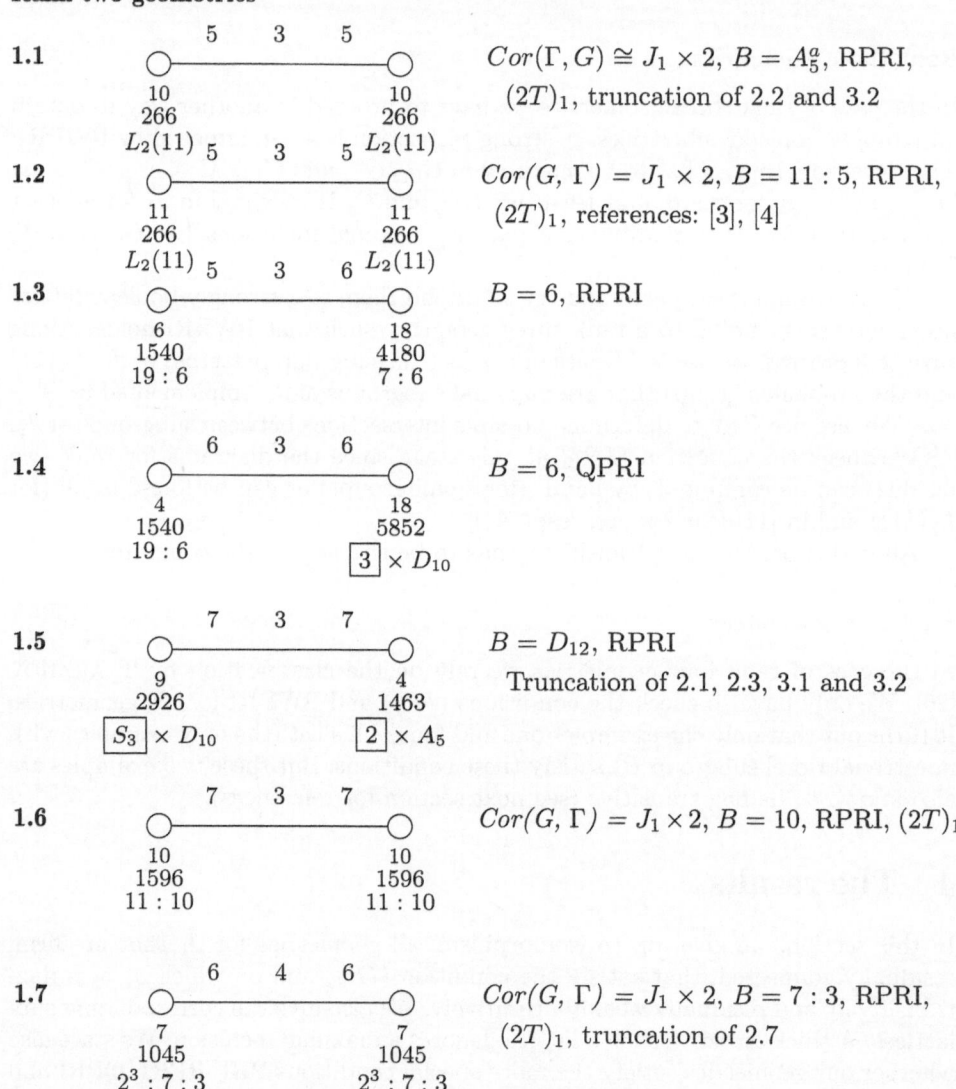

$Cor(G, \Gamma) = J_1 \times 2$, $B = 7 : 3$, RPRI, $(2T)_1$, truncation of 2.7

**1.8**

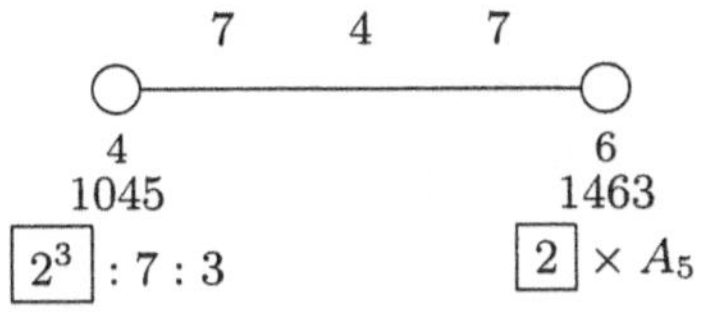

$$\underset{\substack{4 \\ 1045}}{\overset{7 \quad 4 \quad 7}{\bigcirc\!\!-\!\!-\!\!-\!\!-\!\!-\!\!-\!\!-\!\!\bigcirc}}\underset{\substack{6 \\ 1463}}{}$$

$\boxed{2^3}:7:3$        $\boxed{2}\times A_5$

$B = 2 \times A_4$, RPRI

**1.9**

$$\underset{\substack{6 \\ 4180}}{\overset{8 \quad 4 \quad 8}{\bigcirc\!\!-\!\!-\!\!-\!\!-\!\!-\!\!-\!\!-\!\!\bigcirc}}\underset{\substack{6 \\ 4180}}{}$$

$7:6$        $7:6$

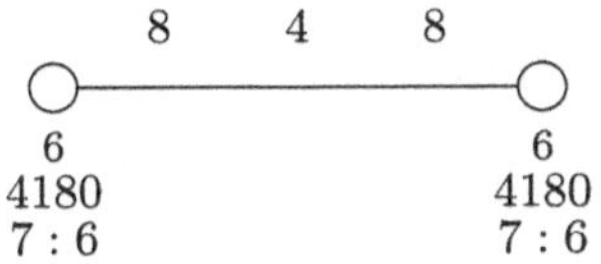

$Cor(G, \Gamma) = J_1 \times 2$, $B = 6$, RPRI, $(2T)_1$

**1.10**

$$\underset{\substack{1 \\ 266}}{\overset{7 \quad 5 \quad 8}{\bigcirc\!\!-\!\!-\!\!-\!\!-\!\!-\!\!-\!\!-\!\!\bigcirc}}\underset{\substack{11 \\ 1596}}{}$$

$L_2(11)$        $11:10 = \boxed{11{:}5}:2$

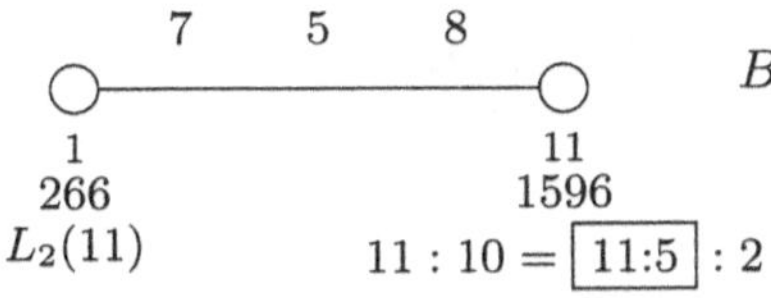

$B = 11:5$, RPRI, $(2T)_1$

References: [3], [4], [1]

**1.11**

$$\underset{\substack{1 \\ 266}}{\overset{8 \quad 5 \quad 8}{\bigcirc\!\!-\!\!-\!\!-\!\!-\!\!-\!\!-\!\!-\!\!\bigcirc}}\underset{\substack{10 \\ 1463}}{}$$

$L_2(11)$        $2 \times \boxed{A_5}$

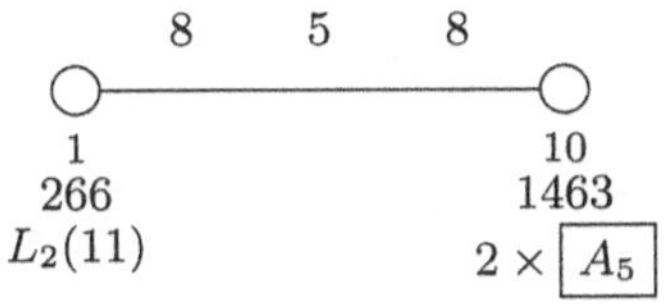

$B = A_5^a$, RPRI, $(2T)_1$

Livingstone graph, Ref.: [20], [1]

Truncation of 2.1 and 3.1

**1.12**

$$\underset{\substack{5 \\ 2926}}{\overset{10 \quad 5 \quad 10}{\bigcirc\!\!-\!\!-\!\!-\!\!-\!\!-\!\!-\!\!-\!\!\bigcirc}}\underset{\substack{2 \\ 1463}}{}$$

$S_3 \times \boxed{D_{10}}$        $\boxed{2}\times A_5$

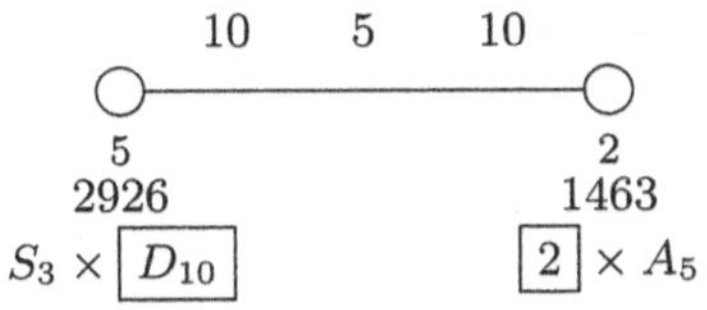

$B = D_{20}$, RPRI, $(2T)_1$

Truncation of 2.3, 3.1 and 3.2

**1.13**

$$\underset{\substack{1 \\ 1045}}{\overset{11 \quad 5 \quad 12}{\bigcirc\!\!-\!\!-\!\!-\!\!-\!\!-\!\!-\!\!-\!\!\bigcirc}}\underset{\substack{7 \\ 4180}}{}$$

$2^3:7:3$        $7:6 = \boxed{7{:}3}:2$

$B = 7:3$, RPRI, $(2T)_1$

Truncation of 2.6

**1.14**

$$\underset{\substack{1 \\ 1540}}{\overset{9 \quad 4 \quad 9}{\bigcirc\!\!-\!\!-\!\!-\!\!-\!\!-\!\!-\!\!-\!\!\bigcirc}}\underset{\substack{18 \\ 14630}}{}$$

$19:6$        $D_{12} = \boxed{6}:2$

$B = 6$

**1.15**

$$\underset{\substack{3 \\ 1540}}{\overset{6 \quad 3 \quad 6}{\bigcirc\!\!-\!\!-\!\!-\!\!-\!\!-\!\!-\!\!-\!\!\bigcirc}}\underset{\substack{18 \\ 7315}}{}$$

$19:6$        $\boxed{2}\times A_4$

$B = 6$

**1.16**

$$\underset{\substack{3 \\ 1540}}{\overset{7 \quad 3 \quad 7}{\bigcirc\!\!-\!\!-\!\!-\!\!-\!\!-\!\!-\!\!-\!\!\bigcirc}}\underset{\substack{18 \\ 7315}}{}$$

$19:6$        $\boxed{2}\times A_4$

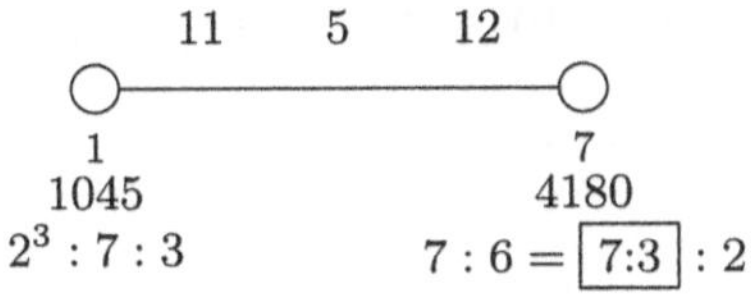

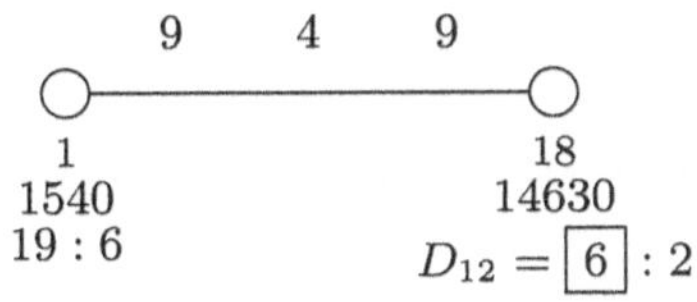

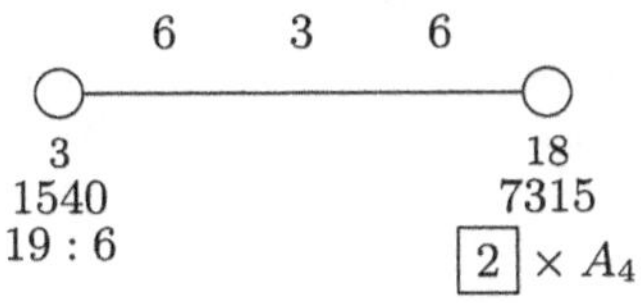

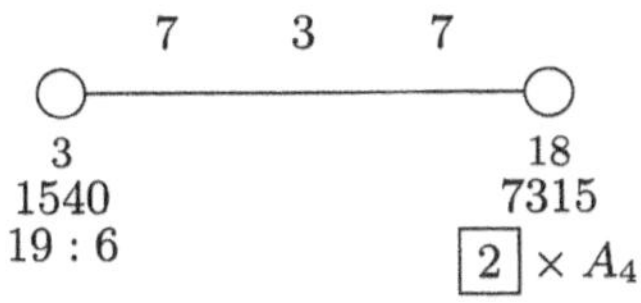

$B = 6$

**1.17** 15 5 16, ○———○, $B = 6$, $(2T)_1$

| | |
|---|---|
| 1 | 6 |
| 4180 | 14630 |
| $7:6$ | $D_{12} = \boxed{6} : 2$ |

**1.18** 9 4 9, ○———○, $B = 6$, $(2T)_1$

| | |
|---|---|
| 3 | 6 |
| 4180 | 7315 |
| $7:6$ | $\boxed{2} \times A_4$ |

**1.19** 9 4 9, ○———○, $B = 6$, $(2T)_1$

| | |
|---|---|
| 3 | 6 |
| 4180 | 7315 |
| $7:6$ | $\boxed{2} \times A_4$ |

**1.20** 9 3 10, ○———○, $B = 6$, QPRI
Truncation of 2.6

| | |
|---|---|
| 4 | 6 |
| 4180 | 5852 |
| $7:6$ | $\boxed{3} \times D_{10}$ |

**1.21** 10 6 10, ○———○, $B = 10$, $(2T)_1$

| | |
|---|---|
| 1 | 10 |
| 1596 | 8778 |
| $11:10$ | $D_{20} = \boxed{10} : 2$ |

**1.22** 9 4 10, ○———○, $B = 10$, QPRI, $(2T)_1$

| | |
|---|---|
| 2 | 10 |
| 1596 | 5852 |
| $11:10$ | $S_3 \times \boxed{5}$ |

**Remarks:**

1) The group-geometry pairs 1.15 and 1.16 (resp. 1.18 and 1.19) are non-isomorphic.

2) There are two non-isomorphic strongly boolean lattices with $G_0$ (resp. $G_1$ and $G_{01}$) isomorphic to $19:6$ (resp. $7:6$ and 6). The first one is geometry number 1.3. The other one does not satisfy the $(IP)_2$ condition.

**Rank three geometries**

**2.1**

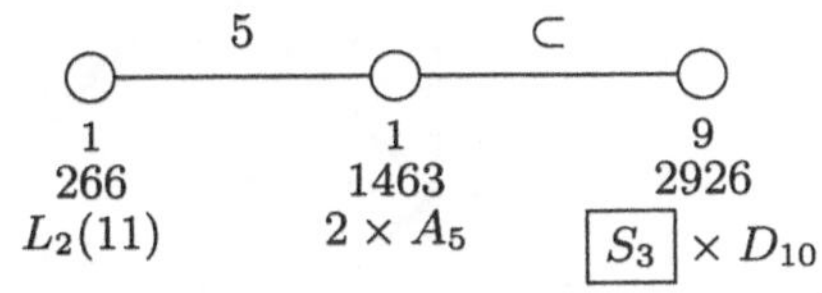


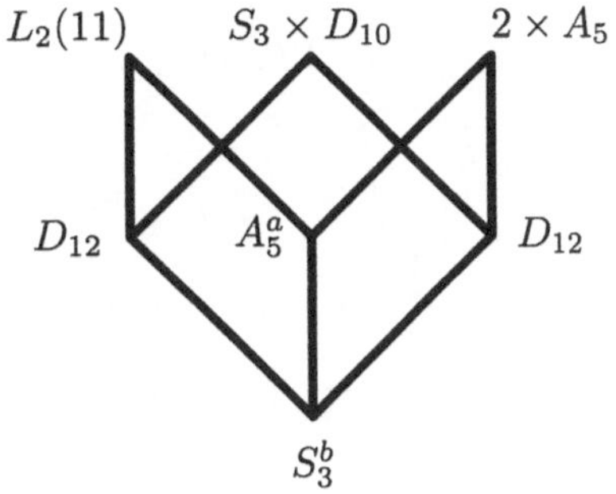


RPRI, References: [2], [1]
Truncation of 3.1

**2.2**

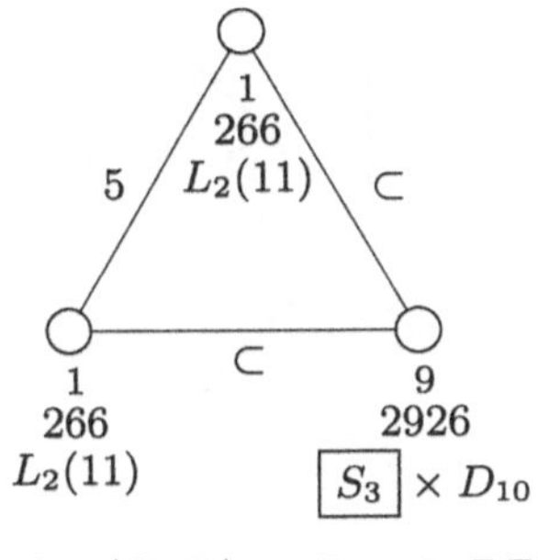


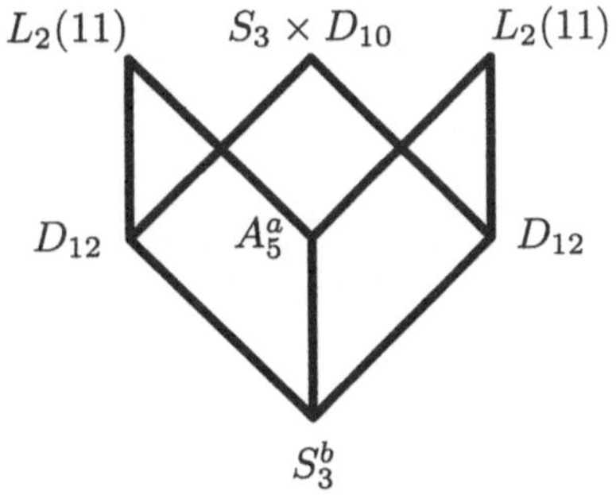


$Cor(G, \Gamma) = J_1 \times 2$, RPRI

**2.3**

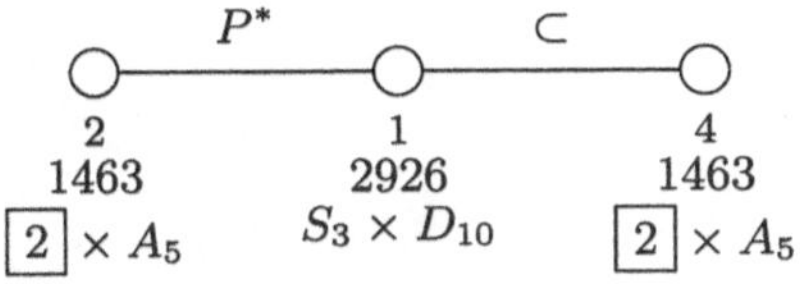


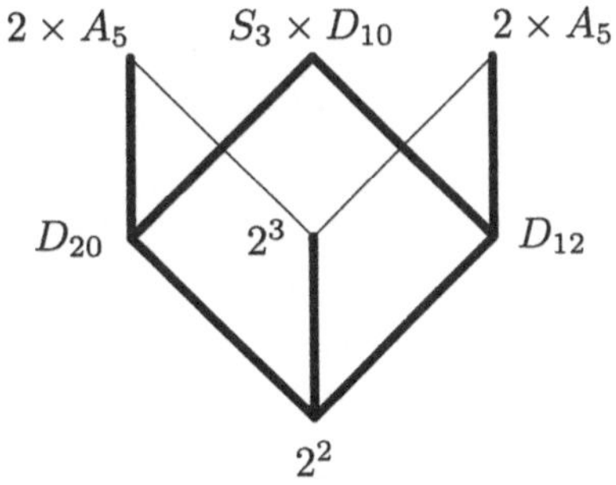


PRI, RQPRI
Truncation of 3.1

**2.4**

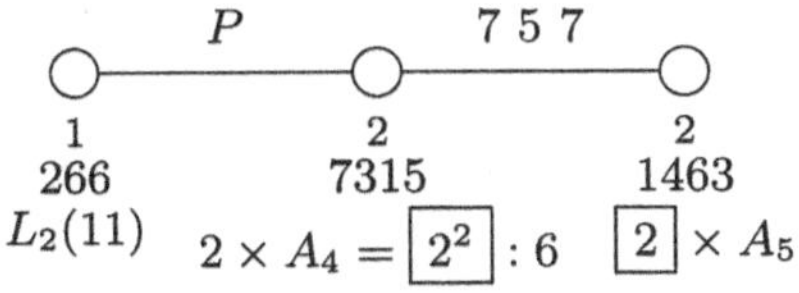


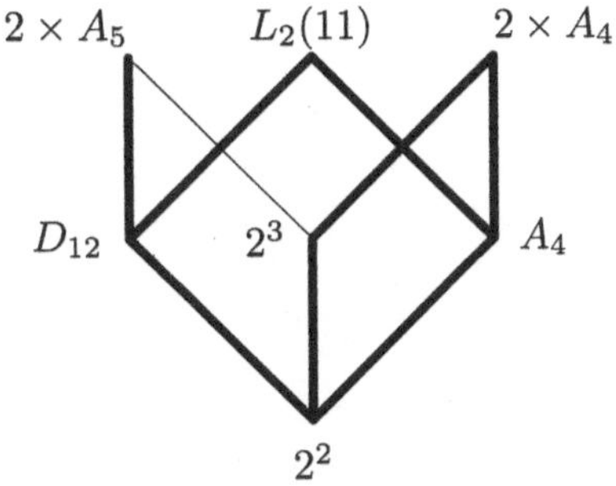


$(2T)_1$

**2.5**

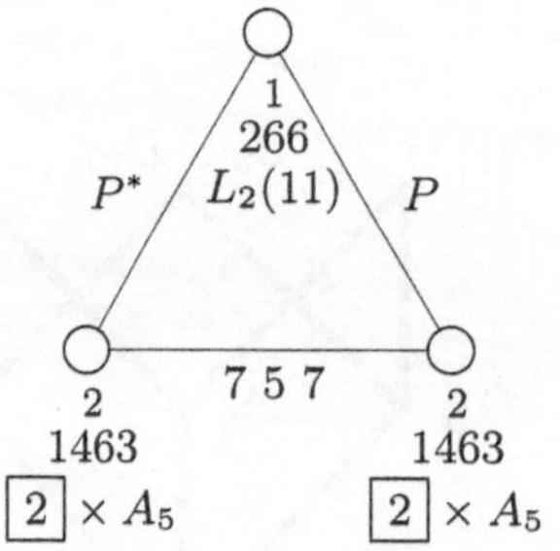


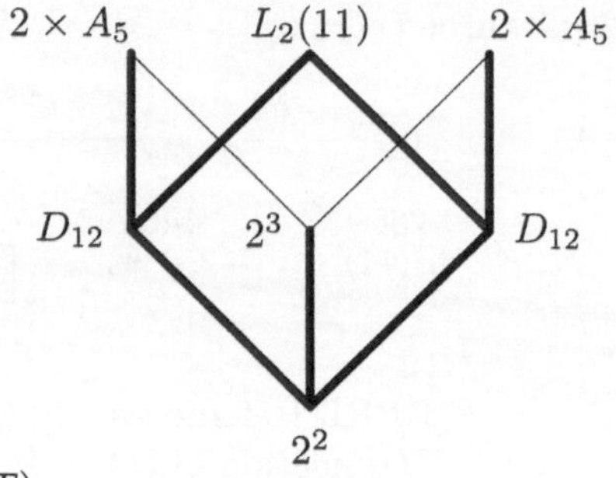


$Cor(G, \Gamma) = J_1 \times 2$, PRI, RQPRI, $(2T)_1$

**2.6**

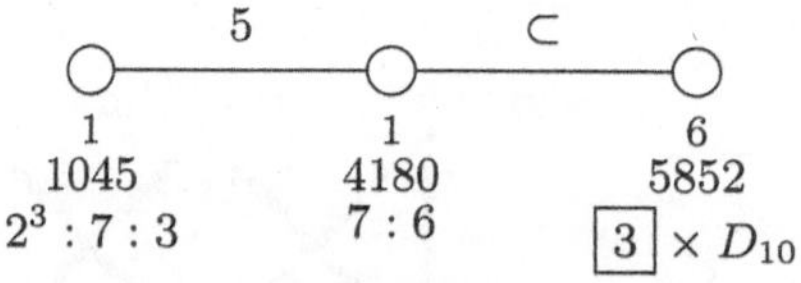


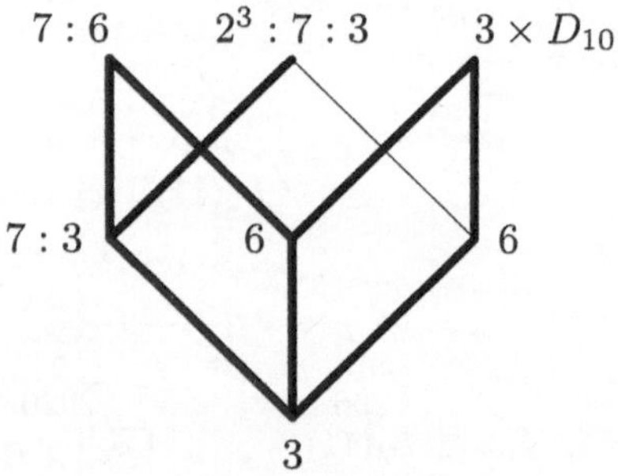


RQPRI, $(2T)_1$, Reference: [26]

**2.7**

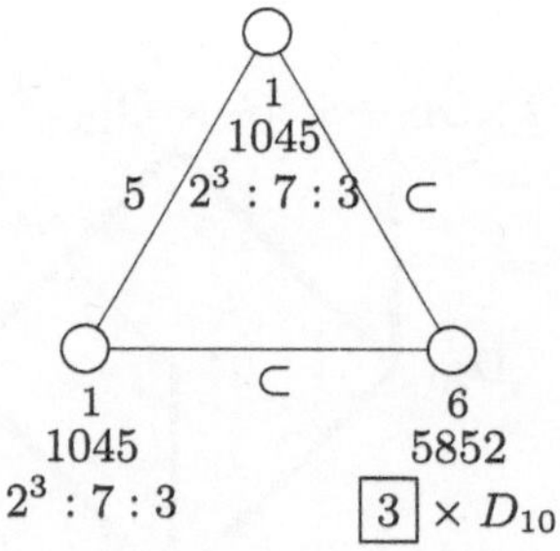


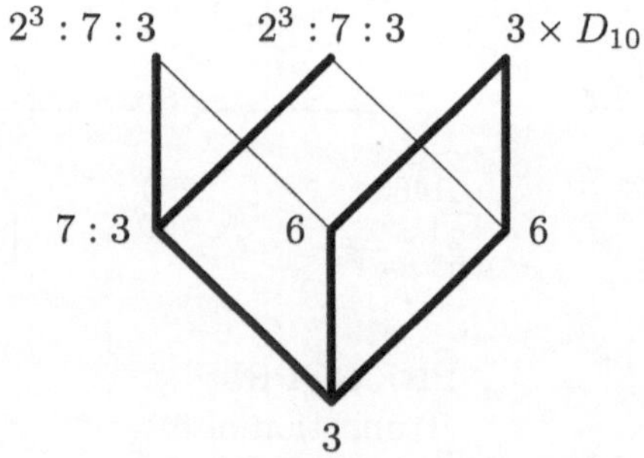


$Cor(G, \Gamma) = J_1 \times 2$, RQPRI

**Rank four geometries**

**3.1**

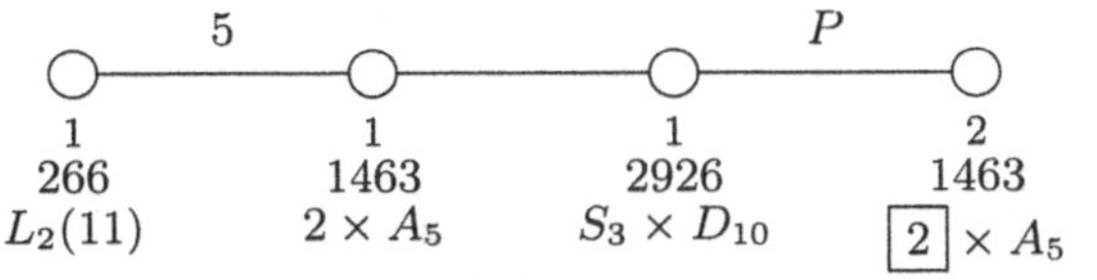


PRI, RQPRI, $(2T)_1$, References: [17], [18], [26], [1], [12]

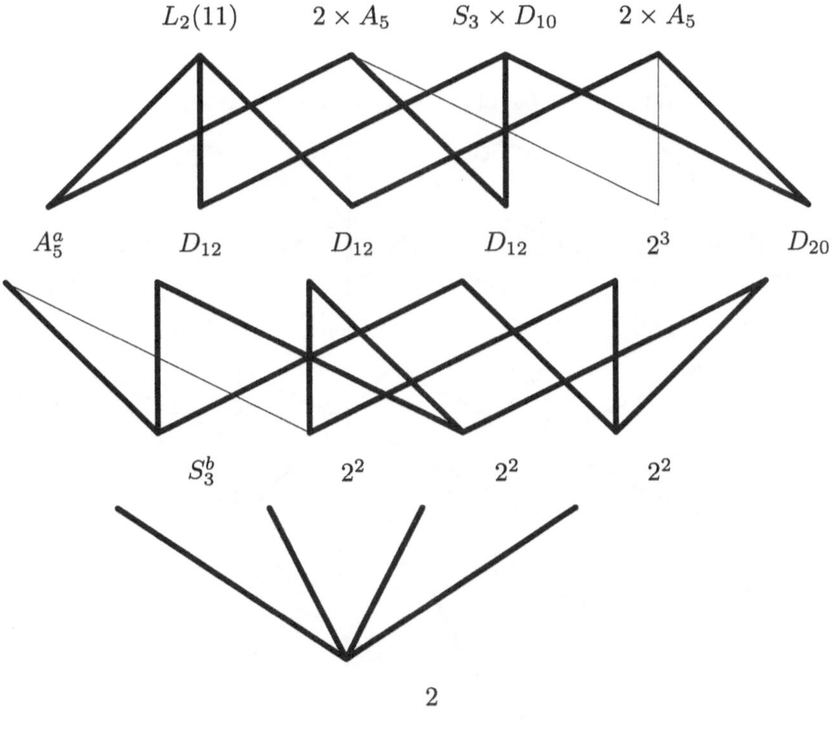

**3.2**

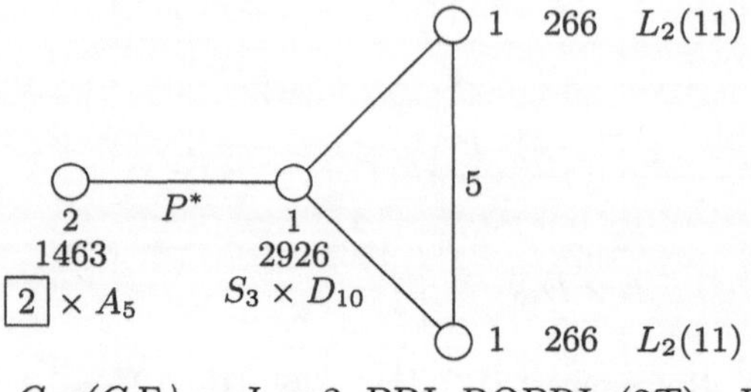


$Cor(G,\Gamma) = J_1 \times 2$, PRI, RQPRI, $(2T)_1$, References: [26], [12]

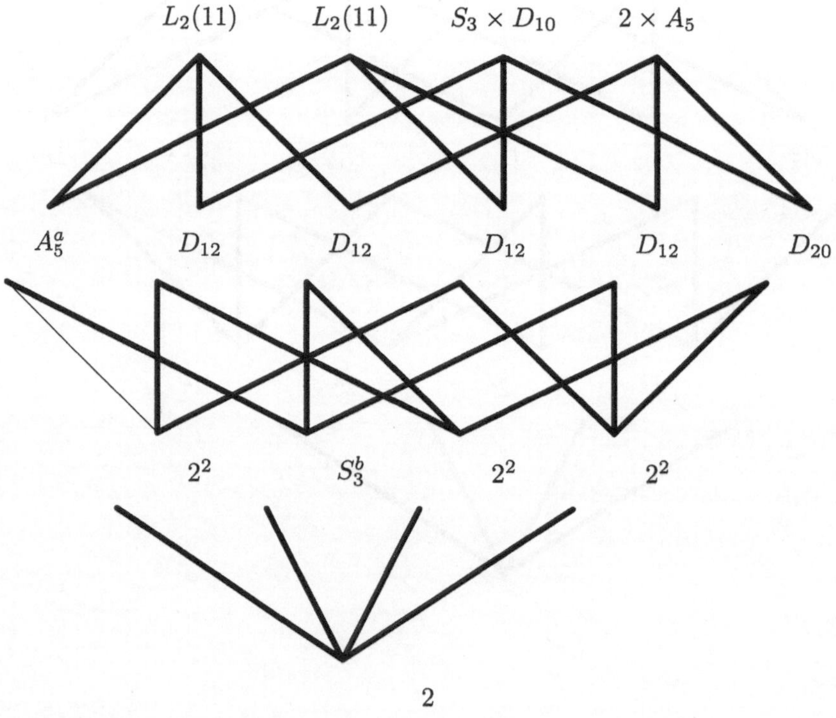

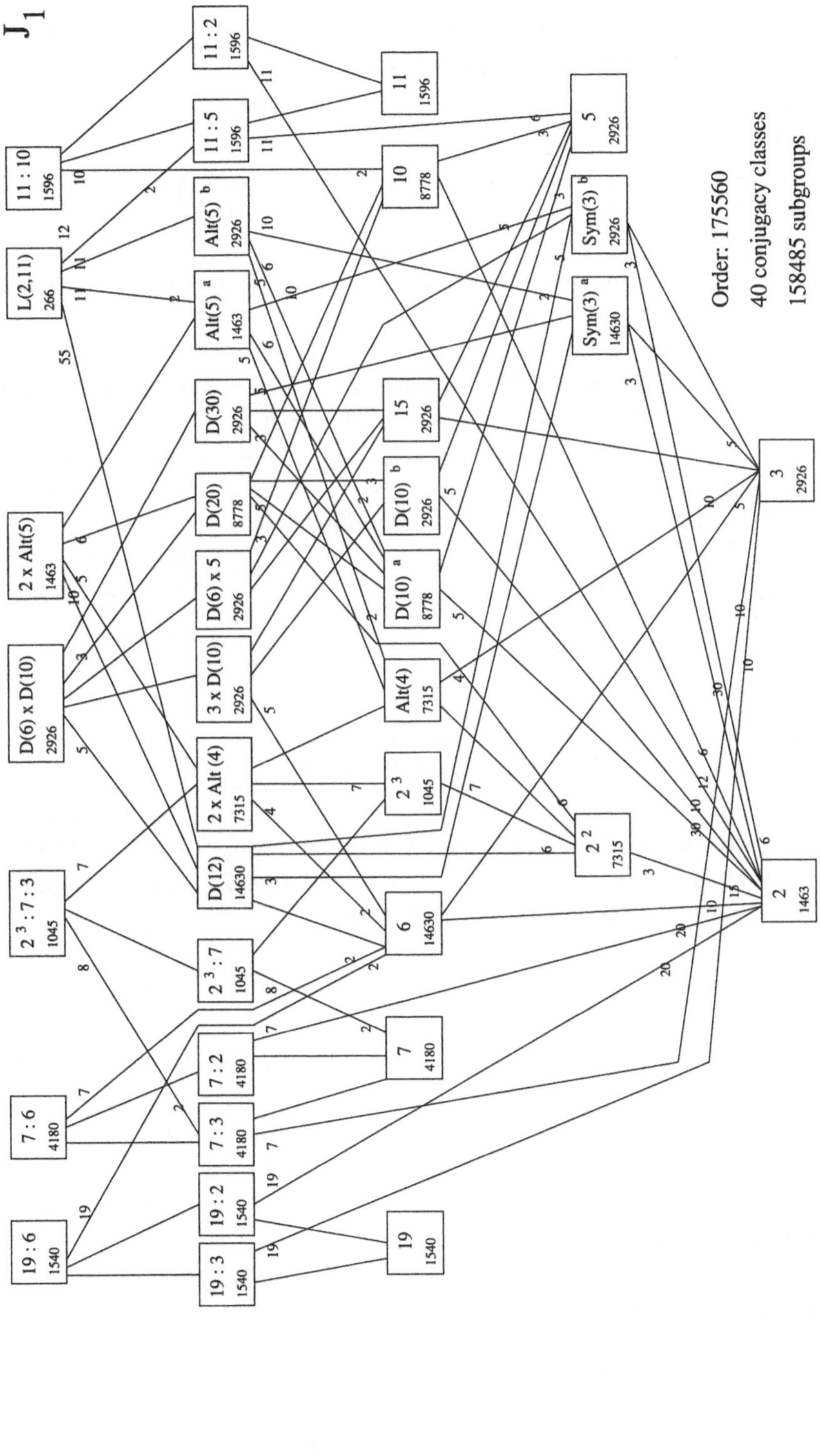
J1
Order: 175560
40 conjugacy classes
158485 subgroups
L(2,11) 266
11 : 10 1596
11 : 5 1596
11 : 2 1596
11 1596
Alt(5) a 1463
Alt(5) b 2926
10 8778
5 2926
Sym(3) a 14630
Sym(3) b 2926
2 x Alt(5) 1463
D(6) x D(10) 2926
D(30) 2926
D(20) 8778
D(6) x 5 2926
3 x D(10) 2926
15 2926
D(10) b 2926
D(10) a 8778
Alt(4) 7315
3 2926
2 3 : 7 : 3 1045
2 x Alt (4) 7315
D(12) 14630
2 3 : 7 1045
2 3 1045
2 2 7315
6 14630
2 1463
7 : 6 4180
7 : 2 4180
7 : 3 4180
7 4180
19 : 6 1540
19 : 2 1540
19 : 3 1540
19 1540

# References

[1] Brouwer, A.E., Cohen, A.M., Neumaier, A. *Distance-Regular Graphs.* Springer-Verlag, Berlin-Heidelberg-New York (1989).

[2] Buekenhout, F. *Diagrams for geometries and groups.* J. Comb. Th. A **27** (1979), 121-151.

[3] Buekenhout, F. *$(g, d^*, d)$-Gons*; in *Finite Geometries* (eds. N.L. Johnson et al.). Lect. Notes in Pure and Appl. Math. **82**, M. Dekker, New York, Basel (1983), 93-111.

[4] Buekenhout, F. *Diagram geometries for sporadic groups*; in *Finite groups coming of age.* AMS series Contemp. Math **45** (1985), 1-32.

[5] Buekenhout, F. *The geometry of the finite simple groups*; in *Buildings and the geometry of diagrams* (ed. L.A. Rosati). Lect. Notes in Math. **1181**, Springer-Verlag, Berlin-Heidelberg-New York (1986), 1-78.

[6] Buekenhout, F. *Finite groups and geometry: A view on the present state and the future*; in *Groups of Lie type and their geometries* (Ed. W.M. Kantor and L. Di Martino), Cambridge U.P. (1995), 35-42.

[7] Buekenhout, F. *Foundations of incidence geometry*; in *Handbook of Incidence Geometry* (ed. F. Buekenhout), Elsevier, Amsterdam (1995), 63-106.

[8] Buekenhout, F., Hermand M. *On flag-transitive geometries and groups.* Travaux de Mathématiques, Université Libre de Bruxelles, Fascicule I (1991), 45-78.

[9] Buekenhout, F., Cara, Ph., Dehon, M. *Geometries of small almost simple groups based on maximal subgroups.* Preprint.

[10] Buekenhout, F., Dehon, M., Leemans, D. *All geometries of the Mathieu group $M_{11}$ based on maximal subgoups.* Experiment. Math. **5** (1996), 101-110.

[11] Buekenhout, F., Dehon, M., Leemans, D. *An atlas of residually weakly primitive geometries for small groups.* Preprint.

[12] Buekenhout, F., Pasini, A. *Finite digram geometries extending buildings*; in *Handbook of Incidence Geometry* (ed. F. Buekenhout). Elsevier, Amsterdam (1995), 1143-1254.

[13] Cannon, J. *A language for group theory.* Departement of pure mathematics, University of Sydney (1982).

[14] Conway J.H., Curtis R.T., Norton S.P., Parker R.A. and Wilson R.A., *Atlas of finite groups.* Clarendon Press, Oxford (1985).

[15] Dehon, M. *Classifying geometries with* CAYLEY. J. Symbolic Computation **17** (1994), 259-276.

[16] Gottschalk, H. *A classification of geometries associated with $PSL(3,4)$.* Diplomarbeit, Gießen (1995).

[17] Ivanov, A.A. *On 2-transitive graphs of girth 5.* Europ. J. Combinatorics **8** (1987), 393-420.

[18] Ivanov, A.A., Shpectorov, S.V. *Geometries for sporadic groups related to the Petersen graph.* Comm. Algebra **16** (1988), 925-953.

[19] Janko, Z. *A new finite simple group with abelian 2-Sylow subgroups and its characterization.* J. Algebra **3** (1966), 147-186.

[20] Livingstone, D. *On a permutation representation of the Janko group.* J. Algebra **6** (1967), 43-55.

[21] Perkel, M. *A charaterization of $J_1$ in terms of its geometry.* Geom. Dedicata **9** (1980), 291-298.

[22] Schönert, M. et al. *GAP: Groups, Algorithms and Programming.* Lehrstuhl D für Mathematik, Aachen (1994).

[23] Tits, J. *Espaces homogènes et groupes de Lie exceptionnels.* Proc. Internat. Congr. Math., Amsterdam, **1** (1954), 495-496.

[24] Tits, J. *Buildings of spherical type and finite $BN$-pairs.* Lect. Notes in Math. **386**, Springer-Verlag, Berlin-Heidelberg-New York (1974).

[25] Tits, J. *Buildings and Buekenhout geometries.* In Ed. C.M.Campbell and L.Robertson. Proc. of Groups-St Andrews 1985. London Math. Soc. Lect. Notes Series **121**, Cambridge University Press, 352-358.

[26] Tsaranov, S.V. *Geometries and amalgams of $J_1$.* Comm. Algebra **18** (1990), 1119-1135.

[27] Whitelaw, T.A. *Janko's group as a collineation group in $PG(6, 11)$.* Proc. Camb. Phil. Soc. **63** (1967), 663-677.

Harald Gottschalk
Universität Halle-Wittenberg
FB Mathematik und Informatik
Institut für Algebra und Geometrie
D-06099 Halle, Germany
e-mail: `harald@coxeter.mathematik.uni-halle.de`

Dimitri Leemans
Département de Mathématique
Service de Géométrie, C.P.216
Université Libre de Bruxelles
B-1050 Bruxelles, Belgium
e-mail: `dleemans@cso.ulb.ac.be`

Trends in Mathematics, © 1998 Birkhäuser Verlag Basel/Switzerland

# The First Cohomology Group and Generation of Simple Groups

Robert M. Guralnick and Corneliu Hoffman

**Abstract**

We obtain an upper bound for the dimension of the first cohomology group of a finite group acting faithfully and irreducibly on a finite dimensional module. We discuss the connection between results of this nature and generation questions for finite simple groups.

## 1 Introduction and statement of results

Let $G$ be a finite group and $k$ a field. The study of $H^1(G,V)$ for $kG$-modules $V$ is intimately connected with questions of generation. Indeed, if $k = F_p$, then the size of $H^1(G,V)$ determines whether the semidirect product $VG$ can be generated by the same number of elements as $G$ or whether certain special types of generating sets for $G$ can be lifted to $VG$. We refer the reader to [Gr], [N], [Sc], [GS], [AG], [G1], and [Lu] for just some results of this nature.

Now assume that $V$ an irreducible $kG$-module which is faithful for $G$ (i.e., $C_G(V) = 1$). Prior to the classification of finite simple groups, there were no known bounds for the size of $H^1$. In [AGu], it was shown that $\dim H^1(G,V) < \dim V$. The method of proof was to show that the problem could be reduced to the case $G$ is simple. The fact that any simple group can be generated by 2 elements (a consequence of the classification) leads to the result that $\dim H^1(G,V) \leq \dim V$. Somewhat better generation results lead to the strict inequality. We will discuss some variations on this theme in the next section.

It seems that the most effective use of generation to bound $\dim H^1$ is to use generating subgroups with a large intersection which has nice cohomological properties relative to the module. This approach was used in [AG] to obtain conditions which force $H^1 = 0$. In [G1], a variation of these conditions was developed to obtain bounds on the size of $H^1$. There the estimate was improved to $\dim H^1(G,V) \leq (2/3)\dim V$. These conditions were also used in [GK] to obtain better bounds for alternating and symmetric groups. In particular, this answered a question of Gruenberg and Roggenkamp ([GR]) about when relation cores for alternating and symmetric groups decompose.

The second author has used these techniques to obtain better bounds for Chevalley groups – particularly in cross characteristic.

See [W] for other types of results relating complexes of subgroups and cohomology.

In this note, we will explore some connections between generation and cohomology. We will also improve the bound obtained in [G1] for sporadic groups. The precise form of the result is stated and proved in the final section.

Combining this result with those in [GK], [H] and the reduction theorem in [AGu] or [G1], we then obtain the following result:

**Theorem 1** *If $G$ is a finite group, $k$ a field with char$k = p$ and $V$ an irreducible $kG$ module which is faithful as a $G$-module, then:*

$$\dim H^1(G,V) \leq \frac{1}{2}\dim V$$

*and the inequality is strict unless $F^*(G) = L_2(2^n), n > 1$ with $V$ the natural module (or a Frobenius twist of it) or $F^*(G) = A_6$ and $V$ the 4-dimensional heart of the natural permutation module in characteristic* 3.

Thus, we see that the inequality is strict for characteristic $> 3$. Of course, $H^1 = 0$ if $k$ has characteristic 0. Indeed, one would suspect a stronger result based on results of Serre, Jantzen and McNinch (see [M], [S]).

**Conjecture 1** *If $V$ is faithful and irreducible and $k$ has positive characteristic $p > 2$, then* $\dim H^1(G,V) \leq 1/(p-2)\dim V$.

A special case of this conjecuture is that for simple algebraic groups $G$ and rational modules $V$ of dimension $< p-1$, then $H^1(G,V) = 0$. This is related to the work of Serre, Jantzen and McNinch mentioned above.

Indeed, there are no known examples where $V$ is absolutely irreducible and $\dim H^1(G,V) > 2$ and indeed, examples where $\dim H^1(G,V) = 2$ are not so common.

Thus, we restate the 1984 conjecture of the first author (see [G2]).

**Conjecture 2** *If $G$ is a finite group and $V$ is faithful and absolutely irreducible, then* $\dim H^1(G,V) \leq 2$.

Since we have a duality between cohomology and homology for finite groups (see [Br]), we obtain the following which has recently been used by Shalev [SS].

**Corollary 1** *If $G$ is a finite group, $k$ a field and $V$ an irreducible $kG$ module which is faithful as a $G$-module, then:*

$$\dim H_1(G,V) \leq \frac{1}{2}\dim V$$

*and the inequality is strict unless $G$ is of the type $L_2(2^n)$ with $V$ the natural module (or a twist of it) or $F^*(G) = A_6$ and $V$ the heart of the natural permutation module in characteristic 3 (in particular $V$ has dimension 4).*

One can ask for similar results for higher cohomology and homology groups. Holt [Ho] used the results of Aschbacher and Guralnick to obtain bounds on the second cohomology group.

We point out another corollary that improves the result in [G1] on minimal relation modules. Recall that a minimal relation module for a finite group $G$ is a module $M = R/[R, R]$ where $R$ is defined by a short exact sequence

$$1 \to R \to F \to G \to 1,$$

where $F$ is a free group with the same number of generators as $G$. It follows (see [Gr]) that $M$ is not uniquely defined but is up to local isomorphism. The proof is exactly the same as the somewhat weaker result in [G1] but using the better bound on cohomology. See [Gr] for many results on relation modules.

**Corollary 2** *If $G$ is a finite simple group and $M$ is a minimal relation module for $G$, then $M \oplus M$ has a free $\mathbb{Z}G$-summand.*

It followed already by [AGu] that some number of copies of $M$ have a free summand. The result in [G1] showed that 3 copies of $M$ have a free summand. Gruenberg (see [Gr]) has shown that $M$ itself will have no nonzero projective summand (using the fact that $G$ can be generated by 2 elements).

# 2 Generation and Cohomology

We first note some weaker consequences of some recent generation results and give a different proof of the Aschbacher-Guralnick result.

**Proposition 1** *Let $V$ be a faithful and irreducible $kG$-module. Then $\dim H^1(G, V) < \dim V$. If $k$ has odd characteristic or $G$ does not contain transvections, then $\dim H^1(G, V) \leq \dim V - 2$. In particular, this last inequality holds unless $k$ has characteristic 2 and $V$ is 2-dimensional.*

Proof. By the reduction theorem of [AGu], we may assume that $G$ is simple. Then by [MSW], $G$ is generated by an involution $x$ and another element $y$. Now any derivation $\delta$ from $G$ into $V$ is determined by its image on $x$ and $y$. Thus $\delta(y) \in V$ and $\delta(x) \in V_1$ where $V_1$ is the $-1$ eigenspace of $x$ (so in characteristic 2, it is the fixed point space of $x$). Clearly, this eigenspace is proper. Moreover, if $x$ is not a transvection (in characteristic 2) or $-x$ is not a reflection (in odd characteristic), we obtain the stronger inequality.

All finite reflection and transvection groups are known and so one can just compute the cohomology for those modules. Alternatively, we use the recent result [GKa] in place of [MSW]. Then the argument above allows us to use any involution. So we conclude the result is true unless every involution in $G$ satisfies the conditions above. It is straightfoward to see that if every involution of $G$ is either a transvection or the negative of a reflection, then $\dim V \leq 3$. One computes the cohomology directly for these small modules.

Since, one would expect (and there are some results of this type) that $V_1$ would be approximately 1/2 the dimension of $V$, this would probabilistically give the theorem. Since in fact most simple groups are generated by an involution and an element of order 3 (see [LM] and [LS]), one would expect a bound more like 1/6. Unfortunately, it seems difficult to guarantee these types of results. The recent result of Guralnick and Kantor [GKa] allows one to choose one of the generators at random – thus, if one element acts nicely (with respect to cohomology), one might expect some better estimates.

In fact, using just two elements to generate does not seem the best way to obtain bounds. Indeed, we shall use (cf.[GK]):

**Lemma 1** *Let $G$ and $V$ as before and suppose $T_1 \leq G$ and $T_2 \leq G$ so that: $G =< T_1, T_2 >$, $B \subset T_1 \cap T_2$, and $T_i =< B, S_i >$ where $B$ is cyclic, $S_i \triangleleft T_i$ and both $S_i$ are $q'$ groups. Then:*

*a)* $\dim Der(G, V) \leq \dim Der(T_1, V) + \dim Der_B(T_2, V)$

*b)* $\dim Der(T_i, V) \leq \dim V$

*c)* $Der_B(T_i, V) \cong [S_i, V]^B$.

# 3 Cohomology of Sporadic Groups

In this section, we will prove the following result. This in combination with the results mentioned in the introduction yield the main theorem.

**Theorem 2** *Let $G$ a sporadic simple group and $k$ a field. Then for any $kG$ module $V$ we have:*

$$\dim H^1(G, V) \leq C_G \dim V,$$

*where $C_G$ is as in Table 1.*

Let $G$ be a simple sporadic group and consider $p$ a particular prime (described below for each group – usually, the largest prime dividing $|G|$) such that $p$ divides $|G|$ but $p^2$ does not divide $|G|$. If the characteristic of the field equals $p$, then we can use Proposition 2.5 of [G1] and see that the dimension of the cohomology group is at most 1. Also we can bound from below the dimensions of the representations for sporadic groups. More precisely, we have that the dimension of the representations are at least $R(G)$ where $R(G)$ is as in Table 1 (cf. Proposition 5.38 of [KL] ).

So we may assume that characteristic of the field is not $p$. Then we will use Lemma 1. Let $P$ be a Sylow $p$-subgroup of $G$ and set $S = N_G(P)$. We know the structure of $S$ (for example [GL] lists all such normalizers). Then $S = P.A$ with $A$ cyclic of order $a$. We will pick $g \in N_G(A)$ and consider $H = \langle S, S^g \rangle$. If we can prove

| G | $C_G$ | R(G) | G | $C_G$ | R(G) |
|---|---|---|---|---|---|
| $M_{11}$ | 1/5 | 5 | He | 1/8 | 18 |
| $M_{12}$ | 1/5 | 6 | Ly | 1/22 | 110 |
| $M_{22}$ | 1/5 | 6 | Ru | 1/14 | 28 |
| $M_{23}$ | 1/11 | 11 | O'N | 1/15 | 31 |
| $M_{24}$ | 1/11 | 11 | Th | 1/15 | 48 |
| $J_1$ | 1/6 | 7 | HN | 1/9 | 56 |
| $J_2$ | 1/6 | 6 | HS | 1/5 | 20 |
| $J_3$ | 1/9 | 9 | $\mathrm{Fi}_{22}$ | 1/5 | 27 |
| $J_4$ | 1/14 | 110 | $\mathrm{Fi}_{23}$ | 1/11 | 234 |
| $\mathrm{Co}_3$ | 1/11 | 22 | $\mathrm{Fi}'_{24}$ | 1/14 | 702 |
| $\mathrm{Co}_2$ | 1/11 | 22 | McL | 1/5 | 21 |
| $\mathrm{Co}_1$ | 1/11 | 24 | B | 1/15 | 234 |
| Sz | 1/10 | 12 | M | 1/29 | 729 |

Table 1:

that $H = G$, then we can use Lemma 1 with $T_1 = S$, $T_2 = S^g$, $B = A$, $S_1 = P$, and $S_2 = P^g$. Note that that $[S_2, V]$ is free $A$ module since $A$ permutes without fixed points the nontrivial eigenspaces for $P$):

$$\dim \mathrm{H}^1(G, V) \leq \frac{1}{a} \dim V.$$

Thus, $C_G \leq 1/a$ in characteristic different from $p$. Combining this with our earlier comments yields $C_G = 1/\min(R(G),\ a)$.

We only need to show that $H = G$ Suppose this is not true, that is $H \subseteq G_0$ with $G_0$ a maximal subgroup of $G$. Then since $S, S^g$ are normalizers of Sylow $p$ subgroups of $G_0$, there is some $y \in G_0$ such that $S^g = S^y$. This means that $gy^{-1} \in N_G(S) = S \Rightarrow g \in yS \subseteq G_0$. In conclusion in order to get that $H = G$ it suffices to pick $g$ such that $g \notin G_0$ for any maximal subgroup $G_0$ that contains $S$. We will examine each sporadic group separately.

If $G = M_{11}$, then let $p = 11$. Then $a = 5$ and $N_G(A) = 5.4$. So we can take $g$ to be an element of order 4 from $N_G(A)$. Then the only possibility for $G_0$ is $L_2(11)$. However $N_{L_2(11)}(A) = 2$ hence the element $g$ cannot be contained in $L_2(11)$ and so we get $C_{M_{11}} = \frac{1}{5}$.

If $G = M_{12}$, let $p = 11$ . As before $S = 11.5$ but $N_G(A) = 5.4$. Again we can pick $g$ an element of order 4 and since the only choice for $G_0$ is $L_2(11)$, we see that $H = G$ and $C_{M_{12}} = \frac{1}{5}$.

If $G = M_{22}$ , again take $p = 11$. Again we have the same choices for $g$ and $G_0$ and so $C_{M_{22}} = \frac{1}{5}$.

If $G = M_{23}$ let $p = 23$,. Then $S = 23.11$ which is already a maximal subgroup. Therefore we only need to pick an element $g \in N_G(11) = 11.5$ to obtain $C_{M_{23}} = \frac{1}{11}$.

If $G = M_{24}$, let $p = 23$. Then $S = 23.11$ and $N_G(A) = 11.10$. There are two choices for $G_0 = L_2(23)$ or $M_{23}$. However, $N_{L_2(23)}(A) = D_{22}$ and $N_{M_{23}}(A) = 11.5$ so if we pick $g \in N_G(A)$ of order 10, $g$ cannot be in any of these two subgroups. The conclusion is that $C_{M_{24}} = \frac{1}{11}$.

If $G = J_1$, let $p = 19$. Then $S = 19.6$ is already a maximal subgroup and so we only need to show that $N_G(A) \neq A$. Since $A \subseteq L_2(11)$ and $N_{L_2(11)}(A) = D_{12}$ we get the desired conclusion. Therefore $C_{J_1} = \frac{1}{6}$.

If $G = J_2$, let $p = 7$. Then $S = 7.6$. Then $C_G(A)$ has order divisible by 12. The choices for $G_0$ are $U_3(3)$ and $L_3(2) : 2$. However, $N_{U_3(3)}(7) = 7.3$ hence $U_3(3)$ cannot contain $S$. Also we get that $C_{L_3(2):2}(A) = A$ hence we can take $g$ to be an element of $C_G(A)$ that will not be in $L_3(2) : 2$ and so $C_{J_2} = \frac{1}{6}$.

If $G = J_3$, let $p = 19$. Then $S = 19.9$. The only choice for $G_0$ is $L_2(19)$. However $|C_G(A)| = 27$ and $\left|C_{L_2(19)}(A)\right| = 9$ so we can find an element $g \in C_G(A) \setminus C_{L_2(19)}(A)$. Therefore $C_{J_3} = \frac{1}{9}$.

If $G = J_4$, let $p = 43$. Then $S = 43.14$ is already maximal. We know $|C_G(14)| = 84$ or 56. In particular, this centralizer is not contained in $S$, whence $C_{J_4} = \frac{1}{14}$.

If $G = McL$, let $p = 11$. Then $S = 11.5$ and $|C_G(A)| = 750$ or 25. The only candidates for $G_0$ are $M_{22}$ or $M_{11}$. However for both groups $|C_{G_0}(A)| = 5$ so we can pick the element $g$ that centralizes $A$ and does not belong to either one of the groups. Therefore $C_{McL} = \frac{1}{5}$.

If $G = HS$, let $p = 11$. Then $S = 11.5$ and $|C_G(5)| = 500$, 300 or 25. Again the possible $G_0$ are of the type $M_{11}$, or $M_{22}$ and in those two groups $A$ is self centralizing. Hence $C_{HS} = \frac{1}{5}$.

If $G = Co_3$, let $p = 23$. Then $S = 23.11$. Then $N_G(A) = 11.5 \times 2$ and $C_G(A) = 22$. The only choice for $G_0$ is $M_{23}$. However $N_{M_{23}}(A) = 11.5$ so we can take an element $g$ of order 2 that normalizes 11 and this is not in $M_{23}$. Therefore $C_{Co_3} = \frac{1}{11}$.

If $G = Co_2$, let $p = 23$. Then $S = 23.11$ and $N_G(11) = 11.10$. Again the only choice for $G_0$ is $M_{23}$ and $N_{M_{23}}(A) = 11.5$ so we can pick $g$ to have order 10 and it will not be contained in $M_{23}$. Therefore $C_{Co_2} = \frac{1}{11}$ .

If $G = Co_1$, let $p = 23$. Then $S = 23.11$ and $N_G(A) = (11.10) \times \Sigma_3$. The possible choices for $G_0$ are $Co_2$, $Co_3$, and $2^{11}M_{24}$. The normalizers of $A$ in these groups are respectively 11.10, $11.5 \times 2$ and 11.10 so it is enough to take $g$ of order 3 and it will not belong to any of these groups. In conclusion $C_{Co_1} = \frac{1}{11}$ .

If $G = Sz$, let $p = 11$. Then $S = 11.10$. Then $|C_G(A)| = 40$ or 20. Also $G_0$ can only be $3^5M_{11}$, $U_5(2)$, or $M_{12} : 2$. The first two groups contain no elements of order 10 so they cannot contain $S$. Also we have that $|C_{M_{12}:2}(A)| = 10$ hence we can find an element $g$ that centralizes $A$ and is not an element of $M_{12} : 2$. This yields $C_{Sz} = \frac{1}{10}$ .

If $G = He$, let $p = 17$. Then $S = 17.8$. Then $|C_G(8)| = 16$. The only candidate for $G_0$ is $Sp_4(4) : 2$. We have $\left|C_{Sp_4(4):2}(8)\right| = 8$, so our choice of $g \notin Sp_4(4) : 2$ is possible and it gives $C_{He} = \frac{1}{8}$ .

If $G = Ly$, let $p = 67$. Then $S = 67.22$ is already maximal so we only need an element $g$ outside of $S$ that normalizes $A$. We can see that $Ly$ has a maximal subgroup of the type $3^5(2 \times M_{11})$, hence a subgroup of the type $2 \times M_{11}$. Now it's easy to see that $N_{2\times M_{11}}(22) = 22.5$, hence we can find an element of order 5 that normalizes $A$. Therefore $C_{Ly} = \frac{1}{22}$ .

If $G = Ru$, let $p = 29$. Then $S = 29.14$. The only possibility for $G_0$ is $L_2(29)$. However $|C_G(A)| = 28$ and $\left|C_{L_2(29)}(A)\right| = 14$ hence we can find an element $g$ that normalizes $A$ and does not belong to $L_2(29)$. So $c_{Ru} = \frac{1}{14}$.

If $G = O'N$, let $p = 31$. Then $S = 31.15$. The only possibility for $G_0$ is $L_2(31)$ but we get that $|C_G(A)| = 45$ and $\left|C_{L_2(31)}(A)\right| = 15$ hence we can pick an element $g$ that centralizes $A$ and is not contained in $L_2(31)$. Therefore $C_{O'N} = \frac{1}{15}$.

If $G = Th$, let $p = 31$. Then $S = 31.15$ is already maximal and $|C_G(A)| = 30$ hence we can pick an element $g$ that centralizes $A$ and get $C_{Th} = \frac{1}{15}$ .

If $G = HN$, let $p = 19$. Then $S = 19.9$. The only choice for $G_0$ is $U_3(8) : 3_1$ (in the notation of [ATL]). Note also that $|C_G(A)| = 27$ and $\left|C_{U_3(8):3_1}(A)\right| = 9$ therefore we can pick an element $g$ that centralizes $A$ and get $C_{HN} = \frac{1}{9}$ .

If $G = Fi_{22}$ let $p = 11$. Then $S = 11.5$. the possibilities for $G_0$ are $2 \cdot U_6(2)$, $2^{10}.M_{22}$ and $M_{12}$. Choose $g \notin A$ of order 5 centralizing $A$. The orders of the centralizers in the three cases are 30, 5, 10 and hence $g$ will not be in any of them. Therefore $C_{Fi_{22}} = \frac{1}{5}$.

If $G = Fi_{23}$, let $p = 17$. Then $S = 17.16$ and the possible choices for $G_0$ are $S_8(2)$ and $S_4(4) : 4$. However $S_8(2)$ will contain no element of order 16 so it cannot contain $S$. Also $C_G(A)$ has order 32 while $C_{S_4(4):4}(A)$ has order 16. Hence we can find an element $g$ that centralizes $A$ and does not belong to $S_4(4) : 4$. Therefore $C_{Fi_{23}} = \frac{1}{16}$.

If $G = Fi'_{24}$, let $p = 29$. Then $S = 29.14$ which is already maximal. So we only need an element $g$ that normalizes $A$ which is not in $A$. We have that $|C_g(A)| \geq 42$, and so $C_{Fi'_{24}} = \frac{1}{14}$.

If $G = B$, let $p = 47$. Then $S = 47.23$ is already maximal. hence we need an element $g \in N_G(A) = (23.11) \times 2$. Using this element we get that $C_B = \frac{1}{23}$.

If $G = M$, let $p = 59$. Then $S = 59.29$ and $N_G(29) = ((29.14) \times 3).2$. Not all the maximal subgroups of the Monster are known, but we do know that the only possible subgroup of $G$ that will contain the 59 local subgroup is of the type $L_2(59)$. If such a subgroup exists then $N_{L_2(59)}(A) = 29.2$. Therefore we can find an element $g$ of order 7 that normalizes $A$ and is not contained in $L_2(59)$ hence $C_M = \frac{1}{29}$.

# References


[AG] J. Alperin and D. Gorenstein, A vanishing theorem for cohomology, Proc. Amer. Math. Soc. 32 (1972), 87–88.

[AJL] H. H. Andersen, J. Jorgensen and P. Landrock The Projective Indecomposable Modules of $SL(2, p^n)$, Proc. London Math. Soc., (3), 46 (1983), 38–52.

[AGu] M. Aschbacher and R. Guralnick, Some applications of the first cohomology group. J. Algebra 90 (1984), 446–460.

[ATL] J.H. Conway, R.T. Curtis, S.P. Norton, R.A. Parker and R.A. Wilson, An ATLAS of Finite Groups, Oxford University Press, Oxford, 1985.

[Br] K. Brown, Cohomology of Groups, Graduate Texts in Mathematics, 87. Springer-Verlag, New York-Berlin, 1982.

[GL] D. Gorenstein and R. Lyons, The local structure of finite groups of characteristic 2 type, Mem. Amer. Math. Soc., 276 (1983).

[Gr] K. W. Gruenberg, Relation modules of finite groups. Conference Board of the Mathematical Sciences Regional Conference Series in Mathematics, No. 25. American Mathematical Society, Providence, R.I., 1976.

[GR] K. W. Gruenberg and K. Roggenkamp, Decomposition of the augmentation ideal and of the relation modules of a finite group. Proc. London Math. Soc. (3) 31 (1975), 149–166; Proc. London Math. Soc. (3) 45 (1982), 89–96.

[G1] R. Guralnick, Generation of simple groups, J. Algebra 103 (1986), 381–401.

[G2] R. Guralnick, The dimension of the first cohomology group, in V. Dlab, P. Gabriel, and G. Michler, eds, Representation theory II, Groups and Orders, vol. 1178, Lecture Notes in Mathematics, 94–97, Springer-Verlag, Berlin, Heidelberg, New York, Tokoyo, 1986.

[GKa] R. Guralnick and W. Kantor, Probabilistic generation of finite simple groups, submitted.

[GK] R. Guralnick and W. Kimmerle, On the cohomology of the alternating and symmetric groups and decomposition of relation module, J. Pure Appl. Algebra 69 (1990), 135–140.

[GS] R. Guralnick and J. Saxl, Generating simple groups by conjugates, preprint.

[H] C. Hoffman, On the cohomology of the finite Chevalley groups, preprint.

[Ho] D. Holt, On the second cohomology group of a finite group. Proc. London Math. Soc. (3) 55 (1987), no. 1, 22–36.

[JP] W. Jones and B. Parshall, On the 1-cohomology of finite groups of Lie type, in Proceedings of the Conference on Finite Groups (Park City, Utah, 1975), pp. 313–328. Academic Press, New York, 1976.

[KL] P. Kleidman and M. Liebeck, The subgroup structure of the finite classical groups Cambridge University Press, (1990).

[L] A. Lucchini, A bound on the number of generators of a finite group, Arch. Math. (Basel) 53 (1989), 313–317.

[LS] M. W. Liebeck and A. Shalev, Classical groups, probabilistic methods, and the $(2,3)$-generation problem, Ann. of Math. 144 (1996) 77–125.

[LM] F. Lübeck and G. Malle, $(2,3)$-generation of exceptional groups, preprint.

[MSW] G. Malle, J. Saxl and T. Weigel, Generation of classical groups. Geom. Ded. 49 (1994) 85–116.

[M] G. McNinch, Dimensional criteria for semisimplicity of representations, Proc. London Math. Soc, to appear.

[N] M. Neubauer, On monodromy groups of fixed genus. J. Algebra 153 (1992), 215–261.

[Sc] L. Scott, Scott, Matrices and cohomology, Ann. of Math. 105 (1977), 473–492.

[S] J.-P. Serre, Sur la semi-simplicité des produits tensoriels de représentations de groupes, Invent. Math. 116(1994), 513–530.

[SS] D. Segal and A. Shalev, On groups with bounded conjugacy classes, preprint.

[W] P. J. Webb, A local method in group cohomology, Comment. Math. Helv. 62 (1987), 135–167.

Robert M. Guralnick,
University of Southern California,
Los Angeles, CA 90089-1113, USA
e-mail: `guralnic@math.usc.edu`

Cornelius Hoffman
University of Southern California,
Los Angeles, CA 90089-1113, USA
e-mail: `choffman@aludra.usc.edu`

Trends in Mathematics, © 1998 Birkhäuser Verlag Basel/Switzerland

# A Characterization of the Hall–Janko Group $J_2$ by a $c.L^*$–Geometry

Cécile Huybrechts*† and Antonio Pasini

**Abstract**

A c.U*–geometry is a geometry over the diagram c.L*, the point residues of which are finite dual unitals. Only one flag–transitive example is known. Its full automorphism group is $Aut(J_2)$, but $J_2$ also acts flag–transitively on it. We shall prove that this geometry is indeed the unique flag-transitive c.U*–geometry, thus obtaining a new geometric characterization of the Hall–Janko group $J_2$.

## 1 Introduction

We recall that a c.L*–*geometry* is a connected rank three incidence geometry of points, lines and planes in which plane–, line– and point–residues are respectively *circles* (namely, linear spaces with all lines of size 2), generalized digons and dual linear spaces. The diagram associated to such a geometry is pictured below.

(c.L*)

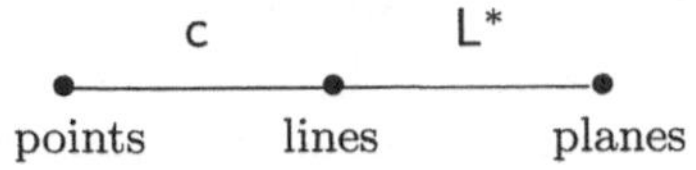


A c.U*–*geometry* (a c.$U_H^*$–*geometry*) is a c.L*–geometry in which point–residues are finite dual (hermitian) unitals. Many c.U*–geometries are known. As noticed in Huybrechts [17], the gluing construction of Buekenhout-Huybrechts-Pasini [7] is the easiest way of producing them. Indeed, a parallelism can be defined in every hermitian or Ree unital; moreover each circle with an even number of points admits a parallelism. Thus, any Ree unital and any hermitian unital of odd order can be glued with such a circle. Infinitely many c.U*–geometries are obtained in this way, but none of them is flag–transitive.

In fact, only one example of flag–transitive c.U*–geometry is known. It has been constructed in 1982 by M. Hermand in an unpublished work [14] supervised by F. Buekenhout and it is mentioned in Buekenhout's "catalogue" [3, (104)]. Following

*Chargé de Recherches du Fonds National Belge de la Recherche Scientifique

†This work was made possible thanks to support of the Fonds National Belge de la Recherche Scientifique

Buekenhout–Huybrechts [6], we call it the *Buekenhout–Hermand geometry.* Its flag–transitive groups of automorphisms are $J_2$ and $Aut(J_2)$, the latter being the full automorphism group. The Buekenhout–Hermand geometry has 100 points and its point–residues are dual hermitian unitals of order 3. Its collinearity graph is one of the natural graphs associated to $J_2$, namely the strongly regular graph on 100 vertices, degree 63 and parameters $(v, k, \lambda, \mu) = (100, 63, 38, 42)$. We refer to [6] for further details. In the present paper, we prove the following:

**Theorem 1.1** *The Buekenhout–Hermand geometry is, up to isomorphism, the unique flag–transitive* $\mathsf{c.U}^*_{\mathsf{H}}$*–geometry.*

The Buekenhout–Hermand geometry is strongly related with the extended generalized hexagon $EGH(J_2)$ for $J_2$ constructed by Buekenhout [2, 3]. Indeed, these two geometries do not only have the same collinearity graph and the same flag-transitive automorphism groups but also, as we will show in Section 3, each of them can be recovered from the other one. This connection will be especially useful to prove Theorem 1.1 since a a purely geometrical characterization of $EGH(J_2)$ has been obtained by Cuypers [10]. Note that, contrary to the Buekenhout-Hermand geometry, which is characterized by our Theorem 1.1, $EGH(J_2)$ is not uniquely determined by its rank 2 residues, even if flag–transitivity is assumed. Indeed, the universal cover of $EGH(J_2)$ is infinite, as its minimal circuit diagram is non-spherical.

As a by–product of the classification of the finite flag-transitive linear spaces by Buekenhout, Delandtsheer, Doyen, Kleidman, Liebeck and Saxl [5], all flag-transitive unitals are known. On this basis, we will also obtain the following improvement of Theorem 1.1:

**Corollary 1.2** *Up to isomorphism, the Buekenhout–Hermand geometry is the unique flag–transitive* $\mathsf{c.U}^*$*–geometry.*

Notice that, contrary to Theorem 1.1, which is independent from the classification of finite simple groups, Corollary 1.2 relies on it. Indeed that classification is used in the result of [5], from which Corollary 1.2 follows. Furthermore, the classification of 2–transitive groups is also used in the proof of Corollary 1.2.

The material of the present paper is part of an attempt to classify flag-transitive $(\mathsf{c} \cdot \mathsf{L}^*)$-geometry with finite point residues. We refer to Huybrechts-Pasini [18] for a survey of the situation.

The paper is organized as follows. In Section 2, we state some lemmas on $\mathsf{c.L}^*$–geometries, we gather some properties needed afterwards on 2-transitive groups and flag-transitive unitals and we recall the characterization of $EGH(J_2)$ by Cuypers. Theorem 1.1 and Corollary 1.2 are proved in Section 3.

# 2 Preliminaries

## 2.1 Some terminology and notation

We assume some knowledge of a few basic definitions and facts from the theory of diagram geometries (see, for instance, Buekenhout [4] or Pasini [22]). As in [4] and [22], we denote by $\Gamma_x$ (resp. $\Gamma_F$) the residue of an element $x$ (a flag $F$) of a geometry $\Gamma$. The words "point", "line" and "plane" we use for the elements of the geometries considered in this paper allude to an ordering of the set of types according to which the residue $\Gamma_a$ of a point $a$ is regarded as a point–line geometry, its points and lines being respectively lines and planes in $\Gamma$. Thus, we may distinguish $\Gamma_a$ from its dual. We denote the latter by $\Gamma_a^*$.

**(Partial) linear spaces and circular spaces.** We recall that a *(partial) linear space* (resp. *linear space*) is a rank 2 geometry consisting of *points* and *lines* such that any two points are incident with at most (resp. exactly) one line, any point is incident with at least two lines and any line with at least two points. A *circle* is a linear space with exactly two points on each line. We also call it *circular space.*

**Collinearity.** Let $\Gamma$ be a geometry with three types of elements, say *points*, *lines* and *planes*, where point–residues are partial linear spaces, plane–residues are circular spaces and line–residues are generalized digons.

Two distinct points of $\Gamma$ are said to be *collinear* if there is a line incident with both of them. If $a, b$ are collinear points of $\Gamma$, then we write $a \perp b$. We also denote by $a^{\perp}$ the set of points collinear with a given point $a$, with the convention that $a \in a^{\perp}$. We say that three points (or a point and a line, or two lines) are *coplanar* if they are incident to the same plane. The *collinearity graph* of $\Gamma$ is the graph $C(\Gamma)$ with the points of $\Gamma$ as vertices and the collinearity relation $\perp$ as the adjacency relation. The *distance* $\delta(a, b)$ of two points $a, b$ of $\Gamma$ is their distance in $C(\Gamma)$. The *distance* $\delta(a, S)$ of a point $a$ from a set of points $S$ is the minimal distance of $a$ from points of $S$. The *diameter* $\delta(\Gamma)$ of $\Gamma$ is the diameter of $C(\Gamma)$.

**The property (LL).** Given a geometry $\Gamma$ as above, let (LL) denote the following condition:
$(LL)$ distinct points of $\Gamma$ are incident with at most one line.
It is easy to see that properties (LL) and (IP) are equivalent for such a geometry $\Gamma$. In particular, when (LL) holds in $\Gamma$, then the following hold.

(i) If the two points of a line $L$ are incident with a plane $\pi$, then $L$ is incident with $\pi$.

(ii) Distinct planes of $\Gamma$ have at most two common points. In particular, the planes of $\Gamma$ may be identified with their point set.

If a point $a$ is incident with a line $l$ (or a plane $A$) then we say that $a$ is on $l$ (belongs to $A$) or that $l$ passes through $a$; accordingly, we write $a \in l$ (resp. $a \in A$). We also identify lines with their point set and write $ab$ for $\{a, b\}$, if $\{a, b\}$ is a line.

**Notation for stabilisers and kernels.** Given an automorphism group $G$ of a geometry $\Gamma$ and a flag $F$ of $\Gamma$, we denote by $G_F$ the stabiliser of $F$ in $G$ and by $K_F$ the *kernel* of $G_F$, namely the elementwise stabiliser of $\Gamma_F$ in $G_F$. We write $\overline{G}_F$ for $G_F/K_F$.

## 2.2 Some lemmas on c.L*–geometries

### 2.2.1 The plane index

Henceforth $\Gamma$ is a c.L*–geometry satisfying $(LL)$ and with orders $(s,t)$ (namely, every plane has $s+2$ points and every line is incident to $t+1$ planes):

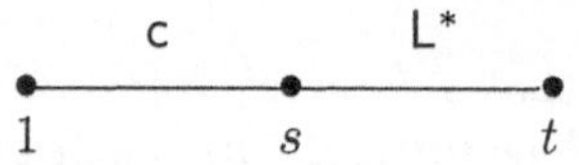


Given a point $a$ of $\Gamma$ and a plane $A$ at distance 1 from $a$, we set $\varphi_{a,A} = |a^\perp \cap A|$ and we denote by $\varphi$ the minimal value of $\varphi_{a,A}$ for $(a,A)$ a point–plane pair at distance 1. As $\Gamma$ has order $(s,t)$, we have $t+2 \leq \varphi_{a,A} \leq s+2$ for any $a$ and $A$ as above.

**Lemma 2.1** *If $\varphi > (s+2)/2$, then $\Gamma$ has diameter $\delta(\Gamma) \leq 2$.*

*Proof.* Let $a, b$ be points at distance 3, if possible. Then there is a plane $A$ with $\delta(a,A) = \delta(b,A) = 1$. However, as $(s+2)/2 < \varphi_{a,A}, \varphi_{b,A}$, at least one of the $s+2$ points of $A$ belongs to $a^\perp \cap b^\perp$; contradiction. □

If $\varphi_{a,A} = \varphi$ for every point $a$ of $\Gamma$ and every plane $A$ at distance 1 from $a$, then we call $\varphi$ the *plane index* of $\Gamma$ and we say that $\Gamma$ *admits plane index*. From now on, we assume that $\Gamma$ admits plane index $\varphi$.

**Lemma 2.2** *Given a point $a$, there are precisely*

$$\frac{st(s+1)(st+t+1)}{(t+1)\varphi}$$

*planes at distance 1 from $a$. Thus, $(t+1)\varphi$ divides $st(s+1)(st+t+1)$.*

*Proof.* Easy, by counting in two ways the number of point–plane flags $\{b, A\}$ with $\delta(a,A) = \delta(a,b) = 1$. □

A *non-planar triangle* of $\Gamma$ is a triple of pairwise collinear points of $\Gamma$ not contained in any plane of $\Gamma$. A *completely non-planar tetrahedron* of $\Gamma$ (or simply a *cnp-tetrahedron*) is a set of four points of $\Gamma$ all 3-subsets of which are non-planar triangles.

Let $a$, $b$ be collinear points and let $l$ be a line through $b$ non coplanar with $a$. If $l \subseteq a^\perp$ then $a$ and $l$ form a non-planar triangle; if otherwise, we say that $l$ is *going away from* $a$. We denote by $A_b^a$ (resp. $T_b^a$) the set of lines through $b$ going away from $a$ (resp. contained in $a^\perp$).

**Lemma 2.3** *Given any two collinear points $a, b$, we have*

$$|A_b^a| \;=\; \frac{st(s+2-\varphi)}{t+1}, \qquad |T_b^a| \;=\; \frac{st(\varphi-t-2)}{t+1}.$$

*Proof.* There are $st$ planes through $b$ not containing $a$ and each of them contains $s+2-\varphi$ points non collinear with $a$ and $\varphi-1$ points collinear with $a$ and distinct from $b$. Furthermore, $t+1$ points out of the latters are coplanar with the line $ab$. The statement easily follows. □

By Lemma 2.3, when $\varphi = t+2$ (resp. $\varphi = s+2$) any three pairwise collinear points of $\Gamma$ are coplanar (resp. any two points of $\Gamma$ are collinear). In the next lemma we consider the case of $\varphi = s+1$.

**Lemma 2.4** *Let $\varphi = s+1$. Then $t+1$ divides $s$ and the relation "being non-collinear" is an equivalence relation on the point-set of $\Gamma$. It has $s+2$ equivalence classes, all of size $s+1-s(t+1)^{-1}$.*

*Proof.* It is clear from Lemma 2.3 that $t+1$ divides $s$. Given a point $a$ and a plane $A$, let $\delta(a, A) > 1$, if possible. We have $\delta(\Gamma) \leq 2$ by Lemma 2.1. Hence $\delta(a, b) = 2$ for any point $b \in A$. Therefore, for every point $b \in A$ there is a plane $B$ through $b$ at distance 1 from $a$. However, as $\varphi = s+1$, $b$ is the unique point of $B$ non-collinear with $a$. On the other hand, $A$ and $B$ meet in a line $l$ containing $b$. By the above, the other point of $l$ is collinear with $a$. Hence $A \cap a^\perp \neq \emptyset$; contradiction. Consequently, $\delta(a, A) \leq 1$.

Let $b$ be a point at distance 2 from $a$, if any such point exists. Given a point $c \in b^\perp$, we have $\delta(a, A) = 1$ for every plane $A$ containing $bc$. Hence $a \perp c$, as $\varphi = s+1$ and $a \not\perp b$. Consequently, $b^\perp = a^\perp$. Therefore, two points at distance 2 from $a$ are never collinear. It follows that the relation "being non-collinear" is an equivalence relation. We denote it by $\not\perp$ and we use the symbol $a^{\not\perp}$ to denote the class of $\not\perp$ containing $a$.

Let $c$ be a point collinear with $a$. As $\not\perp$ is an equivalence relation, $c$ is collinear with every point of $a^{\not\perp}$ and we obtain the following from Lemma 2.3:

$$|a^{\not\perp}| \;=\; 1 + |A_c^a| \;=\; 1 + s - \frac{s}{t+1}.$$

Moreover, every point not on a given plane is not collinear with exactly one point of that plane. Therefore, $\not\perp$ has as many equivalence classes as points in a plane. The statement immediately follows. □

The following is an easy consequence of Lemma 2.4.

**Corollary 2.5** *Let $\varphi = s+1$ and let $G$ be a group of automorphisms of $\Gamma$. Given a point $a$, we say that two lines $ab$, $ac$ through $a$ are 'parallel' if $b \not\perp c$. Then 'being parallel' is an equivalence relation and its classes are permuted by $G_a$. (Hence $G_a$ acts imprimitely on the set of lines of the linear space $\Gamma_a^*$.)*

### 2.2.2 The collinearity index

Let $\Gamma$ be a c.L*–geometry admitting a flag–transitive automorphism group $G$. As the lines of $\Gamma$ have size two, the group $G$ is transitive on the ordered pairs of collinear points. Hence the number of lines through two given collinear points of $\Gamma$ is a constant, say $\nu$. We say that two lines $l$, $m$ of $\Gamma$ are *equivalent*, and we write $l \equiv m$, when $l$ and $m$ have the same points. Clearly, $\equiv$ is an equivalence relation and, for any point $a$ of $\Gamma$, it induces a $G_a$–invariant partition $\mathcal{P}_a$ on the set of lines of the linear space $\Gamma_a^*$. The members of $\mathcal{P}_a$ bijectively correspond to the points of $a^{\perp}$. Hence, $\mathcal{P}_a$ has at least $s+1$ classes. Furthermore, all members of $\mathcal{P}_a$ have size $\nu$.

The relation $\equiv$ is the identity precisely when $(LL)$ holds in $\Gamma$. Therefore,

**Lemma 2.6** *If $(LL)$ fails to hold in $\Gamma$, then the stabiliser $G_a$ of a point $a$ of $\Gamma$ acts imprimitively on the set of lines of the linear space $\Gamma_a^*$.*

(Indeed, if $(LL)$ does not hold in $\Gamma$, then $\mathcal{P}_a$ is a non-trivial $G_a$–invariant partition of the set of lines of $\Gamma_a^*$.)

### 2.2.3 Kernels

Given a c.L*-geometry $\Gamma$, let $G$ be a (possibly non flag–transitive) subgroup of $Aut(\Gamma)$.

**Lemma 2.7** *We have $K_a \leq K_A = K_{a,A}$ for any point–plane flag $\{a, A\}$ of $\Gamma$.*

*Proof.* Clearly, $K_{a,A} \geq K_a, K_A$. However $K_{a,A}$ is itself contained in $K_A$, as $\Gamma_A$ is a circular space. The statement follows. □

**Lemma 2.8** *We have $K_a = 1$ (hence $\overline{G}_a = G_a$) for any point $a$.*

*Proof.* We first show that $K_a = K_b$ for any two collinear points $a$ and $b$. Let $l$ be a line through $a$ and $b$. By Lemma 2.7, we have $K_a \leq K_A$ for any plane $A$ incident to $a$. Therefore $K_a$ fixes every line through $b$ coplanar with $l$. Since $\Gamma_b$ is a dual linear space, $K_a$ also fixes every element of $\Gamma_b$; that is, $K_a \leq K_b$. Similarly, $K_b \leq K_a$. Hence $K_a = K_b$. By connectedness, $K_a = K_b$ for any two (possibly non collinear) points $a$, $b$ of $\Gamma$. Hence, $K_a = 1$. □

## 2.3 Miscellanea on permutation groups

In this section we state some results on permutation groups to be used later. We denote by $E_q$ any *elementary abelian* group of order $q$. As for the rest, we follow the ATLAS notation [9]. Firstly, we recall an old number–theoretic result:

**Proposition 2.9** (Gerono [12]) *If $q, q'$ are prime powers with $q' - q = 1$, then either $(q, q') = (8, 9)$ or one of the numbers $q$, $q'$ is an odd prime and the other one is power of 2.*

In Hering-Kantor-Seitz [13], the 2-transitive permutation groups whose point stabilizers admit a normal regular subgroup are classified. (See Huppert-Blackburn [15], Theorems XI 13.2 and XII 9.1 respectively for the cases involving Ree groups and the sharply 2-transitive groups.) Comparing that result with Proposition 2.9 one easily obtains the following:

**Proposition 2.10** *Let $G$ be a 2–transitive group of permutations of a finite set $\Omega$ such that for every $a \in \Omega$ the stabilizer $G_a$ of $a$ has a normal subgroup regular on $\Omega \backslash \{a\}$. Assume moreover that $|\Omega| = q^2+1$ for some prime power $q \geq 3$ and that $G_a$ is solvable and contains an elementary abelian group of order $q^2$. Then $L_2(q^2) \leq G \leq P\Gamma L_2(q^2)$, $G$ acts on $\Omega$ as on the $q^2+1$ points of $PG(1,q^2)$ and*

$$E_{q^2} : \mathbb{Z}_{\frac{q^2-1}{(2,q^2-1)}} \leq G_a \leq A\Gamma L_1(q^2).$$

The next statement is easy to obtain from the classification of 2–transitive permutation groups, via Proposition 2.9:

**Proposition 2.11** *Let $q$ be an odd prime power. Then there is no 2–transitive permutation group $G$ of degree $q^2+1$ with point–stabilizer $G_a$ satisfying the following: $G_a$ is solvable, it contains an elementary abelian group of order $q^2$ and $q^3$ divides $|G_a|$.*

The following statement can be easily deduced from the Dickson list [11] of subgroups of $L_2(q)$.

**Proposition 2.12** *For any prime power $q > 2$, there are no subgroups of $L_2(q^2)$ of index $q-1$ or $(q-1)/2$.*

## 2.4 Flag–transitive unitals

We recall that a finite *unital of order* $q$ (where $q$ is an integer $> 1$) is a linear space where every line has $q+1$ points and every point is in $q^2$ lines. Such a linear space has $q^3+1$ points. Since the affine plane $AG(2,3)$ is the unique unital of order 2, we will only consider unitals of order $q > 2$ in the sequel.

The next statement is a straightforward application of the classification of finite flag–transitive linear spaces by Buekenhout, Delandtsheer, Doyen, Kleidman, Liebeck and Saxl [5].

**Proposition 2.13** *Let $\mathcal{U}$ be a finite unital of order $q \geq 3$ and assume that $\mathcal{U}$ admits a flag-transitive automorphism group $X$. Then $q$ is a prime power and one of the following holds:*
*(i) $\mathcal{U}$ is hermitian and $U_3(q) \leq X \leq P\Gamma U_3(q)$.*
*(ii) $\mathcal{U}$ is a Ree unital, $q = 3^k$ with $k$ odd and ${}^2G_2(q) \leq X \leq Aut({}^2G_2(q))$.*

(Note that the case called '1–dimensional' in [5] cannot occur here since it happens only for linear spaces having a prime power number of points whereas $q^3+1$ is never a prime power when $q \geq 3$.)

### 2.4.1 Hermitian unitals

We recall that for any prime power $q$, the *hermitian unital* $U_H(q)$ may be defined as follows from a non-degenerate hermitian polarity $\pi$ of $PG(2,q^2)$. The *points* and *lines* of $U_H(q)$ are respectively the absolute points and the non-absolute lines of the polarity $\pi$, with symmetrized inclusion as the incidence relation. We refer to O'Nan [21] for more details.

Henceforth, $\mathcal{U}$ is a hermitian unital of order $q \geq 3$ and $p$ is a given point of $\mathcal{U}$. We denote by $X$ any flag–transitive subgroup of $Aut(\mathcal{U}) = P\Gamma U_3(q)$ (hence $U_3(q) \leq X$, as stated in Proposition 2.13($i$)).

**Proposition 2.14** (O'Nan [21, §1]) *The kernel $K_p$ of $X_p$ is the group of perspectivities of centre $p$ and axis the tangent $p^\pi$ to the unital at $p$. Hence $K_p$ is contained in $U_3(q)$ (in fact, it is the kernel of the stabilizer of $p$ in $U_3(q)$), it is elementary abelian of order $q$ and transitive on $L \setminus \{p\}$, for every line $L$ of $\mathcal{U}$ through $p$.*

**Proposition 2.15** (O'Nan [21, §1]) *We have $X_p = Q{:}Z$, where $Q = Z(Q)^{\cdot}E_{q^2}$ (non-split extension) and $\mathbb{Z}_{q^2-1}/\mathbb{Z}_{(3,q+1)} \leq Z \leq \mathbb{Z}_{q^2-1}{:}Aut(\mathbb{F}_{q^2})$.*

*Moreover, $Z(Q) = K_p$ ($= E_q$ by Proposition 2.14) and $Q$ acts regularly on the point set of $\mathcal{U} \setminus \{p\}$.*

As a consequence, the following holds.

**Corollary 2.16** *Both $X_p$ and $\overline{X}_p$ are solvable and $\overline{X}_p$ contains an elementary abelian group of order $q^2$ acting regularly on the set of lines through $p$.*

**Proposition 2.17** (O'Nan [21, Lemma 1.12]) *If $X_p$ contains the stabilizer of $p$ in $PGU_3(q)$, then $X_p$ acts transitively by conjugacy on the elements of $K_p \setminus \{1\}$.*

As $U_3(q)$ is point–transitive on $\mathcal{U}$, we have $X = X_pU_3(q)$. On the other hand, $K_p \leq U_3(q)$ by Proposition 2.14. Hence the factor group $\overline{X}_p = X_p/K_p$ uniquely determines $X$. In particular,

**Corollary 2.18** *If $X_p$ and the stabilizer of $p$ in $PGU_3(q)$ induce the same group on the set of lines through $p$, then $X = PGU_3(q)$.*

Considering that $A\Gamma L_1(q)$ has a unique subgroup isomorphic to $AGL_1(q)$, we may improve Corollary 2.18 as follows:

**Corollary 2.19** *For any subgroup $Y$ of $P\Gamma U_3(q)$, if $Y_p \geq K_p$ and $\overline{Y}_p \cong AGL_1(q^2)$, then $Y = PGU_3(q)$.*

### 2.4.2 The unital $U_H(3)$ and the generalized hexagon $H(2)$

Let $\pi$ be the polarity of $PG(2,9)$ associated to the hermitian unital $\mathcal{U} = U_H(3)$ of order 3. As above, $X$ is a flag–transitive subgroup of $Aut(\mathcal{U})$. We say that two lines of $\mathcal{U}$ are *$\mathcal{U}$–secant* (resp. *$\mathcal{U}$–disjoint*) if their intersection in $PG(2,9)$ belongs (resp. does not belong) to $\mathcal{U}$. We say that two lines $L$, $M$ of $\mathcal{U}$ are *$\mathcal{U}$–conjugate* (also, *conjugate*) if the pole $L^\pi$ of $L$ belongs to $M$. Note that, if this is the case, the lines $L$ and $M$ are $\mathcal{U}$–disjoint.

**Lemma 2.20** *The stabilizer $X_L$ of a line $L$ of $\mathcal{U}$ has two orbits on the set of lines of $\mathcal{U}$ that are $\mathcal{U}$-disjoint from $L$, one of length 24 (namely the set of lines that are $\mathcal{U}$-disjoint from $L$ but not conjugate to $L$) and the other one of length 6 (namely the set of lines that are conjugate to $L$).*

*Proof.* Clearly each of the two sets of the statement is preserved by $X_L$. Moreover $X_L$ also fixes $L$ as well as the set of lines that are $\mathcal{U}$-secant with $L$. This provides 4 sets stabilized by $X_L$. The transivity of $X_L$ on each of these 4 sets follows from the character table of $U_3(3)$ given in the Atlas [9]. □

It is well-known that the generalized hexagon $\mathcal{H} = H(2)$ associated to the group $G_2(2)$ can be constructed from the polarity $\pi$. We recall that construction here. A *polar triangle* of $\mathcal{U}$ is a set of three pairwise $\mathcal{U}$-conjugate lines. We obtain $\mathcal{H}$ by taking as *points* and *lines* the polar triangles and the lines of $\mathcal{U}$, respectively, with symmetrized inclusion as incidence relation.

Conversely, the unital $\mathcal{U}$ can be recovered from $\mathcal{H}$. We first notice the following:

**Lemma 2.21** *Two lines $L$, $M$ of $\mathcal{U}$ have distance* 1, 2 *or* 3 *in the dual $\mathcal{H}^*$ of $\mathcal{H}$ if and only if they are $\mathcal{U}$-conjugate, $\mathcal{U}$-disjoint but not $\mathcal{U}$-conjugate or, respectively, $\mathcal{U}$-secant.*

(The proof is easy; we leave it for the reader.) Thus, the nine lines through a given point of $\mathcal{U}$ have mutual distance 3 in the collinearity graph of $\mathcal{H}^*$.

**Lemma 2.22** *Every set of nine mutually $\mathcal{U}$-secant lines of $\mathcal{U}$ is the bundle of lines through a point of $\mathcal{U}$.*

*Proof.* Assume the contrary and let $\Lambda$ be a set of nine mutually $\mathcal{U}$-secant lines not through the same point of $\mathcal{U}$. Denote by $P$ the set of intersecting points of lines of $\Lambda$. Define the *degree* of a point $p$ of $P$ as the number of lines of $\Lambda$ through it. By definition, the degree of every point of $P$ is at least two. We now show that it is at most 4. Since $\Lambda$ is not the bundle of lines through $p$, there exists a line $L$ of $\Lambda$ not containing $p$. As a consequence, the lines of $\Lambda$ through $p$ meet $L$ in distinct points of $\mathcal{U}$, and so there are at most 4 such lines.

In accordance with the above restrictions on the degree, a direct computation shows that every line $L$ of $\Lambda$ contains at least two points of $P$ of degree at least three. In particular, $\Lambda$ contains 5 lines forming the configuration pictured below.

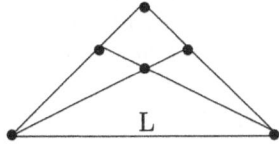

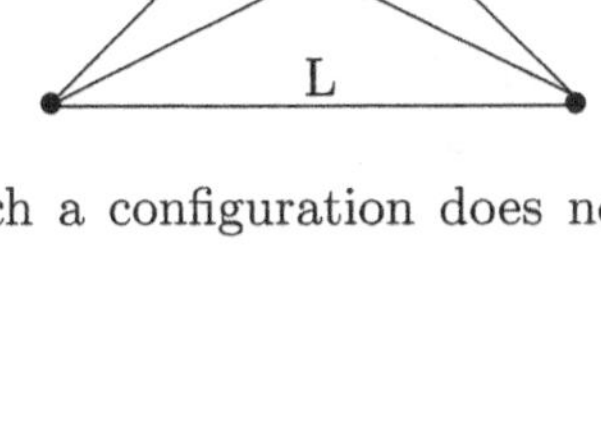


However, such a configuration does not exist as was shown by O'Nan [21, §3]. □

The next proposition assembles the previous two lemmas. It makes it clear that $\mathcal{U}$ can be recovered from $\mathcal{H}$.

**Proposition 2.23** *Any set of nine lines of $\mathcal{H}$ at mutual distance 3 in $\mathcal{H}^*$ is the bundle of lines of $\mathcal{U}$ through a point of $\mathcal{U}$ and conversely.*

### 2.4.3 Ree unitals

In this section, we assume some knowledge of Ree groups (see, for instance, Huppert and Blackburn [15]). The Ree unitals have been found by Lüneburg [20]. Before defining them, we first recall that the Ree group $R(q) = {}^2G_2(q)$ ($q = 3^{2m+1}$, $m$ a nonnegative integer) acts on a set $S$ of $q^3 + 1$ objects. Every involution of $R(q)$ fixes precisely $q + 1$ elements of $S$, any two distinct elements of $S$ are fixed by some involution and distinct involutions have at most one common fixed point. Thus, if we take the elements of $S$ as *points* and the sets of elements fixed by an involution as *lines*, with the natural incidence relation, we obtain a unital of order $q$: the *Ree unital* $U_R(q)$.

Let $X$ be a flag-transitive automorphism group of $U_R(q)$. Then $R(q) \leq X \leq Aut(R(q))$, by Proposition 2.13(i).

**Proposition 2.24** Knapp [19, Theorem 2.4] *We have $X = R(p){:}U$ with $U = Z_\lambda \leq Aut(\mathbb{F}_q)$ normalizing two Sylow 3–subgroups of $R(q)$.*

Let $p$ be a point of $U_R(q)$. Then $X_p$ is the normalizer $N_X(Q)$ in $X$ of a Sylow 3–subroup $Q$ of $R = R(q)$. We may assume that $Q$ is one of the Sylow 3–subgroups of $R$ fixed by the group $U$ mentioned in the above Proposition. Thus, $X_p = R_p{:}U = N_R(Q){:}U$ and, as $U \leq Aut(\mathbb{F}_q)$ has odd order, all involutions of $X_p$ belong to $R_p$.

Given a line $L$ through $p$, the stabilizer $X_L$ of $L$ in $X$ is the centralizer $C_X(t)$ of the (unique) involution $t$ having $L$ as set of fixed points (Lüneburg [20]). There are exactly $q^2$ involutions in $R_p$ (accordingly, exactly $q^2$ lines through $p$).

**Proposition 2.25** Huppert[15, Chapter XI] *The following hold:*
*(i)* $R_p/Q = \mathbb{Z}_{q-1}$.
*(ii) The commutator subgroup $Q'$ of $Q$ is elementary abelian of order $q^2$ and* $|Z(Q)| = q$.
*(iii) $Q$ acts transitively by conjugation on the $q^2$ involutions of $R_p$ and we have $Q' = N_Q(C_Q(t)) = Z(Q) \times C_Q(t)$ (whence $|C_Q(t)| = q$) for any involution $t$ of $R_p$. Thus, exactly $q - 1$ involutions of $R_p$ different from $t$ have the same centralizer as $t$ in $Q$. If $t'$ is any of the remaining $q^2 - q$ involutions, then $C_{R_p}(t) \cap C_{R_p}(t') = 1$.*
*(iv) Every element of $R$ fixing more than two points is an involution; consequently, the stabilizer in $R_p$ of a point $x \neq p$ moves all points of $S \setminus \{p, x\}$.*

**Corollary 2.26** *We have $K_p = 1$.*

*Proof.* The kernel $K_p$ of $X_p$ is the intersection of the centralizers in $X_p$ of the involutions of $R_p$. Hence $K_p \cap R = 1$, by Proposition 2.25(*iii*). Therefore, $\langle K_p, R_p \rangle = K_p \times R_p$. Thus, $K_p$ centralizes $R_p$. Hence $K_p$ centralizes $R_{p,x}$ for every point $x \neq p$. However, $p$ and $x$ are the unique fixed points of $R_{p,x}$, by Proposition 2.25(*iv*). Consequently, $K_p$ fixes $x$. This forces $K_p = 1$, as $x$ is any point other than $p$. □

### 2.5 A characterization of $EGH(J_2)$

We recall that an *extended generalized hexagon of order* $(s, t)$ is a rank 3 geometry, with elements called *points*, *lines* and *circles*, where residues of points are generalized hexagons of order $(s, t)$, residues of circles are circular spaces with $s+2$ points and residues of lines are generalized digons. Its diagram is as follows :

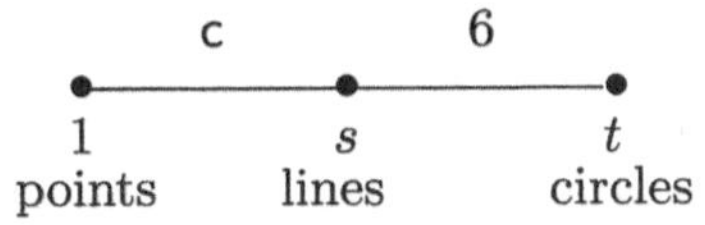


As mentioned in Section 1, the Hall–Janko group $J_2$ acts flag–transitively on an extended generalized hexagon which we called $EGH(J_2)$. It has order $(2, 2)$. A characterization of $EGH(J_2)$ has been obtained by Cuypers [10, Proposition 4.2] by means of a property dealing with triples of mutually collinear points of the collinearity graph not contained in any circle. We call these triples *empty triangles.*

**Proposition 2.27** (Cuypers [10]) *Up to isomorphism, $EGH(J_2)$ is the unique extended generalized hexagon $\Gamma$ of order $(2, 2)$ satisfying both $(LL)$ and the following property:*

$(*)$ *three distinct points $a, b, c$ form an empty triangle of $\Gamma$ if and only if $b, c \in a^{\perp}$ and the lines $ab$ and $ac$ have distance 3 when regarded as points of the generalized hexagon $\Gamma_a$.*

We will exploit this result in the next section, to finish the proof of our Theorem 1.1.

## 3 Proofs

### 3.1 Proof of Theorem 1.1

Henceforth $\Gamma$ is a $c.U_H^*$–geometry and $G$ is a flag–transitive subgroup of $Aut(\Gamma)$. The geometry $\Gamma$ admits order $(q^2-1, q)$, where $q$ is the order of the residual unitals of $\Gamma$.

As noticed in Subsection 2.4, a unital of order 2 is isomorphic to $AG(2, 3)$. Flag-transitive $c.Af^*$-geometries have been classified by Baumeister, Del Fra, Meixner and Pasini [1] and it turns out that there are no flag-transitive extensions of $AG(2,3)$. Thus, henceforth, we assume $q \geq 3$.

The proof is organized as follows. We first show that the residual unitals of $\Gamma$ are of order 3. Next, we determine some geometric properties of $\Gamma$ in view to construct an extended generalized hexagon satisfying the assumptions of Cuypers [10, Proposition 4.2]. This allows us to conclude.

### 3.1.1 Restrictions on the order of the geometry

**Lemma 3.1** *Let $\{a, A\}$ be a point–plane flag of $\Gamma$. Then,*

$$\begin{array}{ll} K_a = 1, & PGU_3(q) \leq \overline{G}_a = G_a \leq P\Gamma U_3(q), \\ K_A = E_q, & PGL_2(q^2) \leq \overline{G}_A \leq P\Gamma L_2(q^2), \\ K_{a,A} = K_A, & AGL_1(q^2) \leq \overline{G}_{a,A} \leq A\Gamma L_1(q^2). \end{array}$$

*Proof.* We have $K_a = 1$ (hence $\overline{G}_a = G_a$) and $K_A = K_{a,A}$ by Lemmas 2.7 and 2.8. The plane-residues of $\Gamma$ are finite since the point-residues of $\Gamma$ the same property. Moreover, in view of its flag-transitivity on $\Gamma$, the group $G$ induces a 2-transitive permutation group on the plane-residues of $\Gamma$. Therefore, by Corollary 2.16, we may apply Proposition 2.10 to deduce that $L_2(q^2) \leq \overline{G}_A \leq P\Gamma L_2(q^2)$ and $\overline{G}_{a,A}$ is a subgroup of $A\Gamma L_1(q^2)$ containing a subgroup

$$\widetilde{H} = E_{q^2}{:}\mathbb{Z}_{\frac{q^2-1}{(2,q^2-1)}}.$$

On the other hand, by Proposition 2.15, $\overline{G}_{a,A}$ also contains a subgroup

$$H \;=\; E_{q^2}{:}\mathbb{Z}_{\frac{q^2-1}{(3,q+1)}}.$$

However, both $H$ and $\widetilde{H}$ are subgroups of $AGL_1(q^2)$ and their indices are relatively prime. Therefore $AGL_1(q^2) \leq \overline{G}_{a,A}$. Hence $PGU_3(q) \leq G_a$ and $PGL_2(q^2) \leq \overline{G}_A$ (Corollary 2.19). Finally, $K_{a,A} = E_q$ by Proposition 2.14. □

**Lemma 3.2** *We have $q = 3$.*

*Proof.* Let $\{a, A\}$ be a point–plane flag of $\Gamma$. By Proposition 3.1, there is a subgroup $H$ of $G_A$ such that $H = K_A.\overline{H}$ with $\overline{H} \cong PGL_2(q^2)$ and $\overline{H}_a \cong AGL_1(q^2)$. Therefore $H_a = K_A.AGL_1(q^2)$. However, $K_a = 1$ and $K_A = K_{a,A}$ by Proposition 3.1. Therefore, $H_a$ is a subgroup of the stabilizer of $A$ in $P\Gamma U_3(q)$. Hence $H_a$ is the stabilizer in $PGU_3(q)$ of the point $A$ of the unital $\Gamma^*_a$. Consequently, $H$ acts transitively by conjugation on $K_A \setminus \{1\}$ (indeed $H_a$ has the same property by Proposition 2.17).

Let $C = C_H(K_A)$. By the above, $H/C$ contains a subgroup of index $q-1$. Since $K_A$ is commutative, $K_A \leq C$. Moreover, $C \trianglelefteq H$ (because $K_A \trianglelefteq H$). Therefore $C/K_A \trianglelefteq \overline{H}$ and $H/C \cong \overline{H}/(C/K_A)$. Since $\overline{H}$ is an almost simple group of socle $L_2(q^2)$, either $C = K_A$ or $L_2(q^2) \trianglelefteq C/K_A$.

Assume $C = K_A$. Then $H/C \cong \overline{H}$ and there is a subgroup of index $q-1$ in $PGL_2(q^2)$. Hence $L_2(q^2)$ has a subgroup of index $q-1$ or $(q-1)/2$, which is impossible by Proposition 2.12. Therefore, $L_2(q^2) \trianglelefteq C/K_A$. Here $H/C$ is smaller than $PGL_2(q^2)/L_2(q^2)$, which is isomorphic to $\mathbb{Z}_{(2,q^2-1)}$. However, we have shown above that $q-1$ divides $|H/C|$. Hence $q = 3$. □

### 3.1.2 Some geometric properties of the case of order 3

**Lemma 3.3** *The property $(LL)$ holds in $\Gamma$.*

*Proof.* Let $a$ be a point of $\Gamma$. As $q = 3$, there are 63 lines through $a$ and $PGU_3(3) \leq G_a \leq P\Gamma U_3(3)$. However, $PGU_3(3)$ has a unique action of degree 63, which is primitive (see [9]). The statement follows from Lemma 2.6. □

**Lemma 3.4** *The geometry $\Gamma$ admits plane index.*

*Proof.* It suffices to show that the stabilizer $G_a$ of a point $a$ is transitive on the set of planes at distance 1 from $a$. Let $A$ be a plane containing $a$. The group $\overline{G}_{a,A}$ contains $AGL_1(9)$, whence it is 2-transitive on the nine lines of $A$ through $a$. Hence $G_{a,A}$ is transitive on the set of lines of $A$ not through $a$. By assumption, $G_a$ is transitive on the set of planes containing $a$. Therefore, $G_a$ is transitive on the set of lines not through $a$ but coplanar with $a$. Let $l$ be one of those lines, contained in $A$ and let $b$ be a point of $l$. Using Proposition 2.14 in $\Gamma_b$ we see that $G_{a,b,l}$ is transitive on the set of planes through $l$ different from $A$. On the other hand, $G_a$ is transitive on the set of planes on $a$ and every plane at distance 1 from $a$ contains some lines coplanar with $a$ but not through $a$. Therefore, $G_a$ is transitive on the set of planes at distance 1 from $a$. □

**Lemma 3.5** *The plane index of $\Gamma$ is 6 and the following holds:*

*$(**)$ A triple $\{a,b,c\}$ of points of $\Gamma$ is a non-planar triangle if and only if $b, c \in a^{\perp}$ and the lines $ab$ and $ac$ are conjugate in the unital $\Gamma_a^*$.*

*Proof.* Let $\varphi$ be the plane index of $\Gamma$. Clearly, $5 \leq \varphi \leq 10$. The extreme cases $\varphi = 5$ and $\varphi = 10$ are excluded by the divisibility condition of Lemma 2.2.

Let $a \perp b$. By Lemma 2.3, we have $|A_b^a| = 6(10-\varphi)$ and $|T_b^a| = 6(\varphi - 5)$. As $5 < \varphi < 10$, none of these sets is empty. Hence $\{A_b^a, T_b^a\}$ is a partition of the set of lines through $b$ that are not coplanar with $a$. Clearly, each of $A_b^a$ and $T_b^a$ is stabilized by $G_{b,ab}$. On the other hand, by Lemma 2.20, $G_{b,ab}$ has two orbits on the set $A_b^a \cup T_b^a$, of respective sizes 6 and 24. Hence those orbits are just $A_b^a$ and $T_b^a$ and $\varphi$ is either 6 or 9 according to whether $|T_b^a| = 6$ or 24.

However the case of $\varphi = 9$ is excluded by Corollary 2.5, since the stabilizer in $G$ of a point of $\Gamma$ acts primitively on the set of lines through that point. Therefore, $\varphi = 6$ and $T_b^a$ is the set of lines of $\Gamma_b$ that are $\Gamma_b^*$-conjugate to the line $ab$. That is, $(**)$ holds. □

**Lemma 3.6** *Given two collinear points of* $\Gamma$ *and a plane containing exactly one of them, there is a unique non-planar triangle containing these two points as well as another point of the plane.*

*Proof.* By (**) of Lemma 3.5 the above is a straightforward consequence of the fact that in a hermitian unital $\mathcal{U}$ every point not on a given line $l$ is contained in a unique line conjugate to $l$. □

**Lemma 3.7** *Every quadruple of mutually collinear points of* $\Gamma$ *containing two non-planar triangles is a cnp–tetrahedron.*

*Proof.* Any two triangles contained in the same 4–clique share an edge. The result immediately follows from this remark and from Lemma 3.6. □

**Lemma 3.8** *Given a point* $a$ *of* $\Gamma$ *and a triple* $S$ *of lines through* $a$*, the lines of* $S$ *belong to the same cnp–tetrahedron if and only if* $S$ *is a polar triangle of the unital* $\Gamma_a^*$.

*Proof.* Straightforward, by (**) of Lemma 3.5 and by Lemma 3.7. □

**Lemma 3.9** *Every non-planar triangle of* $\Gamma$ *is contained in exactly one cnp-tetrahedron.*

*Proof.* Easy, by Lemma 3.8 and by the existence and uniqueness of a polar triangle containing two $\mathcal{U}$–conjugate lines of a hermitian unital $\mathcal{U}$. □

### 3.1.3 Construction of an extended generalized hexagon

We now associate an extended generalized hexagon $H(\Gamma)$ to $\Gamma$ by taking as *points*, *lines* and *circles* the points, the lines and the cnp-tetrahedra of $\Gamma$, respectively, with the natural incidence relation. Lemma 3.8 shows that $H(\Gamma)$ is indeed an extended generalized hexagon and that $(2, 2)$ is its order. Also, $H(\Gamma)$ inherits the $(LL)$ property from $\Gamma$ (Lemma 3.3).

**Lemma 3.10** *The property* $(*)$ *of Proposition* 2.27 *holds in* $H(\Gamma)$.

*Proof.* A triple of points $\{a, b, c\}$ is an empty triangle of $H(\Gamma)$ if and only if it is a clique of the collinearity graph of $\Gamma$ but it is not contained in any cnp–tetrahedron of $\Gamma$. By Lemmas 3.5 and 3.8, the latter holds if and only if $\{a, b, c\}$ is contained in some plane of $\Gamma$. However, by Lemma 2.21, two lines through a point $a$ of $\Gamma$ are coplanar in $\Gamma$ if and only if they are at distance three in the generalized hexagon associated to the unital $\Gamma_a^*$. The conclusion follows. □

Lemma 3.10 and Proposition 2.27 give us the following:

**Corollary 3.11** *We have* $H(\Gamma) \cong EGH(J_2)$.

This finishes the proof of Theorem 1.1. Indeed, $\Gamma$ and $H(\Gamma)$ have the same points and the same lines and, as the points of the unital $U_H(3)$ can be recovered as certain sets of lines from the generalized hexagon $H(2)$ (see Proposition 2.23), we can recover $\Gamma$ from $H(\Gamma)$. However, $H(\Gamma) \cong EGH(J_2)$. Hence $\Gamma$ is uniquely determined, too. Needless to say, $\Gamma$ is the Buekenhout–Hermand geometry.

## 3.2 Proof of Corollary 1.2

Let $\Gamma$ be a c.U*-geometry of order $q \geq 3$ and admitting a flag–transitive automorphism group $G$. Then, by Proposition 2.13(ii), $q$ is a prime power and, for any point $a$ of $\Gamma$, the unital $\Gamma_a^*$ is either hermitian or a Ree unital. Assume the latter. Then, by Corollary 2.26, the stabilizer $G_{a,A}$ of a point–plane flag $\{a, A\}$ acts faithfully on the $q^2$ points of $A \setminus \{a\}$ and, by Proposition 2.25, it is solvable, it contains an elementary abelian group of order $q^2$ and $q^3$ divides $|G_{a,A}|$. However, this goes against Proposition 2.11. Thus, $\Gamma_a^*$ is not a Ree unital, we are driven back to the hermitian case and Theorem 1.1 gives us the conclusion.

**Acknowledgment.** In an earlier version of this paper, we used Burnside's theorem on minimal normal subgroups [8, §27] to prove Theorem 1.1. We thank Hans Cuypers who suggested the present alternative use of [10, Proposition 4.2] and made other useful comments. We also thank Satoshi Yoshiara for useful information.

# References

[1] B. Baumeister, A. Del Fra, T. Meixner and A. Pasini, "Flag-transitive $c.Af^*$ geometries", to appear in *Contributions to Algebra and Geometry.*

[2] F. Buekenhout, "Diagrams for geometries and groups", *J. Comb. Th. A*, **27** (1979), 121–151.

[3] F. Buekenhout, "Diagram geometries for sporadic groups", *Contemp. Math.* **45** (1985), 1–45.

[4] F. Buekenhout, "Foundations of incidence geometry", in *Handbook of Incidence Geometry*, pp. 63–105, ed. F. Buekenhout, Elsevier, Amsterdam, 1995.

[5] F. Buekenhout, A. Delandtsheer, J. Doyen, P. B. Kleidman, M. W. Liebeck and J. Saxl, "Linear spaces with flag-transitive automorphism groups", *Geom. Dedicata* **36** (1990), 89–94.

[6] F. Buekenhout and C. Huybrechts, "A (c · L*)-geometry for the sporadic group $J_2$", submitted.

[7] F. Buekenhout, C. Huybrechts and A. Pasini, "Parallelism in diagram geometry", *Bull. Belg. Math. Soc. Simon Stevin* **1** (1994), no.3, 355–397.

[8] W. Burnside, *Theory of groups of finite order*, Dover publications, 1911.

[9] J. Conway, R. Curtis, S. Norton, R. Parker and R. Wilson, *Atlas of finite groups*, Clarendon Press, Oxford, 1985.

[10] H. Cuypers, "Extended generalized hexagons and the Suzuki chain", preprint.

[11] L. E. Dickson, *Linear groups*, Dover Publ. New-York 1958.

[12] C. G. Gerono, "Note sur la résolution en nombres entiers et positifs de l'équation $x^m = y^n + 1$", *Nouv. Ann. Math (2)* **9** (1870), 469–471 and **10** (1871), 204–206.

[13] C. Hering, W. Kantor and G. M. Seitz, "Finite groups with a split BN-pair of rank 1", *J. Algebra* **20**(1972), 435–475.

[14] M. Hermand, *Du groupe de Hall-Janko aux semidesigns réguliers*, Mémoire de licence, Université Libre de Bruxelles, 1982.

[15] B. Huppert and N. Blackburn, *Finite Groups III*, Springer-Verlag, Berlin Heidelberg New-York, 1982.

[16] C. Huybrechts, *Réductions des géométries de type $L \cdot L^*$*, Thèse de doctorat, Université Libre de Bruxelles,1996.

[17] C. Huybrechts, "An arithmetic reduction of finite rank 3 geometries with linear spaces as plane residues and with dual linear spaces as point residues", *J. Alg. Comb.* **5** (1996), 329–335.

[18] C. Huybrechts and A. Pasini, "Flag-transitive $(\mathsf{c} \cdot \mathsf{L}^*)$-geometries", in preparation.

[19] W. Knapp, "Primitive Permutationsgruppen mit einem Subkonstituent vom grad $2^f$", *Math. Z.* **136** (1974), 261–275.

[20] H. Lüneburg, "Some remarks concerning the Ree groups of type $(G_2)$", *J. Algebra* **3** (1966), 256–259.

[21] M. O'Nan, "Automorphisms of unitary block designs", *J. Algebra* **20** (1972), 495–511.

[22] A. Pasini, *Diagram Geometries*, Oxford U. P., Oxford, 1994.

[23] M. Ronan, "On the second homotopy group of certain simplicial complexes and some combinatorial applications", *Quart. J. Math. Oxford*, **32** (1981), 225–233.

Cecile Huybrechts
Université Libre de Bruxelles, Departement de Mathematique
Campus Plaine, B–1050, Bruxelles, Belgique.
e-mail: `huyb@ulb.ac.be`

Antonio Pasini
University of Siena, Department of Mathematics
Via del Capitano 15, I–53100 Siena, Italy.
e-mail: `pasini@unisi.it`

Trends in Mathematics, © 1998 Birkhäuser Verlag Basel/Switzerland

# Affine Extended Dual Polar Spaces

Alexander A. Ivanov

**Abstract**

An affine extended dual polar space is a geometry with the diagram

in which the residue of an element of the leftmost type is a classical dual polar space and which possesses a flag-transitive automorphism group with a normal subgroup acting regularly on the set of elements of the leftmost type. We reduce the classification of affine extended dual polar spaces to calculation of the universal representation groups of classical dual polar spaces with 3 points on each line.

# 1 Introduction

Concerning diagram geometries we basically follow the terminology of [Pasi]. In particular by the definition all geometries of rank at least 2 are (residually) connected. A geometry $\mathcal{G}$ is said to be an *extended dual polar space* (EDPS for short) if it has diagram of the form

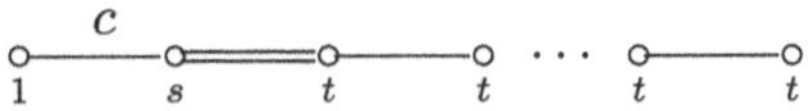


and the residue of an element of the leftmost type is a classical dual polar space. Let $\mathcal{G}$ be an EDPS of rank $m+1 \geq 3$ and assume that the types on the diagram increase rightwards from 1 to $m+1$. If $\mathcal{G}$ possesses a flag-transitive automorphism group $G$ which contains a normal subgroup $T$ whose action on the set of elements of type 1 is regular, then $\mathcal{G}$ is called *affine extended dual polar space* (AEDPS for short). In this case $T$ is called the *translation group* of $\mathcal{G}$ (with respect to $G$). In the present paper and [Ivn2] we address the classification problem of flag-transitive EDPS's. The main result of [Ivn2] is the following theorem (recall that Borel subgroup is the elementwise stabilizer of a maximal flag).

**Theorem 1** *Let $\mathcal{G}$ be an EDPS of rank at least three which possesses a flag-transitive automorphism group whose Borel subgroup is finite. Then one of the following holds:*

(i) $\mathcal{G}$ is isomorphic to one of the 19 exceptional EDPS's whose diagrams and full automorphism groups are given in Table 1;

(ii) there is a 2-covering $\psi : \tilde{\mathcal{G}} \to \mathcal{G}$ where $\tilde{\mathcal{G}}$ is affine.                     □

## Table 1

| The  diagrams | The  groups |
|---|---|
| $\overset{c}{\underset{1 \quad\quad 2 \quad\quad 2}{\circ\!-\!-\!-\!\circ\!=\!=\!=\!\circ}}$ | $Sym_8$, $\quad$ $U_4(2).2$ |
| $\overset{c}{\underset{1 \quad\quad 2 \quad\quad 4}{\circ\!-\!-\!-\!\circ\!=\!=\!=\!\circ}}$ | $Sp_6(2) \times 2$, $\quad$ $Sp_6(2)$ |
| $\overset{c}{\underset{1 \quad\quad 4 \quad\quad 2}{\circ\!-\!-\!-\!\circ\!=\!=\!=\!\circ}}$ | $3 \cdot U_4(3).2^2$, $\quad$ $U_4(3).2^2$ |
| $\overset{c}{\underset{1 \quad\quad 3 \quad\quad 3}{\circ\!-\!-\!-\!\circ\!=\!=\!=\!\circ}}$ | $U_5(2).2$ |
| $\overset{c}{\underset{1 \quad\quad 3 \quad\quad 9}{\circ\!-\!-\!-\!\circ\!=\!=\!=\!\circ}}$ | $McL.2$ |
| $\overset{c}{\underset{1 \quad\quad 9 \quad\quad 3}{\circ\!-\!-\!-\!\circ\!=\!=\!=\!\circ}}$ | $HS.2$, $\quad$ $Suz.2$ |
| $\overset{c}{\underset{1 \quad\quad 2 \quad\quad 2 \quad\quad 2}{\circ\!-\!-\!-\!\circ\!=\!=\!=\!\circ\!-\!-\!-\!\circ}}$ | $Sp_8(2)$, $\quad$ $3 \cdot Fi_{22}.2$, $\quad$ $Fi_{22}.2$ |
| $\overset{c}{\underset{1 \quad\quad 2 \quad\quad 4 \quad\quad 4}{\circ\!-\!-\!-\!\circ\!=\!=\!=\!\circ\!-\!-\!-\!\circ}}$ | $Co_2 \times 2$, $\quad$ $Co_2$ |
| $\overset{c}{\underset{1 \quad\quad 4 \quad\quad 2 \quad\quad 2}{\circ\!-\!-\!-\!\circ\!=\!=\!=\!\circ\!-\!-\!-\!\circ}}$ | $Fi_{24}$ |
| $\overset{c}{\underset{1 \quad\quad 3 \quad\quad 3 \quad\quad 3}{\circ\!-\!-\!-\!\circ\!=\!=\!=\!\circ\!-\!-\!-\!\circ}}$ | $Fi_{24}$ |
| $\overset{c}{\underset{1 \quad\quad 9 \quad\quad 3 \quad\quad 3}{\circ\!-\!-\!-\!\circ\!=\!=\!=\!\circ\!-\!-\!-\!\circ}}$ | $F_1$ |
| $\overset{c}{\underset{1 \quad\quad 2 \quad\quad 2 \quad\quad 2 \quad\quad 2}{\circ\!-\!-\!-\!\circ\!=\!=\!=\!\circ\!-\!-\!-\!\circ\!-\!-\!-\!\circ}}$ | $Fi_{23}$ |

In the present paper we deal with AEDPS's. First let us recall some definitions. Let $\mathcal{S}$ be a geometry one of whose types is called "points" and some other type is called "lines". Let $P$ and $L$ denote the point set and the line set of $\mathcal{S}$ and suppose that *every line is incident to exactly three points*. We will always assume that every line is uniquely determined by the set of points it is incident to. A group $T$ is said to be a *representation group* of $\mathcal{S}$ if there is a mapping $\varphi : p \mapsto t_p$ of $P$ into $T$ such that

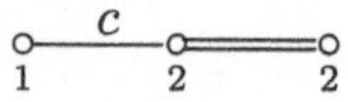

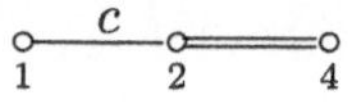

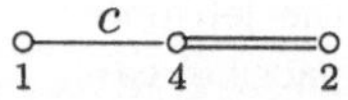

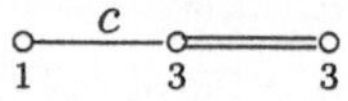

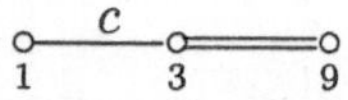

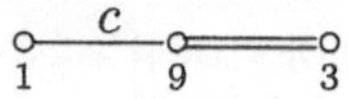

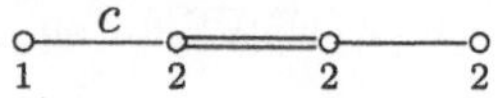

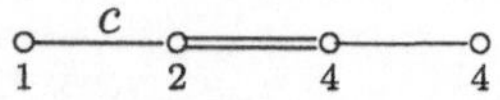

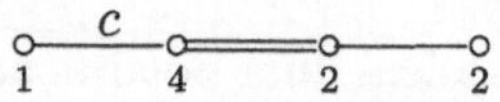

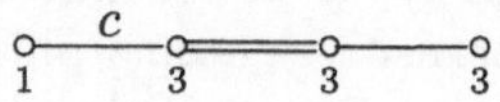

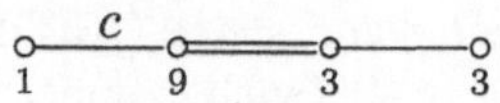

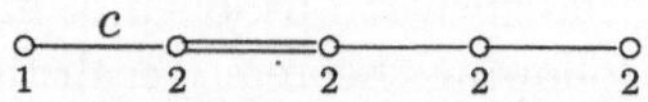

(a) $T$ is generated by the $t_p$ for all $p \in P$;

(b) $t_p^2 = 1$ for $p \in P$;

(c) $t_p t_q t_r = 1$ whenever $\{p, q, r\} \in L$.

For an element $\sigma$ of $\mathcal{S}$ of any type (not necessary point or line) let $T_\sigma$ denote the subgroup of $T$ generated by the elements $t_p$ for all points $p$ which are incident to $\sigma$ in $\mathcal{S}$. The representation group $T$ is called *separable* if $T_\sigma \neq T_\nu$ whenever $\sigma \neq \nu$. In the separable case we will say that $T_\sigma$ *represents* $\sigma$ in $T$. Let $H$ be an automorphism group of $\mathcal{S}$. Then $T$ is said to be *$H$-admissible* if for every $h \in H$ the mapping $t_p \mapsto t_{p^h}$, $p \in P$ defines an automorphism of $T$ (here $p^h$ denotes the image of $p$ under $h$). A representation group is *admissible* if it is $S$-admissible for the full automorphism group $S$ of $\mathcal{S}$. Let $T$ and $\tilde{T}$ be representation groups of $\mathcal{S}$ with the generating sets $\{t_p \mid p \in P\}$ and $\{\tilde{t}_p \mid p \in P\}$, respectively. Suppose that there is a homomorphism $\chi$ of $\tilde{T}$ onto $T$ such that $\chi : \tilde{t}_p \mapsto t_p$ for every $p \in P$. Then $\chi$ is called a *representation homomorphism*. The *universal representation group* $\mathcal{R}(\mathcal{S})$ of $\mathcal{S}$ is defined by the following presentation

$$R(\mathcal{S}) = \langle z_p,\ p \in P \mid z_p^2 = 1,\ z_p z_q z_r = 1 \text{ for } \{p, q, r\} \in L \rangle$$

It is clear that $\mathcal{R}(\mathcal{S})$ is a representation group of $\mathcal{S}$ with respect to the mapping $p \mapsto z_p$. Moreover, for any representation group $T$ of $\mathcal{S}$ the mapping $z_p \mapsto t_p$ defines a representation homomorphism of $\mathcal{R}(\mathcal{S})$ onto $T$ and $\mathcal{R}(\mathcal{S})$ is admissible. Finally, if $\mathcal{S}$ possesses a separable representation group then $\mathcal{R}(\mathcal{S})$ is separable.

Let $\mathcal{D}$ be a classical dual polar space of rank $m \geq 2$ with three points on every line, *i.e.* $\mathcal{D}$ is either the dual polar space $\mathcal{D}(Sp_{2m}(2))$ of symplectic type with the diagram

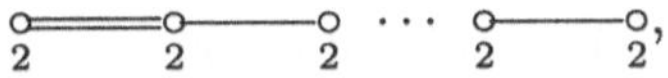


or the dual polar space $\mathcal{D}(U_{2m}(2))$ of unitary type with the diagram

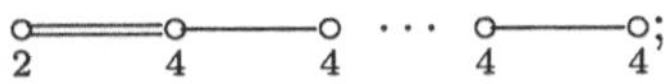


and let $D$ be a flag-transitive automorphism group of $\mathcal{D}$.

We will often use the following *basic property of classical dual polar spaces*: every element can be identified with the set of points (maximal totally isotropic subspaces) it is incident to, so that the incidence relation is given via inclusion.

Let $T$ be a $D$-admissible separable representation group of $\mathcal{D}$ and for an element $\sigma$ in $\mathcal{D}$ let $T_\sigma$ be the subgroup representing $\sigma$ in $T$. Let $\mathcal{A}(\mathcal{D}, T)$ be the incidence system of rank $m+1$ whose elements of type 1 are the elements of $T$ (right cosets of the identity subgroup) and for $2 \leq i \leq m+1$ the elements of type $i$ are all the right cosets of the subgroups $T_\sigma$ for all elements $\sigma$ of type $i-1$ in $\mathcal{D}$; two elements are incident if one of the corresponding cosets contains the other one.

**Lemma 1.1** *The following assertions hold:*

(i) *in the above notation $\mathcal{A}(\mathcal{D},T)$ is an AEDPS;*

(ii) *the semidirect product $G = T : D$ acts flag-transitively on $\mathcal{A}(\mathcal{D},T)$ and $T$ is the translation group of $\mathcal{A}(\mathcal{D},T)$ with respect to $G$;*

(iii) *if $\tilde{T}$ is another $D$-admissible separable representation group of $\mathcal{D}$ and $\chi : \tilde{T} \to T$ is a representation homomorphism then $\chi$ induces a 2-covering of $\mathcal{A}(\mathcal{D},\tilde{T})$ onto $\mathcal{A}(\mathcal{D},T)$.*

*Proof:* Let $\alpha$ be the element of type 1 in $\mathcal{A} = \mathcal{A}(\mathcal{D},T)$ which is the identity element of $T$. Then the elements incident to $\alpha$ are exactly the subgroups $T_\sigma$ representing in $T$ the elements of $\mathcal{D}$. By the basic property of the classical dual polar spaces since $T$ is separable this means that $\mathrm{res}_{\mathcal{A}}(\alpha) \cong \mathcal{D}$. Clearly $G$ acts transitively on the set of elements of type 1 in $\mathcal{A}$ and hence the residue of any such element is isomorphic to $\mathcal{D}$. It follows from the definition that if $K_i$ and $K_j$ are incident elements of type $i$ and $j$ respectively with $i < j$ then $K_i \subset K_j$. This shows that every maximal flag in $\mathcal{A}$ contains an element of type 1 and also that $\mathcal{A}$ belongs to a string diagram. Finally let $\gamma$ be an element of type 3 in $\mathcal{A}$. Without loss of generality we can assume that $\gamma = T_l$ where $l$ is a line in $\mathcal{D}$. Since $T$ is a separable representation group of $\mathcal{D}$, $T_l$ is elementary abelian of order 4. Now the elements of type 1 and 2 incident to $\gamma$ are the elements of $T_l$ and the cosets of all subgroup of order 2 in $T_l$, respectively. This is obviously a $c$-geometry with $s = 2$ and (i) follows. Now (ii) follows directly from the construction of $\mathcal{A}(\mathcal{D},T)$. For a homomorphism $\chi$ as in (iii) we define a morphism $\varphi$ of $\mathcal{A}(\mathcal{D},\tilde{T})$ onto $\mathcal{A}(\mathcal{D},T)$ by $\varphi(\tilde{T}_\sigma\tilde{t}) = T_\sigma\chi(\tilde{t})$ where $\sigma \in \mathcal{D}$ and $\tilde{t} \in \tilde{T}$. It is easy to deduce from the proof of (i) that $\varphi$ is a 2-covering. □

In the present paper we prove the following.

**Theorem 2** *Let $\mathcal{G}$ be an AEDPS of rank $m+1$ where $m \geq 2$. Let $G$ be a flag-transitive automorphism group of $\mathcal{G}$ and $T$ be the translation group of $\mathcal{G}$ with respect to $G$. Let $\alpha$ be an element of type 1 in $\mathcal{G}$, $\mathcal{D} = \mathrm{res}_{\mathcal{G}}(\alpha)$ and $D$ be the action induced on $\mathcal{D}$ by the stabilizer of $\alpha$ in $G$. Then*

(i) *$\mathcal{D}$ has three points on every line, that is $\mathcal{D} \cong \mathcal{D}(Sp_{2m}(2))$ or $\mathcal{D} \cong \mathcal{D}(U_{2m}(2))$;*

(ii) *$T$ is separable $D$-admissible representation group of $\mathcal{D}$ and $\mathcal{G}$ is isomorphic to $\mathcal{A}(\mathcal{D},T)$;*

(iii) *$\mathcal{A}(\mathcal{D},\mathcal{R}(\mathcal{D}))$ is the universal 2-cover of $\mathcal{G}$.*

This result reduces the classification of AEDPS's to the study of the groups $\mathcal{R}(\mathcal{D})$ for classical dual polar spaces $\mathcal{D}$ with three points on every line. The current knowledge about these universal representation groups is summarized in the following theorem.

**Theorem 3** *Let $\mathcal{D}$ be a classical dual polar space of rank $m$ with 3 points on every line and $\mathcal{R} = \mathcal{R}(\mathcal{D})$ be the universal representation group of $\mathcal{D}$. Let $f_m$ denote the multiplicity of the smallest eigenvalue of the collinearity graph of $\mathcal{D}$, that is $f_m = (2^m+1)(2^{m-1}+1)/3$ or $(4^m+2)/3$ depending on whether $\mathcal{D}$ is of symplectic or unitary type. Then*

(i) *$\mathcal{R}$ is separable;*

(ii) *$\mathcal{R}$ has a factor group $\bar{\mathcal{R}}$ which is elementary abelian 2-group of rank $f_m$ and if $m \leq 3$ then $\bar{\mathcal{R}}$ is the maximal abelian factor group of $\mathcal{R}$;*

(iii) *the commutator subgroup of $\mathcal{R}$ is trivial if $m = 2$ and of order 2 if $m = 3$;*

(iv) *if $\mathcal{D}$ is of symplectic type and $m \geq 3$ then $\mathcal{R}$ is non-abelian.*

The above two theorems imply the following result which we need in [Ivn2].

**Corollary 4** *Let $\mathcal{D}$ be a classical dual polar space with 3 points on every line and $D$ be a flag-transitive automorphism group of $\mathcal{D}$. Then there exist a 2-simply connected AEDPS $\mathcal{G}$ and a flag-transitive automorphism group $G$ of $\mathcal{G}$ so that for an element $\alpha$ of type 1 in $\mathcal{G}$ we have $\mathrm{res}_{\mathcal{G}}(\alpha) \cong \mathcal{D}$ and the stabilizer of $\alpha$ in $G$ is isomorphic to $D$.* □

In the above corollary we can put $\mathcal{G} = \mathcal{A}(\mathcal{D}, \mathcal{R}(\mathcal{D}))$ and $G = \mathcal{R}(\mathcal{D}) : D$.

## 2 Preliminaries

In this section we summarize some known facts about classical dual polar spaces and their extensions. The proofs can be found in [BCN] and [Pasi].

Let $\mathcal{D}$ be a classical dual polar space of rank $m \geq 2$ with the following diagram:

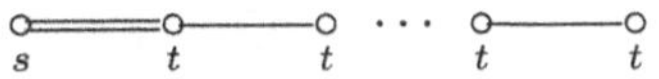


Then $\mathcal{D} = \mathcal{D}(X)$ where $X \cong Sp_{2m}(t)$, $\Omega_{2m+1}(t)$, $\Omega^-_{2m+2}(t)$, $U_{2m}(t^{\frac{1}{2}})$ or $U_{2m+1}(t^{\frac{1}{2}})$ and $s = t$, $t$, $t^2$, $t^{\frac{1}{2}}$ or $t^{\frac{3}{2}}$, respectively. The elements corresponding to the nodes from the left to the right on the diagram will be called *points*, *lines*, *planes*, ..., *hyperplanes*.

Let $\mathcal{G}$ be an EDPS of rank $m+1 \geq 3$ in which the residue of an element of type 1 is isomorphic to $\mathcal{D}$ and let $G$ be a flag-transitive automorphism group of $\mathcal{G}$. For an element $\alpha$ of type 1 in $\mathcal{G}$ let $G(\alpha)$ denote the stabilizer of $\alpha$ in $G$.

**Lemma 2.1** *In the above notation $G(\alpha)$ acts primitively on the point set of the dual polar space $\mathrm{res}_{\mathcal{G}}(\alpha) \cong \mathcal{D}$.*

*Proof:* Clearly $G(\alpha)$ induces a flag-transitive action on the residual dual polar space $\text{res}_{\mathcal{G}}(\alpha)$. If $m \geq 3$ then a flag-transitive automorphism group of $\mathcal{D}(X)$ contains $X$ and acts primitively on the point set of $\mathcal{D}(X)$ (compare Chapter 9 in [Pasi]). In the case $m = 2$ we can apply the classification of flag-transitive EDPS's of rank 3 [DGMP]. □

With $\mathcal{D}$ as above let $\Delta$ be the collinearity graph of $\mathcal{D}$ that is a graph on the point set of $\mathcal{D}$ in which two points are adjacent if there is a line incident to both of them. For a point $p$ of $\mathcal{D}$ and $0 \leq i \leq m$ let $\Delta_i(p)$ denote the set of vertices whose distance from $p$ in $\Delta$ is $i$. The properties of $\Delta$ stated below are well known (cf. Section 9.4 in [BCN]).

**Lemma 2.2** *The following assertions hold:*

(i) *$\Delta$ is a distance-regular graph of diameter $m$ with parameters*

$$c_i = \left[{i \atop 1}\right]_t, \quad b_i = st^i\left[{m-i \atop 1}\right]_t \quad for \quad 0 \leq i \leq m,$$

*in particular the valency of $\Delta$ is $k = b_0 = s\left[{m \atop 1}\right]_t$;*

(ii) *the points incident to a line form a clique of size $s + 1$ and all cliques in $\Delta$ are of this type;*

(iii) *the points incident to a plane $\mu$ induce a subgraph $Q(\mu)$ which is the point graph of a generalized quadrangle of order $(s,t)$ (a quad); any two vertices at distance 2 in $\Delta$ are in a unique quad;*

(iv) *the subgraph induced on $\Delta_m(p)$ is connected.* □

Let $Y$ denote the space of functions defined on the vertex set of $\Delta$ and taking values in the field of rational numbers. For $p \in \Delta$ let $y_p$ be the function such that $y_p(q) = \delta_{p,q}$ for every $q \in \Delta$. Clearly $\{y_p \mid p \in \Delta\}$ is a basis of $Y$ and we define the inner product on $Y$ with respect to which this basis becomes orthonormal. Let $A$ be the *adjacency operator* on $Y$ which is the linear transformation defined by

$$A : y_p \mapsto \sum_{q \in \Delta_1(p)} y_q.$$

Then $A$ has exactly $m + 1$ distinct eigenvalues: $\lambda_0 = k > \lambda_1 > ... > \lambda_m$ and $U$ splits into the orthogonal sum

$$Y = Y_0 \oplus Y_1 \oplus ... \oplus Y_m$$

where $Y_i$ is the subspace of eigenvectors with eigenvalue $\lambda_i$. (Notice that all the eigenvalues $\lambda_i$ are rational.) Let $y_p^i$ denote the orthogonal projection of $y_p$ into $Y_i$ multiplied by a positive scalar which makes it unit and let $f_i = \dim Y_i$. The following lemma comes from the general theory of Bose–Mesner algebras of association schemes applied to the dual polar spaces [BCN].

**Lemma 2.3** *In the above notation if $p$ and $q$ are adjacent, then $\langle y_p^i, y_q^i \rangle = \frac{\lambda_i}{k}$. If $s = 2$ then $\frac{\lambda_m}{k} = -\frac{1}{2}$ and $f_m = (2^m + 1)(2^{m-1} + 1)/3$ if $X = Sp_{2m}(2)$ while $f_m = (4^m + 2)/3$ if $X = U_{2m}(2)$.* □

The final result of this section can be proved independently and also follows from the classification of flag-transitive EDPS's of rank 3 [DGMP].

**Lemma 2.4** *Let $\mathcal{G}$ be an EDPS of rank $m + 1 \geq 3$ and $G$ be a flag-transitive automorphism group of $\mathcal{G}$. Let $\gamma$ be an element of type 3 in $\mathcal{G}$ and $\bar{G}(\gamma)$ be the action induced by the stabilizer of $\gamma$ in $G$ on the set of elements of type 1 incident to $\gamma$. Then one of the following holds:*

(i) *$s = 2$ and $\bar{G}(\gamma) \cong Sym_4$;*

(ii) *$s = 3$ and $Alt_5 \leq \bar{G}(\gamma) \leq Sym_5$;*

(iii) *$s = 4$ and $Alt_6 \leq \bar{G}(\gamma) \leq Sym_6$;*

(iv) *$s = 9$ and $\bar{G}(\gamma) \cong Mat_{11}$.* □

# 3 The graph of EDPS

In many circumstances it is convenient to study an EDPS $\mathcal{G}$ via its graph $\Gamma = \Gamma(\mathcal{G})$ whose vertices are the elements of type 1 and two vertices are adjacent if there is an element of type 2 incident to both of them. We will often refer to elements of type 1 in $\mathcal{G}$ as *vertices*. An element of type 2 in $\mathcal{G}$ is incident to exactly two vertices which are adjacent. Suppose that $G$ is a flag-transitive automorphism group of $\mathcal{G}$ and $G(\alpha)$ is the stabilizer in $G$ of a vertex $\alpha$. Then by Lemma 2.1 $G(\alpha)$ acts primitively on the set of elements of type 2 incident to $\alpha$. Hence either different elements of type 2 in $\mathcal{G}$ correspond to different edges in $\Gamma$ or there is only one edge. It is obvious that the latter is impossible. So in the flag-transitive case the elements of type 2 in $\mathcal{G}$ can be identified with the *edges* of $\Gamma$. Notice that $\Gamma$ is connected and the action of $G$ on $\Gamma$ is faithful.

Let $\mathcal{G}$ be a flag-transitive EDPS, let $\Gamma = \Gamma(\mathcal{G})$ and let $\delta$ be an element of type $i$ in $\mathcal{G}$. Let us associate with $\delta$ a subgraph $\Gamma[\delta]$. If $i = 1$ then $\Gamma[\delta]$ consists of the single vertex $\delta$ and has no edges. Otherwise $\Gamma[\delta]$ is the subgraph in $\Gamma$ formed by the vertices and edges incident to $\delta$ (notice that in general $\Gamma[\delta]$ is not an induced subgraph). As we have observed, if $i = 2$ then $\Gamma[\delta]$ consists of two vertices and an edge joining these vertices; if $i = 3$ then $\Gamma[\delta]$ is a complete subgraph on $s + 2$ vertices. Suppose that $i \geq 4$. Then the elements of type less than $i$ incident to $\delta$ form an EDPS $\mathcal{G}_\delta$ of rank $i - 1$ and it is easy to see that $\Gamma[\delta]$ is isomorphic to $\Gamma(\mathcal{G}_\delta)$, in particular it is connected.

The following two lemmas easily follow from the basic property of the classical dual polar spaces.

**Lemma 3.1** *Let $\mathcal{G}$ be a flag-transitive EDPS of rank $m+1$. Then the incidence system whose elements of type $i$, $1 \leq i \leq m+1$ are the subgraphs $\Gamma[\delta]$ taken for all elements $\delta$ of type $i$ in $\mathcal{G}$ with respect to the incidence relation given by inclusion, is isomorphic to $\mathcal{G}$.* □

**Lemma 3.2** *Let $\alpha$ be a vertex of $\Gamma$, $\delta_1$ and $\delta_2$ be distinct elements from $\mathrm{res}_{\mathcal{G}}(\alpha)$ which are not incident to each other. Then there is an edge incident to $\alpha$ which is contained in $\Gamma[\delta_1]$ but not in $\Gamma[\delta_2]$.* □

As above let $\mathcal{G}$ be a flag-transitive EDPS of rank at least 3 and $\Gamma = \Gamma(\mathcal{G})$. Suppose that $\hat{\Gamma}$ is a graph possessing a morphism $\chi : \hat{\Gamma} \to \Gamma$ onto $\Gamma$. By this we mean that for every $\hat{x} \in \hat{\Gamma}$ the restriction of $\chi$ to $\hat{\Gamma}(\hat{x})$ is a bijection onto $\Gamma(\chi(\hat{x}))$. Suppose in addition that for every element $\varrho$ of type 3 in $\mathcal{G}$ the preimage $\chi^{-1}(\Gamma[\varrho])$ is a disjoint union of subgraphs each of which is isomorphic to $\Gamma[\varrho]$.

**Lemma 3.3** *Under the above assumptions there exists an EDPS $\hat{\mathcal{G}}$ possessing a 2-covering onto $\mathcal{G}$ such that $\hat{\Gamma} = \Gamma(\hat{\mathcal{G}})$.*

*Proof:* Define $\hat{\mathcal{G}}$ to be the geometry whose elements are the connected components of the subgraphs $\chi^{-1}(\Gamma[\delta])$ for all elements $\delta$ of $\mathcal{G}$. The type function is inherited from $\mathcal{G}$ and the incidence relation is defined by inclusion. Then the result follows from Lemmas 3.1, 3.2 and from the properties of $\chi$. □

**Lemma 3.4** *Let $\mathcal{G}^t$ denote the rank 3 truncation of $\mathcal{G}$ containing elements of type 1, 2 and 3. Let $\psi : \mathcal{H} \to \mathcal{G}^t$ be a covering. Then there exists an EDPS $\hat{\mathcal{G}}$ possessing a 2-covering onto $\mathcal{G}$ such that $\mathcal{H} = \hat{\mathcal{G}}^t$.*

*Proof:* Let $\hat{\Gamma}$ be the graph on the set of elements of type 1 in $\mathcal{H}$ in which two elements are adjacent if they are incident to a common element of type 2. Then $\psi$ induces a morphism $\chi : \hat{\Gamma} \to \Gamma(\mathcal{G})$ and since $\psi$ is a covering, for every element $\varrho$ of type 3 in $\mathcal{G}^t$ all connected components of $\chi^{-1}(\Gamma[\varrho])$ are isomorphic to $\Gamma[\varrho]$ (which is a complete graph on $s+2$ vertices). Hence the result follows from Lemma 3.3. □

# 4 Affine geometries

In this section we prove Theorem 2. Let $\mathcal{G}$ be an AEDPS. Let $G$ be a flag-transitive automorphism group of $\mathcal{G}$ and $T$ be the translation group of $\mathcal{G}$ with respect to $G$. Let $\alpha$ be a vertex of $\Gamma = \Gamma(\mathcal{G})$ and $\mathcal{D} = \mathrm{res}_{\mathcal{G}}(\alpha)$. Then the set $\Gamma(\alpha)$ of vertices adjacent to $\alpha$ in $\Gamma$ can be identified with the vertex set of the collinearity graph $\Delta$ of $\mathcal{D}$. By Lemma 2.1 the action of $D := G(\alpha)$ on $\Gamma(\alpha)$ is primitive. For a point $p$ of $\mathcal{D}$ let $t_p$ be the unique element from $T$ which maps $\alpha$ onto $p$ (where the latter is considered as a vertex from $\Gamma(\alpha)$).

**Lemma 4.1** $t_p^2 = 1$.

*Proof:* It is clear that $H := G(\alpha) \cap G(p)$ centralizes $t_p$. Since the action of $G(\alpha)$ on $\Gamma(\alpha)$ is primitive, $H$ does not stabilize in $\Gamma(\alpha)$ points other than $p$ but it certainly stabilizes the preimage $q$ of $\alpha$ under $t_p$ and $q$ is in $\Gamma(\alpha)$ since $\alpha$ and $p$ are adjacent. So $q$ must be equal to $p$ *i.e.* the orbit of $t_p$ containing $\alpha$ has length 2. Since $t_p \in T$ and the latter acts regularly on the vertex set, the result follows. □

**Lemma 4.2** *Let $\beta$ be the edge $\{\alpha, p\}$. Then*

(i) *every line of $\mathcal{D}$ contains exactly* 3 *points;*

(ii) $t_p$ *stabilizes* $\mathrm{res}_{\mathcal{G}}(\{\alpha, \beta\})$ *elementwise;*

(iii) *if* $\{p, q, r\}$ *is a line of $\mathcal{D}$ then* $t_p t_q t_r = 1$.

*Proof:* Let $\gamma$ be an element of type 3 and $\bar{G}(\gamma)$ be the action induced by the stabilizer of $\gamma$ in $G$ on the set of vertices incident to $\gamma$. Let $\beta$ be an edge incident to $\gamma$. Then there is a unique element $t_\beta \in T$ which stabilizes $\beta$ as a whole (for instance if $\beta = \{\alpha, p\}$ then $t_\beta = t_p$). The elements $t_\beta$ taken for all edges in $\gamma$ generate a subgroup which is normal in $G(\gamma)$ and whose image in $\bar{G}(\gamma)$ is a regular normal subgroup. By Lemma 2.4 $s = 2$ and (i) follows.

By the flag-transitivity $H := G(\alpha) \cap G(p)$ acts transitively on the of elements of type 3 incident to $\beta$. By Lemma 4.1 and its proof $t_p$ is an involution and it commutes with $H$. Since the lines in $\mathcal{D}$ have size 3, the number of elements of type 3 incident to $\beta$ is odd. Hence $t_p$ stabilizes them elementwise. Finally, since $\mathrm{res}_{\mathcal{G}}(\{\alpha, \beta\})$ is a projective geometry, (ii) follows.

Let $\{p, q, r\}$ denote the line of $\mathcal{D}$ which corresponds to $\gamma$. Then $\langle t_p,\ t_q,\ t_r \rangle$ induces an action of order 4 on the set of vertices incident to $\gamma$ so that $t_p t_q t_r$ stabilizes each of these vertices. Since $T$ acts regularly on the vertex set of $\Gamma$, we obtain (iii). □

By Lemmas 4.1 and 4.2 (iii) $T$ is a representation group of $\mathcal{D}$. In addition $T$ is $D$-admissible and $D$ acts flag-transitively on $\mathcal{D}$.

**Lemma 4.3** *$T$ is separable.*

*Proof:* If $m = 2$ then we can apply the classification of the flag-transitive EDPS's [DGMP]. If $m \geq 3$ and $\mathcal{D} = \mathcal{D}(X)$ then $D$ contains $X$ and the stabilizers in $D$ of different elements of $\mathcal{D}$ are different maximal subgroups in $D$. Hence if $T_{\delta_1} = T_{\delta_2}$ then $T_{\delta_1}$ is normalized by two different maximal subgroups in $D$, which means that it is normalized by the whole $D$ and $T_{\delta_1} = T$ since $D$ is transitive on the point-set of $\mathcal{D}$. Let $h$ be a hyperplane in $\mathcal{D}$ containing $\delta_1$. Then there is a non-trivial element in $D$ which fixes $h$ (and $\delta_1$) point-wise. This shows that $T_{\delta_1}$ and $T$ have different centralizers in $D$, a contradiction. □

Now to complete the proof of Theorem 2 (ii) it remains to establish the following.

**Lemma 4.4** *$\mathcal{G}$ is isomorphic to $\mathcal{A}(\mathcal{D}, T)$.*

*Proof:* As above let $\alpha$ be a vertex of $\Gamma = \Gamma(\mathcal{G})$. For an element $\delta$ of $\mathcal{G}$ define

$$T(\delta) = \{t \mid t \in T, \quad \alpha^t \in \Gamma[\delta]\}$$

If $\delta$ is a vertex then $T(\delta)$ consists of the unique element of $T$ which maps $\alpha$ onto $\delta$. In this case $T(\delta)$ can also be considered as a right coset in $T$ of the identity subgroup. Suppose that $\delta$ is an element of type $i \geq 2$ incident to $\alpha$. Then $T(\delta)$ is the stabilizer of $\delta$ in $T$ and $\delta$ can be considered as an element from $\mathcal{D} = \text{res}_{\mathcal{G}}(\alpha)$. We claim that in this case $T(\delta) = T_\delta$. If $i = 2$ then $\delta$ is just an edge containing $\alpha$ and the claim is obvious. Suppose that $i \geq 3$ and let $p$ be a point of $\mathcal{D}$ incident to $\delta$. By Lemma 4.2 (ii) $t_p$ stabilizes $\delta$ and hence $T_\delta \subseteq T(\delta)$. Since the subgraph $\Gamma[\delta]$ is connected, it is easy to see that $T_\delta$ acts transitively on the vertex set of $\Gamma[\delta]$. This means that if $T(\delta)$ would contain $T_\delta$ properly then there would be an element in $T(\delta) \setminus T_\delta$ stabilizing $\alpha$ which is impossible since the action of $T$ on the vertex set of $\Gamma$ is regular. Hence $T(\delta) = T_\delta$. Now suppose that $\delta$ is an arbitrary element in $\mathcal{G}$ of type $i \geq 2$ (non-necessary incident to $\alpha$). Since $T$ acts transitively on the vertex set of $\Gamma$, the set $T(\delta)$ is non-empty. Let $t \in T(\delta)$ and put $\sigma = \delta^{t^{-1}}$ so that $\alpha \in \Gamma[\sigma]$. Every element $t' \in T(\sigma)t$ maps $\sigma$ onto $\delta$ and hence it maps $\alpha$ into $\Gamma[\delta]$ which means that $t' \in T(\delta)$. If $t_1$ is an element from $T(\delta)$, then $\alpha^{t_1 t^{-1}} \in \Gamma[\sigma]$ and there is a (unique) element $r \in T(\delta)$ such that $\alpha^r = \alpha^{t_1 t^{-1}}$. Hence $t_1 t^{-1} r$ is an element of $T$ which stabilizes $\alpha$. This element must be the identity since $T$ acts regularly on the vertex set of $\Gamma$. This shows that $t_1 \in T(\sigma)t$ and hence $T(\delta)$ is a right coset of $T(\sigma)$. Since $T(\sigma) = T_\sigma$ we conclude that $T(\delta)$ is a right coset of $T_\sigma$ in $T$.

Since $T$ is separable, every element $\sigma$ of $\mathcal{D}$ can be identified with the subgroup $T_\sigma$ and under this identification the incidence relation is given by inclusion. This implies that two elements $\delta_1$ and $\delta_2$ are incident in $\mathcal{G}$ if and only if either $T(\delta_1) \subseteq T(\delta_2)$ or $T(\delta_2) \subseteq T(\delta_1)$. Hence the mapping $\delta \mapsto T(\delta)$ establishes an isomorphism of $\mathcal{G}$ onto $\mathcal{A}(\mathcal{D}, T)$. □

By the above lemma and Lemma 1.1 (iii) there is a 2-covering of $\mathcal{A}(\mathcal{D}, \mathcal{R}(\mathcal{D}))$ onto $\mathcal{G}$, and to complete the proof of Theorem 2 (iii) it is sufficient to show that $\mathcal{A}(\mathcal{D}, \mathcal{R}(\mathcal{D}))$ is 2-simply connected.

As in the previous section for an EDPS $\mathcal{G}$ let $\mathcal{G}^t$ denote the rank 3 truncation of $\mathcal{G}$ containing elements of type 1, 2 and 3. Let $\mathcal{H} \to \mathcal{G}^t$ be the universal covering of $\mathcal{G}^t$. Then by Lemma 3.4 $\mathcal{H} = \hat{\mathcal{G}}^t$ for an EDPS $\hat{\mathcal{G}}$ possessing a 2-covering onto $\mathcal{G}$.

**Lemma 4.5** *In the above notation suppose that $\mathcal{G}$ is affine, $G$ is an automorphism group of $\mathcal{G}$ and $T$ is the translation group of $\mathcal{G}$ with respect to $G$.*

(i) *$\hat{\mathcal{G}}$ is affine, there is a flag-transitive automorphism group $\hat{G}$ of $\hat{\mathcal{G}}$ such that the translation group $\hat{T}$ of $\hat{\mathcal{G}}$ with respect to $\hat{G}$ possesses a representation homomorphism onto $T$;*

(ii) *if* $\mathcal{G} = \mathcal{A}(\mathcal{D}, R(\mathcal{D}))$ *then* $\hat{\mathcal{G}} = \mathcal{G}$.

*Proof:* Every representation homomorphism onto $R(\mathcal{D})$ is an isomorphism and $R(\mathcal{D})$ is the translation group of $\mathcal{A}(\mathcal{D}, R(\mathcal{D}))$ with respect to its automorphism group, hence (ii) is implied by (i). Let $\hat{G}$ be the group of all liftings of elements from $G$ to automorphisms of $\hat{\mathcal{G}}$. Let $\psi$ be the natural homomorphism of $\hat{G}$ onto $G$, so that the kernel $K$ of $\psi$ is the group of deck transformations. Since $\chi : \hat{\mathcal{G}}^t \to \mathcal{G}^t$ is the universal cover, $K$ acts regularly on $\chi^{-1}(\delta)$ for every $\delta \in \mathcal{G}^t$. Let $\hat{T}$ be the full preimage of $T$ with respect to $\psi$. Since $T$ is normal in $G$, $\hat{T}$ is normal in $\hat{G}$. We claim that $\hat{T}$ acts regularly on the vertex set of $\hat{\mathcal{G}}$. In fact, $K$ acts regularly on $\chi^{-1}(\alpha)$ for every vertex $\alpha$ of $\mathcal{G}$ and $T = \hat{T}/K$ acts regularly on the vertex set of $\mathcal{G}$, or equivalently on the set of orbits of $K$ on the vertex set of $\hat{\mathcal{G}}$. Hence the claim follows and $\hat{\mathcal{G}}$ is affine. Let $\hat{\alpha}$ be a vertex of $\hat{\mathcal{G}}$ and let $\mathcal{D}$ denote $\mathrm{res}_{\hat{\mathcal{G}}}(\hat{\alpha})$ isomorphic to $\mathrm{res}_{\mathcal{G}}(\chi(\hat{\alpha}))$. Then by Lemma 4.4 $\mathcal{G} \cong \mathcal{A}(\mathcal{D}, T)$, $\hat{\mathcal{G}} \cong \mathcal{A}(\mathcal{D}, \hat{T})$. Hence the 2-cover $\hat{\mathcal{G}} \to \mathcal{G}$ induces a representation homomorphism of $\hat{T}$ onto $T$ and (i) follows. □

Let $\mathcal{A} = \mathcal{A}(\mathcal{D}, \mathcal{R}(\mathcal{D}))$ and $G$ be a flag-transitive automorphism group of $\mathcal{A}$. Let $\Phi = \{\alpha_i \mid 1 \leq i \leq m+1\}$ be a maximal flag in $\mathcal{A}$ where $\alpha_i$ is of type $i$ and let $G_i$ be the stabilizer of $\alpha_i$ in $G$. Then by a standard principle [Pasi] Lemma 4.5 imply that $G$ is the universal completion of the amalgam consisting of the subgroups $G_1$, $G_2$ and $G_3$. For $1 \leq i < j \leq m+1$ let $P_{ij}$ denote the elementwise stabilizer in $G$ of the flag $\Phi \setminus \{\alpha_i, \alpha_j\}$. Then by the same principle $\mathcal{A}$ is 2-simply connected if and only if $G$ is the universal completion of the amalgam consisting of all the $P_{ij}$. On the other hand if $k \in \{1, 2, 3\}$ then $\mathrm{res}_{\mathcal{A}}(\alpha_k)$ is 2-simply connected and hence $G_k$ is the universal completion of the amalgam consisting of the $P_{ij}$ for $k \notin \{i, j\}$. This shows that $\mathcal{A}$ is 2-simply connected.

# 5 On the universal representation groups

By Theorem 2 the structure of the affine EDPS's is determined by that of the universal representation groups of the classical dual polar spaces with three points per a line. In this section we summarize our knowledge about these representation groups (compare Theorem 3).

Let $\mathcal{D}$ be a classical dual polar space of rank $m$ with 3 points on every line (that is $\mathcal{D} = \mathcal{D}(Sp_{2m}(2))$ or $\mathcal{D} = \mathcal{D}(U_{2m}(2))$) and $\mathcal{R} = \mathcal{R}(\mathcal{D})$ be the universal representation group of $\mathcal{D}$. Let $f_m$ denote the multiplicity of the smallest eigenvalue of the collinearity graph of $\mathcal{D}$ (compare Lemma 2.3): $f_m = (2^m + 1)(2^{m-1} + 1)/3$ or $(4^m + 2)/3$ depending on whether $\mathcal{D}$ is of symplectic or unitary type.

It is well known and follows from the classification of the flag-transitive EDPS's of rank 3 [DGMP] that the AEDPS's with automorphism groups $2^5 : Sp_4(2)$ and $2^6 : O_6^-(2)$ are simply connected. By Theorem 2 (iii) this implies the following.

**Lemma 5.1** *If* $m = 2$ *then* $\mathcal{R}$ *is elementary abelian* 2*-group of rank* $f_2$. □

Now we are going to construct an abelian factor group $\bar{\mathcal{R}}$ of $\mathcal{R}$. Let $P$ and $L$ be the point set and the line set of the dual polar space under consideration. Let $\mathcal{V} = \mathcal{R}/\mathcal{R}'$ be the largest abelian factor group of $\mathcal{R}$. Then $\mathcal{V}$ has the set $\{z_p \mid p \in P\}$ of generators and we split the set of its relations into the following three families:

$R_1$ : $[z_p, z_q] = 1$ for $p, q \in P$;

$R_2$ : $z_p z_q z_r = 1$ for $\{p, q, r\} \in L$;

$R_3$ : $z_p^2 = 1$ for $p \in P$.

If we drop the relations $R_1$ we obtain a presentation for $\mathcal{R}$. On the other hand the group subject only to the relations $R_1$ is a free abelian group whose rank is $|P|$. Let $\mathcal{A}$ denote the group generated by the $z_p$ subject to the relations $R_1$ and $R_2$. Then $\mathcal{V} = \mathcal{A}/2\mathcal{A}$ and we are going to produce a lower bound on the rank of the torsion-free part of $\mathcal{A}$.

Let $\Delta$ be the collinearity graph of $\mathcal{D}$. Then $P$ and $L$ can be identified with the set of vertices and the set of triangles in $\Delta$, respectively. Let $Y$ be the space of rational valued functions on $\Delta$ and let $Y_m$ be the $m$-th eigenspace of the adjacency operator as defined before Lemma 2.3. If $\{p, q, r\} \in L$ then $y_p^m + y_q^m + y_r^m$ is the zero vector since $y_p^m$ is a unit vector for $p \in P$ and $\langle y_p^m, y_q^m \rangle = -\frac{1}{2}$ whenever $p$ and $q$ are adjacent (cf. Lemma 2.3). This means that the mapping

$$\varrho : z_p \mapsto y_p^m$$

defines a homomorphism of $\mathcal{A}$ into the additive group of $Y_m$. It is easy to see that $\mathrm{Im}(\varrho)$ is a free abelian group whose rank is $f_m$ that is the dimension of $Y_m$. Hence $\bar{\mathcal{R}}$ defined as $\mathrm{Im}(\varrho)/2\,\mathrm{Im}(\varrho)$ is a homomorphic image of $\mathcal{V}$ and this image is an elementary abelian 2-group of rank $f_m$. So we have the following.

**Lemma 5.2** *The group $\mathcal{R}$ has a factor group $\bar{\mathcal{R}}$ which is an elementary abelian 2-group of rank $f_m$.* □

It was conjectured in [Bro] that $\bar{\mathcal{R}} = \mathcal{V}$ in the symplectic case and it is natural to include the unitary case in this conjecture.

**Conjecture 5.3** *The group $\mathcal{V} = \mathcal{R}/\mathcal{R}'$ is an elementary abelian 2-group of rank $f_m$.*

Lemma 5.1 proves Conjecture 5.3 for $m = 2$. The conjecture was proved in the symplectic case for $m = 3$ in [CS] and for $m = 4$, 5 in [Coo] and in the unitary case for $m = 3$ in [Yosh2]. Earlier some of these results were established by A.E. Brouwer (unpublished) using computer calculations.

**Lemma 5.4** *If $m = 3$ then the commutator subgroup of $\mathcal{R}$ has order 2.*

*Proof:* In the considered situation we have $\mathcal{D} = \mathcal{D}(Sp_6(2))$ or $\mathcal{D} = \mathcal{D}(U_6(2))$. First we are going to show that $\mathcal{R}$ is non-abelian.

Let $\mathcal{B}$ be a building with the diagram

in which the residue of a point is isomorphic to $\mathcal{D}$ and $B$ be the full automorphism group of $\mathcal{B}$. This means that $B \cong F_4(2)$ if $t = 2$ and $B \cong {}^2E_6(2) : Sym_3$ if $t = 4$. The elements of $\mathcal{B}$ which correspond to the nodes from the left to the right on the diagram will be called, respectively points, lines, planes and symplecta. Every element will be identified with the set of points it is incident to, so that the incidence relation is by inclusion. Let $P$ denote the point set of $\mathcal{B}$. For $p \in P$ let $H_p$ denote the set of points which are *not* at the maximal distance from $p$ in the collinearity graph of $\mathcal{B}$. Then $H_p$ is a geometrical hyperplane. Let $\mathcal{G}$ be a geometry whose elements are all elements of $\mathcal{B}$ which are not contained in $H_p$ with the incidence relation being as in $\mathcal{B}$. A plane $\pi$ in $\mathcal{B}$ contains 7 points and 7 lines. It is easy to see that either $\pi \subseteq H_p$ or $\pi \cap H_p$ is a line. In the latter case $\pi \setminus H_p$ is the affine plane of order 2 which one can easily identify with the geometry of vertices and edges of a complete graph on 4 vertices. If $q$ is a point outside $H_p$ then every element which contains $q$ is not contained in $H_p$ and hence the residues of $q$ in $\mathcal{B}$ and $\mathcal{G}$ are the same. This shows that $\mathcal{G}$ is an EDPS. Let $G$ be the stabilizer of $p$ in $B$. It is a standard fact about Lie type groups that $G \cong T : D$ where $T = O_2(G)$, $T$ acts regularly on $P \setminus H_p$ and $D$ is the full automorphism group of $\mathcal{D} \cong \mathrm{res}_{\mathcal{B}}(p)$. So $\mathcal{G}$ is flag-transitive and affine. In addition $T$ is non-abelian (its commutator subgroup has order 2). Namely, $G \cong 2^{1+8}.2^6 : Sp_6(2)$ in the symplectic case and $G \cong 2^{1+20} : U_6(2).Sym_3$ in the unitary case. By Theorem 2 (ii) this shows that $\mathcal{R}$ is non-abelian.

Thus to complete the proof we have to show that the commutator subgroup of $\mathcal{R}$ has order at most 2. To prove this we will make use of the following result (cf. Lemma 2.2 in [IPS]).

**Lemma 5.5** *Let $\mathcal{H}$ be a geometry with three points on each line. Let $H$ be a representation group of $\mathcal{H}$ and let $p \mapsto h_p$, $p \in P$ be the corresponding mapping. Suppose that for every point $p \in P$ there is a partition of $P$ into two disjoint subsets $A_p$ and $B_p$ such that the following three conditions hold:*

(i) *if $q \in A_p$ then $[h_p, h_q] = 1$;*

(ii) *let $\Sigma_p$ be the graph on $B_p$ in which $x, y \in B_p$ are adjacent if and only if there is $z \in A_p$ such that $\{x, y, z\}$ is a line, then $\Sigma_p$ is connected for every $p \in P$;*

(iii) *if $q \in B_p$ then $p \in B_q$ and the graph on $P$ in which every $p \in P$ is adjacent to all points from $B_p$, is connected.*

*Then the order of $H'$ is at most 2.* □

*Proof of Lemma 5.4 (cont.)* Let $\Delta$ be the collinearity graph of $\mathcal{D}$. Then $\Delta$ is distance-regular of diameter 3 with the parameters given in Lemma 2.2. For $p \in \Delta$ put $B_p = \Delta_3(p)$ and $A_p = \Delta \setminus B_p$. If $q \in A_p$ then by Lemma 2.2 (iii) $p$ and $q$ are incident to a common plane $\mu$ and hence they are contained in a common quad $Q(\mu)$. It is clear that the elements $z_r$ for $r \in Q(\mu)$ generate in $\mathcal{R}$ a representation group of $\mathrm{res}_{\mathcal{D}}(\mu)$. By Lemma 5.1 this means that $[z_p, z_q] = 1$ and condition (i) in Lemma 5.5 holds. The condition (ii) follows from Lemma 2.2 (iv). Finally (iii) holds since the automorphism group of $\Delta$ acts primitively on its vertex-set. □

In [Yosh2] for $\mathcal{D} = \mathcal{D}(Sp_6(2))$ the geometry $\mathcal{A}(\mathcal{D}, \mathcal{R}(\mathcal{D}))$ was constructed and its simple connectedness was established using computer calculations.

**Lemma 5.6** *Let $\tau$ be an element of type $m$ in $\mathcal{D} = \mathcal{D}(Sp_{2m}(2))$, $m \geq 3$. Then the subgroup in $\mathcal{R}$ generated by the elements $z_p$ for all points $p$ incident to $\tau$ is the universal representation group of* $\mathrm{res}_{\mathcal{D}}(\tau)$.

*Proof:* Let $\mathcal{R}(\tau) = \langle z_p \mid p \in \mathrm{res}_{\mathcal{D}}(\tau)\rangle$. Since $\mathrm{res}_{\mathcal{D}}(\tau) \cong \mathcal{D}(Sp_{2m-2}(2))$, clearly $\mathcal{R}(\tau)$ is a quotient of the universal representation group of $\mathrm{res}_{\mathcal{D}}(\tau)$. An elegant construction of $\mathcal{D}$ in terms of its residue $\mathcal{D}(Sp_{2m-2}(2))$ is given in [CS]. Based on this construction it was shown in [CS] that for every abelian representation group $V$ of $\mathcal{D}(Sp_{2m-2}(2))$ there exists an abelian representation group $W$ of $\mathcal{D}$ such that the submodule in $W$ generated by the images of the points from $\mathrm{res}_{\mathcal{D}}(\tau)$ is isomorphic to $V$. This result can be almost literary extended to the case of non-abelian representations and this gives the claim of the lemma. □

Now by Lemmas 5.4 and 5.6 we see that if $m \geq 3$ and $\mathcal{D}$ is of symplectic type then $\mathcal{R}$ is non-abelian.

We conclude the paper by posing two conjectures concerning the structure of $\mathcal{R}$.

**Conjecture 5.7** *The analogue of Lemma* 5.6 *holds in the unitary case.*

The fixed-point set in $\mathcal{D}(U_{2m}(2))$ of a non-trivial automorphism of $GF(4)$ forms a subgeometry isomorphic to $\mathcal{D}(Sp_{2m}(2))$.

**Conjecture 5.8** *If $\mathcal{D}$ is of unitary type then the subgroup in $\mathcal{R}$ generated by the elements $z_p$ for all points $p \in \mathcal{D}(Sp_{2m}(2))$ is the universal representation group of $\mathcal{D}(Sp_{2m}(2))$.*

A possible way of proving Conjectures 5.7 and 5.8 might be to produce a construction of $\mathcal{D}(U_{2m}(2))$ in terms of $\mathcal{D}(U_{2m-2}(2))$ and a construction of $\mathcal{D}(U_{2m}(2))$ in terms of $\mathcal{D}(Sp_{2m}(2))$ analogous to the construction of $\mathcal{D}(Sp_{2m}(2))$ given in [CS].

It would be very important to find out whether the group $\mathcal{R}$ is finite or infinite for $m \geq 4$.

## References

[Bro] A.E. Brouwer, Dimensions of embeddings of near polygons, In: Algebraic Combinatorics, Abstracts, p. 2, Fukuoka, 1993.

[BCN] A.E. Brouwer, A.M. Cohen and A. Neumaier, *Distance Regular Graphs*, Springer Verlag, Berlin, 1989.

[Coo] B.N. Cooperstein, On the generation of some dual near polygons of symplectic type over $GF(2)$, *Europ. J. Combin.* (to appear)

[CS] B.N. Cooperstein and E.E. Shult, Combinatorial construction of some near polygons, Preprint, 1993.

[DGMP] A. Del Fra, D. Ghinelli, T. Meixner and A. Pasini, Flag-transitive extensions of $C_n$ geometries, *Geom. Dedic.* **37** (1991), 253–273.

[Ivn1] A.A. Ivanov, On flag-transitive extended dual polar spaces, In: Combinatorics 94, Abstracts, pp. 55–60, Pescara, 1994.

[Ivn2] A.A. Ivanov, Exceptional extended dual polar spaces, *Europ. J. Combin.* (submitted)

[IPS] A.A. Ivanov, D.V. Pasechnik and S.V. Shpectorov, Non-abelian representations of some sporadic geometries, *J. Algebra* **181** (1996), 523–557.

[Pasi] A. Pasini, *Diagram Geometries*, Clarendon Press, Oxford, 1994.

[Yosh1] S. Yoshiara, On some extended dual polar spaces I, *Europ. J. Combin.* **15** (1994), 73–86.

[Yosh2] S. Yoshiara, Codes associated with some near polygons having three points on each line, *Algebra Colloq.* **2** (1995), 79–96.

Alexander A. Ivanov,
Department of Mathematics,
Imperial College,
180 Queen's Gate,
London, SW7 2BZ, England, UK
e-mail: a.ivanov@ic.ac.uk

# Derivable Nets May be Embedded in Nonderivable Planes

Norman L. Johnson*

**Abstract**

A construction is given that provides examples of infinite dual affine translation planes which contain derivable nets $D$ but which are not derivable by the derivation of $D$.

## 1 Introduction

This article is concerned with the resolution of certain fundamental questions concerning the concept of "derivation". In order to somewhat appreciate why such a study might be interesting, we shall provide a short history of the problems associated with this construction procedure. The reader may be interested in a somewhat more detailed history of derivation given in the survey article [19].

The concept of a finite derivable affine plane was conceived by T.G. Ostrom in the early $1960's$ (see [22] and [24]) and has been arguably the most important construction procedure of affine planes invented in the last thirty years. A finite affine plane may be "derived" to produce another affine plane of the same order. For example, the Hall planes of order $q^2$ originally constructed by Marshall Hall Jr. [9] by coordinate methods were shown by Albert [1] to be constructible from any Desarguesian affine plane of order $q^2$ by the method of derivation. The Hughes planes [10] of order $q^2$ were shown to be derivable and the projective planes constructed were the first examples of finite projective planes of Lenz-Barlotti class II-1 (there is a single, incident, point-line transitivity). The planes obtained were independently discovered by T.G. Ostrom [23] and L.A. Rosati [26] and are called the Ostrom-Rosati planes.

The description of a finite derivable affine plane is as follows:

**Definition 1.1** *Let $\pi$ denote a finite affine plane of order $q^2$ and let $\pi^E$ denote the projective extension of $\pi$ by the adjunction of the set $L_\infty$ of parallel classes as a line.*

*The author gratefully acknowledges the helpful comments of the referee in the writing of this article.

*Let $D_\infty$ denote a subset of $q+1$ points of $L_\infty$. $D_\infty$ is said to be a* ***derivation set*** *if and only it satisfies the following property:*

*If $a$ and $b$ are any two distinct points of $\pi$ whose join in $\pi^E$ intersects $L_\infty$ in $D_\infty$ then there is an affine subplane $\pi_{a,b}$ of order $q$ containing $a$ and $b$ and whose $q+1$ infinite points are exactly those of $D_\infty$.*

*Given any derivation set $D_\infty$, there is a corresponding set $B$ of $q^2(q+1)$ affine subplanes of order $q$ each of which has $D_\infty$ as its set of infinite points.*

*$\pi$ is said to be* ***derivable*** *if it contains a derivation set.*

The main result of Ostrom is

**Theorem 1.2** *(Ostrom [22]). Let $\pi$ be a finite derivable affine plane of order $q^2$ with derivation set $D_\infty$. Let $B$ denote the associated set of $q^2(q+1)$ subplanes of order $q$ each of which has $D_\infty$ as its set of infinite points.*

*Form the following incidence structure $\pi(D_\infty)$: The points of $\pi(D_\infty)$ are the points of $\pi$ and the lines of $\pi(D_\infty)$ are the lines of $\pi$ which do not intersect $D_\infty$ in the projective extension and the subplanes of $B$.*

*Then $\pi(D_\infty)$ is an affine plane of order $q^2$.*

**Definition 1.3** *The affine plane $\pi(D_\infty)$ is called* ***the plane derived from $\pi$ by the derivation of $D_\infty$*** *or merely* ***the plane derived from $\pi$***.

Every Desarguesian affine plane of order $q^2$ is derivable and if $GF(q^2)$ is a field coordinatizing the plane then $GF(q) \cup \{\infty\}$ considered as a subset of the set of points on the line at infinity is a derivation set. Furthermore, given any affine point $O$, the set of lines incident with $O$ and intersecting $GF(q) \cup \{\infty\}$ projectively defines a regulus $R$ in the associated three-dimensional projective space $PG(3,q)$. The set of subplanes of order $q$ incident with $O$ each of whose set of infinite points is $GF(q) \cup \{\infty\}$ is the opposite regulus $R^{opp}$ to $R$. It is immediate that the subplanes of $R$ are Desarguesian subplanes of order $q$.

Early attempts to recognize when a finite affine plane is derivable concentrated on the coordinate systems of the planes. For Desarguesian affine planes, the coordinatizing field $GF(q^2)$ is a right 2-dimensional vector space over $GF(q)$. More generally, assume that $(Q,+,\cdot)$ is a coordinate system for an affine plane $\pi$ of order $q^2$. The points of $\pi$ are the pairs $(x,y)$ for $x,y$ in $Q$ . If we assume that the lines of $\pi$ have the form $x=a, y=x\cdot m+b$ for $a,m,b$ in $Q$ then $\pi$ is derivable provided $(Q,+)$ is a right 2-dimensional vector space over a field $F$ isomorphic to $GF(q)$ ([24]).

It was noticed in Johnson [14] that the construction process of derivation was not dependent upon finiteness but rather on the existence of a class of Baer subplanes.

**Definition 1.4** *Let $\Sigma$ be a projective plane and $\Sigma_o$ a subplane of $\Sigma$.*

*$\Sigma_o$ shall be said to be a point-Baer subplane if and only if every point of the plane is incident with a line of $\Sigma_o$.*

*$\Sigma_o$ shall be said to be a line-Baer subplane if and only if every line of the plane is incident with a point of $\Sigma_o$.*

*$\Sigma_o$ shall be said to be a Baer subplane if and only if the subplane is both point-Baer and line-Baer.*

A counting argument establishes that within a projective plane of order $q^2$ any subplane of order $q$ is both a point-Baer and a line-Baer subplane so that any finite plane of order $q$ is a Baer subplane.

In the general case, an affine plane is derivable if there is an associated derivation set $D_\infty$ which is a subset of the set of infinite points and such that for each two distinct points $a$, $b$ whose join is in $D_\infty$ projectively then there is a Baer subplane $\pi_{a,b}$ containing $a$ and $b$ whose infinite points are exactly those of $D_\infty$.

It is possible to consider incidence structures which are analogous to the subnets of a derivable affine plane defined by a derivation set. The definitions of point-Baer and line-Baer subplanes within a net instead of within a plane are the same as in the plane case with the condition on the lines restricted to the lines of the net.

**Definition 1.5** *A derivable net $N = (P, L, B, C, I)$ is an incidence structure with a set $P$ of points, a set $L$ of lines, a set $B$ of Baer subplanes of the net, a set $C$ of parallel classes of lines and a set $I$ which is called the incidence set such that the following properties hold:*

*(i) Every point is incident with exactly one line from each parallel class, each parallel class is a cover of the points and each line of $L$ is incident with exactly one of the classes of $C$.*

*(ii) Two distinct points are incident with exactly one line of $L$ or are not incident.*

*(iii) If we refer to the set $C$ as the set of infinite points then the subplanes of $B$ are affine planes with infinite points exactly those of the set $C$.*

*(iv) Given any two distinct points $a$ and $b$ of $P$ which are incident with a line of $L$, there is a Baer subplane $\pi_{a,b}$ of $B$ containing (incident with) $a$ and $b$.*

**Remark 1** *Given a derivable net $N = (P, L, B, C, I)$, we may define a parallelism relation on the set $B$ of subplanes. Two subplanes $\rho$ and $\tau$ of $B$ are defined to be parallel if and only if they are disjoint on points. Define $N^* = (P, B, L, C^*, I)$ $= (P^*, L^*, B^*, C^*, I^*)$ as the incidence structure where $P^* = P$ is the set of points, $L^* = B$ is the set of lines, $B^* = L$ is the set of Baer subplanes, $C^*$ is the parallelism on the set of lines $L^*$ defined on the set of Baer subplanes $B$ and $I^*$ is the incidence set $I$.*

*Then $N^*$ is also a derivable net which is called the* ***derived net****. The transfer from $N$ to $N^*$ is called the* ***derivation of*** *$N$.*

Since in this paper we are trying to complete the outstanding problems on derivable affine planes and derivable nets, we shall list some of the original questions.

(1) If $(Q,+,\cdot)$ is a coordinate system for a derivable affine plane $\pi$, is there a skewfield $K$ such that $(Q,+)$ is a right 2-dimensional vector space over $K$ when the lines of the plane are written in the form $y = x \cdot m + b, x = c$ for $m, b, c$ in $Q$ ?

(2) Are the subplanes of any finite derivable net Pappian? Are the subplanes of any derivable net Desarguesian?

(4) Given a finite derivable affine plane with derivation set $D_\infty$, the net of points of the plane and lines which projectively intersect $D_\infty$ is a derivable net $N$. Does there exist a three-dimensional projective space $\Sigma$ such that the set of lines of $N$ incident with a given point is a regulus in $\Sigma$?

(5) What is the structure of a derivable net and is it possible to completely determine the full collineation group?

(6) Can the derivation procedure be considered as a geometric process?

Question (1) has been studied by Gründhofer [8], Lunardon [21], Ostrom [24], among others and was settled affirmatively provided the plane is a translation plane or a cartesian plane. More generally, this was shown to be true in any coordinate system for a net containing a derivable net by Johnson [17].

Prohaska [25] showed (2) to be true in the finite case and Cofman [7] extended the work of Prohaska in the infinite or general case by showing that the subplanes of any derivable net are Desarguesian but not always Pappian.

Furthermore, using certain ideas of Cofman [7], the structure of any derivable net, finite or infinite is completely determined in Johnson [15] and [18]. It is proved that any derivable net can be realized combinatorially within some $PG(3, K)$ where $K$ is a skewfield as follows:

**Theorem 1.6** *(Johnson [15]). Given a derivable net $N = (P, L, C, B, I)$, there is a 3-dimensional projective space $\Pi$ and a fixed line $R$ of $\Pi$ such that the set $P$ of points, the set $L$ of lines, the set $C$ of parallel classes, the set $B$ of Baer subplanes of the net $N$ are the set of lines of $\Pi$ skew to $R$, the set of points of $\Pi - R$, the set of planes of $\Pi$ that contain $R$, and the set of planes of $\Pi$ which do not contain $R$ respectively, where incidence $I$ is the natural incidence of $\Pi$.*

Note that it now follows that the collineation group of any derivable net is isomorphic to the full collineation group $P\Gamma L(4, K)_R$ of the 3-dimensional projective space $\Sigma$ which leaves the line $R$ invariant. Furthermore, with the use of this chacterization result and realization of the group of the derivable net, it is possible to give a complete structure theory for derivable nets.

**Theorem 1.7** *(Johnson [18]). A net is derivable if and only there is a skewfield $K$ and a left 4-dimensional vector space $V$ over $K$ such that the points of the net are the vectors of $V$ and the lines of the net are translates of the following set of $Z(K)$-subspaces:*

*The points are $(x_1, x_2, y_1, y_2) \equiv (x, y)$ for all $x_i$ , $y_i$ in $K$, $i = 1, 2$, and the lines are translates of the $Z(K)$ subspaces $x = 0, y = \delta x$ where $\delta x = \delta(x_1, x_2) = (\delta x_1, \delta x_2)$ for all $\delta$ in $K$.*

We call such nets **pseudo-regulus** nets to emphasize the connection to regulus nets. In the finite case, we obtain the following corollary.

**Corollary 1.8** *Any finite derivable net is a regulus net.*

Hence, questions (1) through (5) have been answered. To consider problem (6), given a derivable net $N$, realize $N$ combinatorially within a 3-dimensional projective space $\Sigma$ with distinguished line $R$ as above. The derivation process produces the derived net $N^*$ from $N$ as noted. It is also true that $N^*$ may also be realized combinatorially within a 3-dimensional projective space $\Sigma^*$ with distinguished line $R^*$. The transfer from $N$ to $N^*$ may be obtained by the action of a polarity $\delta$ of $\Sigma$ which leaves $R$ invariant. In this case, $R = R^*$ and if $N = (P, L, B, C, I)$ and since $L$ is the set of points of $\Sigma - R$ and $B$ is the set of hyperplanes of $\Sigma$ which do not contain the line $R$ then $L\delta = B = L^*$ and $B\delta = L = B^*$. Therefore, to answer the question posed in (6), we may describe the process of derivation geometrically as a polarity or as the action of a polarity. The reader is referred to Johnson [16] for further details.

## 1.1 Extension and embedding problems

There are some remaining problems or questions to be resolved involving the embedding of a derivable net in an affine plane or equivalently the extension of a derivable net to an affine plane.

Any finite derivable net with $q^2$ points per line and $q+1$ parallel classes may be coordinatized by a field isomorphic to $GF(q)$. Hence, any finite regulus net may be embedded in a Pappian affine plane since there is always a quadratic extension field $GF(q^2)$ of any finite field $GF(q)$.

We have mentioned that Ostrom proved that when any finite derivable net is embedded in any affine plane then the affine plane is forced to be derivable.

However, the proof of Ostrom's theorem is only valid for finite derivable nets. To be clear on what is meant by a derivable plane, we formulate the following definition.

**Definition 1.9** *Let $\pi$ be an affine plane which contains a derivable net $N$. Form the following structure $\pi(N)$: The points of $\pi(N)$ are the points of $\pi$ and the lines of $\pi(N)$ are the Baer subplanes of $N$ and the lines of $\pi$ which are not in $N$. If $\pi(N)$ is an affine plane then $\pi$ is said to be derivable.*

The two major unanswered questions concerning derivable nets are:
(7) Can any derivable net be extended to an affine plane ?
(8) If a derivable net is embedded in an affine plane, is the affine plane derivable ?

In this article, both of these questions are resolved.

It is shown that any derivable net can be embedded in a dual translation plane although not always in a Desarguesian or Pappian plane.

We have previously defined point-Baer and line-Baer subplanes. We shall see that the two concepts of point-Baer and line-Baer are independent. Furthermore, we shall use point-Baer subplanes which are not line-Baer subplanes in order to construct derivable nets which are embeddable in an affine plane in a somewhat non-standard manner. It is shown that any infinite derivable net can be embedded in a dual translation plane which is not derivable.

## 2 Spreads, dual spreads and Baer subplanes

**Definition 2.1** *A* ***spread*** *$S$ in $PG(2k-1, K)$, for $K$ a skewfield, and $k$ a positive integer $> 1$ is a set of skew $(k-1)$-dimensional subspaces such that each point of $PG(2k-1, K)$ is incident with one of the subspaces of $S$.*

*A* ***dual spread*** *$S^*$ is a set of skew $(k-1)$-dimensional subspaces such that each hyperplane of $PG(2k-1, K)$ contains a line of $S^*$.*

When $K$ is finite any spread is a dual spread and conversely. However, for any infinite skewfield, there are spreads which are not dual spreads and dual spreads which are not spreads. The known examples are in are in $PG(3, K)$ and the reader is referred to Cameron [6] (p.41, 4.1.6) for more details. The spreads mentioned in Cameron are generalizations of those given in Bruen and Fisher [5] for spreads over countable fields. The reader is also referred to Barlotti [2] and Bernardi [3] for additional background and details.

We shall see that there is an intimate connection between spreads which are dual spreads and Baer subplanes. We have previously defined point-Baer and line-Baer sublanes of a projective plane and noted that when the projective plane is finite then every point-Baer subplane is also a line-Baer subplane. However, in the infinite case, this is no longer the case as is shown by Barlotti [2].

Barlotti shows that there exist line-Baer subplanes which are not point Baer. It is probable that Barlotti was aware that his methods were applicable to the problems that we here discuss. However, where Barlotti would use the Bruck-Bose model for his analysis, we shall use the André representation to show that one may actually obtain derivable nets embedded into affine planes which are not derivable. Furthermore, we observe this fact also using vector spaces instead of projective spaces. Similarly, as in the work of Barlotti, we are interested in what occurs in the affine duals to the projective extensions of certain affine translation planes.

There are certain critical situations when it can always be guaranteed that a given subplane is a line-Baer subplane. We following standard practice of referring to an affine subplane $\pi_o$ of an affine plane $\pi$ as point-Baer or line-Baer if the projective extension subplane $\pi_o^E$ is point-Baer or line-Baer respectively in the projective extension $\pi^E$ of the affine plane.

**Theorem 2.2** *Let $S$ be a spread in $PG(2k-1, K)$. Let $\pi_S$ denote the associated affine translation plane obtained by taking vectors in the $2k$-dimensional vector* *space over $K$ as points and translates of the elements of $S$ (taken as vector subspaces) as lines.*

*Let $\pi_o$ be a $k$-dimensional $K$-vector subspace of $\pi_S$. Assume that $k$ is even and assume that nontrivial intersections with the spread elements with $\pi_o$ are always $k/2$-dimensional $K$-subspaces of $\pi_o$.*

*(1) Then $\pi_o$ is an affine plane which is a line-Baer subplane.*

*(2) If $k = 2$ then a spread $S$ is a dual spread if and only if any 2-dimensional $K$-subspace which is not a line of $\pi_S$ is a point-Baer subplane of $\pi_S$.*

*Proof.* Note that the intersections of $\pi_o$ with elements of $S$ determine a spread in $PG(k-1, K)$ as the lattice of $K$-subspaces of $\pi_o$. Clearly, this forces $\pi_o$ to be an affine subplane of $\pi_S$. Let $N_{\pi_o}$ denote the subnet of $\pi_S$ which is defined by the elements of $S$ which nontrivially intersect $\pi_o$.

Let $L$ be any line of $\pi_S$. If $L$ is a line of $N_{\pi_o}$ then the projective extension of $L$ intersects the projective extension of $\pi_o$. Thus, assume that $L$ is not a line of $N_{\pi_o}$.

By the construction of $\pi_S$, $L = V_\alpha + c$ where $V_\alpha$ is in $S$ (as a vector subspace) and $c$ is a vector of $\pi_S$. Thus, $V_\alpha$ and $\pi_o$ are mutually disjoint $k$-dimensional $K$-subspaces and since $k$ is finite, $V_\alpha \oplus \pi_o = \pi_S$ as a $K$ -vector space.

In order to show that $L$ contains a point of $\pi_o$, it suffices to show that $c$ may be taken in $\pi_o$. Since $c = v + c^*$ where $v$ in $V_\alpha$ and $c^*$ in $\pi_o$, then $L = V_\alpha + c = V_\alpha + v + c^* = V_\alpha + c^*$. This proves (1).

To prove (2), assume $k = 2$ and note similarly as above that any 2-dimensional $K$-subspace which is not a line has a spread of 1-dimensional subspaces automatically induced upon it so there is an induced affine subplane.

By (1), the subplane constructed from any 2-dimensional $K$-subspace is a line-Baer subplane of the projective extension.

Assume that $S$ is a dual spread. Suppose that $\pi_o$ is an affine plane which is a 2-dimensional vector subspace with spread induced from $S$ as above. Let $b$ be any point of the projective extension of $\pi_S$. Since the line at infinity is a line of the projective extension of the subplane $\pi_o$, $b$ may be assumed to be in $\pi_S$. Let $\langle b, \pi_o\rangle$ denote the 3-dimensional subspace generated by $b$ and $\pi_o$. Then, by assumption, there is a spread line $V_\alpha$ of $S$ (as a 2-dimensional vector subspace) contained in $\langle b, \pi_o\rangle$. Hence, $V_\alpha \cap \pi_o$ is a 1-dimensional $K$-subspace. So, for $d$ a point of $\pi_o - V_\alpha \cap \pi_o$ then $\langle b, \pi_o\rangle = V_\alpha \oplus \langle d\rangle$. Thus, $b$ is either in $V_\alpha$ or $V_\alpha + d$ both of which are lines of $\pi_o$. Therefore, $\pi_o$ is a point-Baer subplane and, as observed above, also a line-Baer subplane so is a Baer subplane.

Conversely, assume that every 2-dimensional $K$-subspace which is not a line defines a point-Baer subplane and hence a Baer subplane. Let $\Sigma$ be any plane of the projective space $PG(3, K)$ and consider $\Sigma$ as a 3-dimensional vector subspace of $\pi_S$. Let $\pi_o$ be any 2-dimensional $K$-subspace of $\Sigma$. If $\pi_o$ is a line of $S$ then certainly $\Sigma$ contains a line of the spread. Thus, assume that $\pi_o$ is not a line of $S$ and so defines a point-Baer subplane by assumption. If $b$ is a point of $\Sigma - \pi_o$ then $\Sigma = \langle b, \pi_o\rangle$. Since $\pi_o$ is a point-Baer subplane, there is a line of $\pi_o$, $V_\alpha + d$ for $d$ in $\pi_o$, which nontrivially intersects $b$. However, $\pi_o \cap V_\alpha \neq 0$, so it follows that $V_\alpha = \pi_o \cap V_\alpha \oplus \langle b - d \rangle \subset \Sigma$ . We have thus shown that any plane of the three-dimensional projective space contains a line of the spread so that $S$ is a dual spread. This completes the proof of (2).

We now obtain the following corollary:

**Corollary 2.3** *Let $S$ be any spread of $PG(3, K)$ for $K$ a skewfield. Let $\pi_S$ denote the corresponding translation plane. Let $\pi_S^E$ denote the projective extension and let $D(\pi_S^E)$ denote the dual projective translation plane.*

*Then any 2-dimensional $K$-vector subspace of $\pi_S$ which is not a line defines a point-Baer subplane in $D(\pi_S^E)$.*

*Proof.* Note that any 2-dimensional $K$-vector defines a line-Baer subplane which when taken projectively and dualized becomes a point-Baer subplane of the dual translation plane.

# 3 The construction

In this section, we provide a construction which produces a nonderivable or derivable affine planes depending upon the nature of a given spread. We shall state the theorem only when the spread is not a dual spread whereas a more general construction procedure is possible.

**Theorem 3.1** *Let $S$ be a spread of $PG(3, K)$ which is not a dual spread and $K$ is an infinite skewfield. Choose a plane $\Sigma$ of $PG(3, K)$ which does not contain a line of $S$. Let $\pi_o$ be a line of $\Sigma$ and write $\pi_o = X_o \oplus Y_o$ where $X_o$ and $Y_o$ are 1-dimensional vector $K$-subspaces.*

(1) *Then $\pi_o$ is a line-Baer subplane of the associated translation plane $\pi_S$.*

*Let $\pi_S^E$ denote the projective extension of $\pi_S$ and $D(\pi_S^E)$ the dual of $\pi_S^E$.*

*Denote the parallel classes of $\pi_S$ containing the lines of $S$ containing $X_o$ and $Y_o$ respectively by $(\infty)$ and $(0)$. Form the lines $(\infty)\alpha$ for all $\alpha$ in $Y_o$. Dualize the translation plane, sending $(\infty)$ to the line at infinity of $D(\pi_o^E)$ and the line at infinity of the translation plane to $(\infty)^*$.*

(2) *The net $N$ with infinite points $(\infty)^*, (\infty)\alpha$ for all $\alpha$ in $Y_o$ is a derivable net in the affine dual translation plane $\delta_S$ obtained by deleting $(\infty)$ from $D(\pi_S^E)$.*

(3) *The affine dual translation plane $\delta_S$ containing $N$ is not derivable by the derivation of $N$.*

*Proof.* (1) is immediate from the previous section as $\Sigma$ does not contain a line of $S$.

In order to prove (2), we need to show that given any two affine points $a$, $b$ of the affine dual translation plane such that the join $ab$ is a line of a parallel class either $(\infty)^*$ or $(\infty)\alpha$ for some $\alpha$ in $Y_o$ then there is a Baer subplane $\pi_{a,b}$ of the net $N$ which contains $a$ and $b$. Note that this simply says that the subplane has the same parallel classes as $N$.

For an affine subplane with the same parallel classes as $N$ to be a Baer subplane of the net, it is required only that the subplane be a point-Baer subplane as it is automatically a line-Baer subplane with respect to the net $N$ since every line of the net contains one of the infinite points of the affine subplane.

Considered projectively, it is required that the dual subplane of the corresponding translation plane be a line-Baer subplane. We have seen that any subplane of the translation plane which originates as a 2-dimensional vector subspace is always a line-Baer subplane.

Thus, to show that the net $N$ is a derivable net, it suffices to show that there is a set of line-Baer subplanes of the translation plane which projectively contain the lines $(\infty)\alpha$ for all $\alpha$ in $Y_o$ and which dualize to appropriate subplanes of $N$.

Given points $a$ and $b$ of the affine dual translation plane such that the join $ab$ is a line of $N$, there is a Baer subplane $\pi_{a,b}$, of $N$ containing $a$ and $b$ if and only if there is a projective subplane of $\pi_S^E$ which contains the lines $(\infty)\alpha$ for all $\alpha$ of $Y_o$, contains the line at infinity $l_\infty$ of $\pi_S$ and contains the lines $a$ and $b$ of $\pi_S^E$

Suppose that $a$ and $b$ are (necessarily affine) lines of $\pi_S$ which are concurrent either on the line at infinity of $\pi_S$ or at a line $(\infty)\alpha_o$ for some $\alpha_o$. Hence, it is required to construct a line-Baer affine subplane of $\pi_S$ which contains the lines $a$ and $b$ as well as the lines $(\infty)\beta$ for all $\beta$ in $Y_o$.

Note that the translation group with center $(\infty)$, $G_{(\infty)}$, of the translation plane acts as a translation group of the dual affine translation plane and so fixes each parallel class of $N$. Moreover, the group $G_{(\infty)}$ acts regularly on the lines of each parallel class of $\pi_S$ distinct from $(\infty)$. Since the lines $a$ and $b$ are concurrent at $(\infty)$ if and only if they are infinite points of the corresponding affine dual translation plane, it may be assumed that there is a translation with center $(\infty)$ which maps $b$ onto a line which intersects the zero vector of the line $(\infty)0$.

There are two cases depending on whether $a$ and $b$ are concurrent at an affine line or at the line at infinity of $\pi_S$. The arguments are similar but we shall take these separately.

If $a$ and $b$ are concurrent at the line at infinity, they belong to the same parallel class say $\delta$. Assuming that $b$ is $0\delta$ , take any point $W$ on $b \cap (\infty)\beta$ for $\beta \neq 0$ in $Y_o$. There is a kernel homology group $H$ of $\pi_S$ which acts transitively on the nonzero vectors of any 1-dimensional $K$-vector subspace and fixes each infinite point of $\pi_S$. It follows that that $H$ fixes $(\infty)0$ and acts regularly on the set of lines $(\infty)\alpha$ for all $\alpha$ in $Y_o - \{0\}$. Take $a \cap (\infty)0 = A$ and form the 2-dimensional $K$-subspace $\langle A, W\rangle$ and note that this subspace becomes a line-Baer affine subplane $\pi_1$ of $\pi_S$. Moreover, the above remark shows that the subplane $\pi_1$ contains the lines $(\infty)\alpha$ for all $\alpha$ in $Y_o$ and as a projective plane contains the line at infinity. Therefore, the plane contains 0, $W$ and $A$ as points and $0W = b$ projectively intersects the line at infinity at $\delta$. Hence, it follows that $\pi_1$ contains the line $A\delta = a$,

Thus, in the dual translation plane, there is a point-Baer affine plane of the net $N$ which contains the points $a$ and $b$.

Now assume that the lines $a$ and $b$ are concurrent on a line $(\infty)\alpha_o$. First assume that $\alpha_o \neq 0$ of $Y_o$.

Then, $a \cap b = W$ is a nonzero vector and assuming again that $b \cap (\infty)0 = 0$ and letting $a \cap (\infty)0 = A$, again it follows that $\langle A, W \rangle$ is a line-Baer subplane $\pi_1$ which contains the points $0, W, A$ and hence the lines $0A = a$ and $0W = b$.

Now assume that the lines $a$ and $b$ are concurrent on $(\infty)0$ and without loss of generality, assume that $a \cap b$ is 0. Let $W = b \cap (\infty)\alpha$, $A = a \cap (\infty)\alpha$ for some $\alpha \neq 0$ of $Y_o$. Form the 2-dimensional $K$ space $\langle A, W \rangle$. Since the group $H$ is regular on the lines $(\infty)\beta$ for all $\beta \neq 0$ of $Y_o$ , it follows that as a line-Baer affine subplane, $\langle A, W \rangle$ contains the lines $(\infty)\gamma$ for all $\gamma$ of $Y_o$.

Moreover, $\langle A, W \rangle$ contains the lines $0A = a$ and $0W = b$.

Hence, in all case, given affine points $a$ and $b$ of the affine dual translation plane such that the join $ab$ is a line of $N$ then there is a point-Baer subplane of the associated dual translation plane containing $a$ and $b$. Since any point-Baer subplane of the net $N$ is also line-Baer with respect to the net, it now follows that $N$ is a derivable net. This proves (2).

To prove (3), we need to show that the subplanes of $N$ are not all Baer subplanes of the affine dual translation plane. This is equivalent to showing that not all subplanes of $N$ are line-Baer in the dual translation plane.

To construct the set of lines $(\infty)\alpha$, we used a 2-dimensional $K$-subspace $\pi_o$. When extended and dualized to say $D(\pi_o^E)$, this subplane becomes a point-Baer subplane of the net $N$ considered projectively. We assert that $D(\pi_o^E)$ is not a line-Baer subplane of the dual translation plane. This assertion is equivalent to the assertion that $\pi_o$ is not a point-Baer subplane of the translation plane. So, suppose that $\pi_o$, is a point-Baer subplane of $\pi_S$. Recall that we obtained $\pi_o$ from a 3-dimensional vector subspace $\Sigma$ which did not contain a line of the spread (as a set of 2-dimensional $K$-subspaces). Take any point $A$ of $\Sigma - \pi_o$. If there is a line of $\pi_o$ as a subplane which contains the point $A$ then the arguments of the previous section show that there is a line of the spread contained in $\Sigma$. Hence, $\pi_o$ is not a point-Baer subplane and thus, $D(\pi_o^E)$ is not a line-Baer subplane. So, although $N$ is a derivable net which is contained in the affine dual translation plane, the affine dual translation plane cannot be derivable by derivation of $N$.

Note that an attempt at derivation would result in a structure whose lines are the lines of the affine dual translation plane which are not in $N$ together with the Baer subplanes of $N$. However, since $D(\pi_o^E)$ is not line-Baer, there exists a line of the affine dual translation plane which does not intersect $D(\pi_o^E)$ and would not be considered in the same parallel class as $D(\pi_o^E)$ so that the derived structure would fail to be an affine plane. This completes the proof of the theorem.

**Corollary 3.2** *Derivation of $N$ in the dual translation plane produces a nontrivial linear space which may be considered a union of two nets on the same points but which is not an affine plane. Note that Playfair's axiom is not valid in this structure.*

*So, there is a partition of the lines into parallel classes such that each point is incident with a unique line from each parallel class but two lines from distinct parallel classes do not necessarily intersect.*

*Proof.* We consider the projective version of the constructed incidence structure of points of the dual translation plane and lines as subplanes of $N$ and lines not in $N$.

Since any two points are incident with a unique line, the structure is a linear space.

Let $\pi_1$ be a subplane of $N$ and $b$ a line of the dual translation plane such that $\pi_1$ and $b$ do not intersect projectively. Let $b$ belong to the parallel class $\alpha$. Choose two points $W$ and $A$ of $\pi_1$ such that there exist lines $w$ and $a$ of the dual translation plane which contain $W$ and $A$ respectively but are not lines of $\pi_1$ and such that $w$ and $a$ are in the same parallel class $\delta \neq \alpha$. Then $W, A, \delta$ form a triangle with corresponding lines $W\delta$, $A\delta$ and $\pi_1$. The first two lines are lines of the dual translation plane so that these intersect $b$ as they are not parallel to $b$. However, b but does not intersect the third line $\pi_1$ so that Playfair's axiom cannot be valid.

**Remark 2** *The general construction above also is valid for spreads which are dual spreads producing derivable dual translation planes.*

# 4 The algebraic construction and the extension question

Let $K$ be any skewfield and let $V$ be a 4-dimensional left $K$-vector space. Let $PG(3, K)$ denote the lattice of left vector $K$-subspaces of $V$. Since a spread $S$ considered vectorially is a set of mutually skew left 2-dimensional $K$-subspaces, if we choose any three lines $a, b, c$ of a spread, we may choose a basis for $a$ and for $b$, such that we represent the elements by $(x, y)$ where $x$ is in $a$ and $y$ is in $b$, such that we may consider $V$ as $(K \oplus K) \oplus (K \oplus K)$ where $a$ is identified with $(K \oplus K) \oplus (0 \oplus 0)$, $b$ is identified with $(0 \oplus 0) \oplus (K \oplus K)$. With this basis change, we may also the basis is taken so the $c$ is $(x, x)$. That is, $a$,$b$, $c$ have the equations $y = 0, x = 0, y = x$ respectively.

It is easy to verify that the other spread lines have the general form $y = xM$ where $M$ is a $2 \times 2$ $K$-matrix. Furthermore, as a set, we obtain the spread in the following form:

$$x = 0, \;\; y = x \begin{bmatrix} g(t,u) & f(t,u) \\ t & u \end{bmatrix} \text{ for all elements } u, t \text{ of } K$$

where $g$ and $f$ are functions on $K \times K$ such that $g(0,0) = f(0,0) = 0$ and $g(0,1) = 1$, $f(0,1) = 0$.

In the above construction, we took two specific parallel classes $(\infty)$, $(0)$ of a translation plane with spread in $PG(3, K)$ and took a particular 1-dimensional $K$-subspace $Y_o$ incident with a line of the parallel class $(0)$. It was shown that $(\infty)\alpha$ for all $\alpha$ in $Y_o$ and the line at infinity define a derivable net in the associated dual translation plane where $(\infty)$ is taken as the line at infinity. Furthermore, it was shown that any given left 2-dimensional $K$-space $\pi_o$ which is not a spread 2-space (line) may be chosen so that the dual of the projective extension of $\pi_o$, is a subplane of a derivable net.

We define a multiplication $*$ on $K \oplus K$ as follows: For each element $(t, u)$, there is an associated matrix with the second row equal to $\{t, u\}$. Here, we identify $u$ of $K$ with $(0, u)$.

Define $(s, v) * (t, u) = (s, v) \begin{bmatrix} g(t,u) & f(t,u) \\ t & u \end{bmatrix} = (sg(t, u) + vt, sf(t, u) + vu)$.

It is well known that $(K \oplus K, +, *)$ is a right quasifield. Furthermore, we have $u * (a * b) = (u * a) * b$ for all $u$ in $K$ and $a, b$ in $K \oplus K$.

Furthermore, by defining a multiplication $\circ$ by $a \circ b = b * a$, we obtain that $(K \oplus K, +, \circ)$ is a left quasifield coordinatizing the affine dual translation plane which has the parallel class $(\infty)$ as the line at infinity.

We agree that a generator for $Y_o$ is taken within the basis which means that the set of lines $x = \delta$ for all $\delta$ in $K$ union the line at infinity $(\infty)^*$ becomes the derivable net $N$.

In other words, we obtain that $\{(\infty)^*, (\delta) \text{ for } \delta \in K\}$ is the set of parallel classes for a derivable net $N$.

We consider this algebraically. Notice that the lines incident with $(0, 0)$ of the net $N$ in the dual translation plane have the form $y = x \circ \delta$, $x = 0$ for all $\delta$ in $K$.

In terms of multiplication:

$$\begin{aligned} x \circ \delta = \delta * x = (0, \delta) * (x_1, x_2) &= (0, \delta) \begin{bmatrix} g(x_1, x_2) & f(x_1, x_2) \\ x_1 & x_2 \end{bmatrix} \\ &= (\delta x_1, \delta x_2) = \delta(x_1, x_2). \end{aligned}$$

Hence, the lines incident with $(0, 0)$ of the net $N$ have the form $x = 0, y = \delta x$ for all $\delta$ in $K$.

**We shall use the notation $N_\delta$ when a derivable net has the above form.**

We may now provide our main results. First, we note the following:

**Remark 3** *Given any skewfield $K$, there exist a spread in $PG(3, K)$.*

**Theorem 4.1** (1) *Let $N$ be any derivable net. Then there exists a skewfield $K$ such that $N$ is isomorphic to a derivable net constructible from a spread in $PG(3, K)$.*

(2) *Any derivable net $N$ may be embedded in a dual translation plane.*

*Proof.* By the results of Johnson [18], the net $N$ may be represented in the form $x = 0, y = \delta x$ for all $\delta$ in $K$ some skewfield as above.

Now consider any spread in $PG(3, K)$. We consider the above construction process from a right quasifield $(Q, +, *)$ considered as a left two dimensional $K$-space to produce a derivable net $N_\delta$ which is isomorphic to $N$ . This proves (1).

We now consider the situation described in (2). In the notation, We note that there is a mapping $\Gamma$ from the points of $N_\delta$ to the points of $N$ which maps lines of $N_\delta$ to lines of $N$ and preserves collinearity. However, $N_\delta$ is embedded in a dual translation plane $\pi$. Since the points of $\pi$ and the points of $N_\delta$ are the same, we define the lines of $\pi\Gamma$ to be the $\Gamma$ images of lines of $\pi$. It follows that $\pi\Gamma$ is a dual translation plane containing $N$.

**Theorem 4.2** *Let $N$ be any infinite derivable net. Then exists a dual translation plane containing $N$ which is not derivable with respect to $N$.*

*Proof.* Any derivable net $N$ may be embedded in a dual translation plane originating from a spread in $PG(3,K)$ for some skewfield $K$. However, there exists a spread in $PG(3,K)$ which is not a dual spread which produces a nonderivable dual translation plane $\pi$ containing a derivable net $N_\delta$ . It follows from the above arguments that $N$ is isomorphic to $N_\delta$. Similarly as above, there is a dual translation plane $\pi\Gamma$ containing $N$ and which is isomorphic to $\pi$.

Assume that $\pi\Gamma$ is derivable with respect to the net $N$. Since $\pi$ is not derivable with respect to $N_\delta$, there exists a line $b$ of $\pi$ which projectively does not intersect a projective Baer subplane $\rho_o$ of $N_\delta$. Since $N_\delta$ is isomorphic to $N$, it follows that all subplanes of $N_\delta$ map under $\Gamma$ to subplanes of $N$. We have deliberately forced collinearity by the definition of $\pi\Gamma$ so that it follows that $b\Gamma$ cannot projectively intersect a point of the Baer subplane $\rho_o\Gamma$ of the net $N$. Hence, $\pi$ is a non-derivable dual translation plane containing the derivable net $N$.

**Remark 4** *Although the above results show that any derivable net may be embedded in a dual translation plane, it is not necessarily the case that any derivable net may be embedded in a Desarguesian affine plane.*

*In fact, a derivable net defined by as a regulus net in $PG(3,K)$ for an algebraically closed field $K$ cannot be properly embedded in any net defined by a partial spread in $PG(3,K)$.*

*Proof.* (Due to V. Jha). Any regulus net may be represented in the form $x = 0, y = \delta x$ for all $\delta$ in $K$. Any partial spread extending the net has a transversal of the form $y = xT$ for some $2 \times 2$ $K$-matrix. Since $T$ has an eigenvalue $\rho$ in $K$, it follows that $T - \rho I$ is singular so that $y = xT$ and $y = \rho x$ have two points in common.

# 5 Spreads which are dual spreads

In the previous section, we have seen that there exist derivable nets in nonderivable affine planes. The affine planes in question are dual translation planes. Could such a plane be a translation plane? Note that when $K$ is a field in the above context then a derivable net corresponds to a regulus in some projective space isomorphic to $PG(3,K)$. So, we have an extension plane of a regulus net which is not a derivable net. However, the plane does not necessarily correspond to a spread of $PG(3,K)$. Thus, we arrive at the following question:

*Given a regulus $R$ in $PG(3,K)$ for $K$ a field, assume that there is a spread in $PG(3,K)$ which contains $R$. Is the corresponding translation plane derivable ?*

Another related question is:

*If a spread in $PG(3, K)$ for $K$ a field, is a dual spread and contains a regulus and if the translation plane is derivable, is the derived spread also a dual spread ?*

In Bruen and Fisher [5], it is shown that any subregular spread in $PG(3, K)$ for $K$ a field is a dual spread. More precisely, the proof taken in the context of the results on Baer subplanes and duals spreads allows the following extension.

**Theorem 5.1** *Let $S$ be any spread in $PG(3, K)$ which contains a regulus net $R$.*

*Then the translation plane is derivable with derivable net $N_R$ and the spread obtained by derivation of the regulus net is a dual spread if and only if the original spread is a dual spread.*

*Proof.* First of all, we see that any regulus net defines a derivable net so that any 2-dimensional $K$-subspace which is a subplane of the net automatically becomes a point-Baer subplane of the net and hence also of the plane since the points of the net and the points of the plane are the same. But, we have also noted that such subplanes are also line-Baer. Hence, any such affine plane incident with the zero vector becomes a Baer subplane. Clearly, any translate of a Baer subplane is also Baer. Hence, the translation plane is derivable.

Now derive the plane to $\pi^*$ using the regulus net $N_R$. We need to show that every 3-dimensional $K$-subspace $\Sigma$ contains a line of the new spread. We give a slightly modified version of the proof of [5] (Theorem 2 , p. 802).

Let $\pi_o$ be a subplane of the net $N_R$ which is in $\Sigma$ and let $\pi_1$ and $\pi_2$ be two other distinct subplanes of the net which are incident with the zero vector. Then $\Sigma \cap \pi_i = \langle P_i \rangle \neq 0$ for $i = 1, 2$. Thus, $\langle P_1, P_2 \rangle$ is a 2-space of $\Sigma$ (line of the plane $\Sigma$) and this 2-space nontrivially intersects $\pi_o$ as both are 2-spaces within a 3-space. Hence, the space $\langle P_1, P_2 \rangle$ is a line of the opposite regulus to $R$. Since this is a component by assumption, then $\Sigma$ contains a line of the new spread.

We state the equivalency as a corollary.

**Corollary 5.2** *Let $S$ be a spread in $PG(3, K)$ for $K$ a field and let $R$ be a regulus of $S$.*

*Then any plane of $PG(3, K)$ which contains a line of the opposite regulus to $R$ contains a line of the spread.*

We may use the above results to show that any spread which corresponds to a flock of a quadratic cone or to a flock of a hyperbolic quadric in $PG(3, K)$ must be a dual spread. Our proof depends on the structure of the spread based on functions acting on $K$. J.A. Thas pointed out to the author that it is possible to use the Klein quadric construction associated with such structures to give a proof. Here we give a purely algebraic proof which does not use the Klein quadric and which possibly may extend more naturally to spreads in $PG(3, K)$ for $K$ a skewfield which are unions of pseudo-reguli as opposed to unions of reguli as the key to the proof which we provide does not depend upon the multiplicative commutativity of the field.

The reader is referred to Jha-Johnson [12] for examples of such spreads which are unions of pseudo-reguli in $PG(3, K)$ when $K$ is a non-commutative skewfield.

Hence, we first state the required results.

**Theorem 5.3** *(Jha-Johnson [11]) Let $S$ be a spread in $PG(3, K)$ for $K$ a field which is a union of reguli that share exactly one common line. The reguli of this set are called the base reguli.*

*Then there exists an elation group $E$ of the corresponding translation plane such that each $E$-orbit of lines of $S$ union the axis of $E$ is a base regulus.*

*Furthermore, coordinates may be chosen so that the spread $S$ has the following form:*

$$x = 0, \; y = x \begin{bmatrix} u + g(t) & f(t) \\ t & u \end{bmatrix}$$

*for all $u, t$ in $K$ and $g, f$ functions on $K$.*

*The group $E$ may be represented as follows:*

$$E = \langle \begin{bmatrix} 1 & 0 & u & 0 \\ 0 & 1 & 0 & u \\ 0 & 0 & 1 & 0 \\ 0 & 0 & 0 & 1 \end{bmatrix} \ni u \in K \rangle.$$

**Theorem 5.4** *(Johnson [20]) Let $S$ be a spread in $PG(3, K)$ for $K$ a field which is a union of reguli that share two common lines. The reguli of this set are called the base reguli.*

*Then there exists a homology group $H$ of the corresponding translation plane such that each $H$-orbit of lines of $S$ union the axis and coaxis of $H$ is a base regulus.*

*Furthermore, coordinates may be chosen so that the spread $S$ has the following form:*

$$x = 0, \; y = x \begin{bmatrix} v & 0 \\ 0 & v \end{bmatrix}, \; y = x \begin{bmatrix} uf(t) & ug(t) \\ u & ut \end{bmatrix}$$

*for all $u \neq 0, v, t \in K$ and $f, g$ functions on $K$ where $f$ is bijective.*

*In this form, the group $H$ has the following representation:*

$$H = \langle \begin{bmatrix} 1 & 0 & 0 & 0 \\ 0 & 1 & 0 & 0 \\ 0 & 0 & u & 0 \\ 0 & 0 & 0 & u \end{bmatrix} \ni u \in K - \{0\} \rangle.$$

We also need a technical lemma for the proof that such spreads are always dual spreads.

**Lemma 5.5** *Let $R$ be any regulus in $PG(3,K)$. Choose any two lines $L$, $M$ of $R$ and any two lines $L^*, M^*$ of the opposite regulus.*

*Then coordinates may be chosen so that writing vectors as $(x_1, x_2, y_1, y_2)$ then $L$ is $x = 0$, $M$, is $y = 0$, $L^*$ is $\langle(x_1,0,y_1,0)$ for all $x_1, y_1$ of $K\rangle$ and $M^*$ is $\langle(0,x_2,0,y_2)$ for all $x_2, y_2$ of $K\rangle$.*

*In addition, with this choice of coordinates, the regulus $R$ has the form*

$$x = 0, \; y = x\begin{bmatrix} v & 0 \\ 0 & v \end{bmatrix} \text{ for all } v \text{ in } K$$

*Proof.* From the results of Johnson [18], it follows that any regulus net admits a collineation group $G_1G_2$ where $G_i \simeq GL(2,K)$ for $i = 1,2$ and $G_1 \cap G_2 = Z(G_1) = Z(G_2) \simeq Z(GL(2,q))$. Furthermore, $G_1$ fixes each Baer subplane sharing the zero vector and acts 2-transitively on the components of the net and $G_2$ fixes each components and acts 2-transitively on the Baer subplanes incident with the zero vector. Hence, the choice of coordinates follows directly.

**Theorem 5.6** *Let $S$ be a spread in $PG(3,K)$ for $K$ a field. Then $S$ is a dual spread in either of the following two situations:*

(1) *$S$ is a union of reguli that share exactly one line.*

(2) *$S$ is a union of reguli that mutually share two lines*

*Proof.* In either of the two situations, let $\Sigma$ be any 3-dimensional $K$-subspace. It is asserted that either $\Sigma$ contains a line of the standard regulus net

$$x = 0, \; y = x\begin{bmatrix} u & 0 \\ 0 & u \end{bmatrix} \text{ for all } u \text{ in } K$$

or coordinates may be chosen so that $\Sigma \cap (x = 0) = \langle(0,0,0,1)\rangle$ and $\Sigma \cap (y = 0) = \langle(1,0,0,0)\rangle$. To see this, we note from above that if $\Sigma$ contains a Baer subplane incident with the zero vector of the standard regulus net $N$ then $\Sigma$ contains a line of $N$. So, assume that $\Sigma$ does not contain a line of $N$. Then $\Sigma$ intersects $x = 0$ and $y = 0$ in distinct Baer subplanes of $N$. By the above lemma, we may choose coordinates so that $N$ retains the form and the two Baer subplanes have the form listed in the lemma. Now since $N$ retains its form and since we either have an elation group or homology group whose line or component orbits union the axis or axis and coaxis respectively determine the base regulus nets, this means that the forms of the elation group and/or homology group have not changed. Furthermore, although we allow that perhaps the functions describing the spread have changed, the basic properties listed in the theorems above are still valid for the possibly new functions.

Let $\Sigma \cap (y = x) = \langle(1, \alpha_o, 1, \alpha_o)\rangle$. We note that $\alpha_o \neq 0$ since otherwise, $\Sigma$ would contain the line $x = 0$. It is noted that the choice of the 1-space is made possible by the fact that $\Sigma$ does not contain the line $y = 0$ by assumption.

Hence, we may assume that $\Sigma = \langle(0,0,0,1), (1,0,0,0), (1, \alpha_o, 1, \alpha_o)\rangle$.

Now assume the conditions of (1). It suffices to show that there is a Baer subplane of one of the regulus nets which is also in $\Sigma$. Since $\Sigma$ does not contain $x = 0$ but intersects $\Sigma$ at $\langle(0,0,0,1)\rangle$, it follows that the only way that $\Sigma$ could contain one of the Baer subplanes in question is if it contains an image of the Baer subplane $\langle(0, x_2, 0, y_2)\rangle$ under a mapping

$$g_t = \begin{bmatrix} 1 & 0 & g(t) & f(t) \\ 0 & 1 & t & 0 \\ 0 & 0 & 1 & 0 \\ 0 & 0 & 0 & 1 \end{bmatrix} \quad \text{for } t \in K.$$

The reader is directed back to the statement of the result for the significance and properties of the functions $g$ and $f$. Thus, the question now becomes whether $\Sigma$ contains a Baer subplane of the form $\langle(0, x_2, x_2 t, y_2)\rangle = \pi_t$.

Notice that $(y_2 - x_2)(0,0,0,1) - (x_2/\alpha_o)(1,0,0,0) + (x_2/\alpha_o)(1,\alpha_o,1,\alpha_o)$ are vectors in both $\Sigma$ and $\pi_{1/\alpha_o}$. Furthermore, we see that, for each $t$ in $K$, the regulus nets $S_t$ are given by the partial spread:

$$\{x = 0,\ y = x \begin{bmatrix} u + g(t) & f(t) \\ t & u \end{bmatrix} \text{ for all } u \in K\}.$$

Hence, $\Sigma$ contains a Baer subplane of one of the regulus nets, namely $S_{1/\alpha_o}$ and so contains a line of one of the reguli and hence contains a line of the spread.

Now assume (2). In this case, we show that $\langle(0,0,0,1),(1,0,0,0)\rangle$ is actually a Baer subplane of one of the regulus nets that share both $x = 0$ and $y = 0$. We consider the image of $\pi_o = \langle(1,0,0,0),(0,0,1,0)\rangle$ under mappings

$$h_t = \begin{bmatrix} 1 & 0 & 0 & 0 \\ 0 & 1 & 0 & 0 \\ 0 & 0 & f(t) & g(t) \\ 0 & 0 & 1 & t \end{bmatrix} \quad \text{for all } t \in K.$$

Again, we direct the reader to the statement for the significance and properties of the functions $f$ and $g$ but point out, in particular, that $f$ is bijective on $K$.

The subplane $\pi_o$maps under $h_t$ to $\langle(1,0,0,0),\ (0,0,f(t),g(t))\rangle$. Since $f$ is bijective, there exists an element $t_o$ such that $f(t_o) = 0$.

Furthermore, since

$$\begin{bmatrix} f(t_o) = 0 & g(t_o) \\ 1 & t_o \end{bmatrix}$$

is nonsingular, it follows that $g(t_o) \neq 0$. Note that, for each $t$ in $K$, the regulus nets $R_t$ other than $N$ are

$$\{x = 0, y = 0, y = x \begin{bmatrix} uf(t) & g(t)u \\ u & ut \end{bmatrix} \text{ for all } u \neq 0 \text{ of } K.$$

Hence, $\Sigma$ contains a Baer subplane of $R_{t_o}$ and hence a line of $R_{t_o}$ and thus contains a line of the spread.

This completes the proof of the theorem.

**Corollary 5.7** *Any spread in $PG(3, K)$ that corresponds to a flock of a quadratic cone in $PG(3, K)$ or to a flock of an elliptic or hyperbolic quadric in $PG(3, K)$ is a dual spread.*

**Corollary 5.8** *Let $S$ denote any spread corresponding to a quadric set in $PG(3, K)$ where $K$ is a field. Let $N$ be any regulus in the spread and let $N^*$ denote the opposite regulus.*

*Then $(S - N) \cup N^*$ is a spread which is a dual spread.*

*Proof.* By Jha and Johnson [11], a flock of a quadratic cone corresponds to a spread in $PG(3, K)$ with property (1).

By Johnson [20], a flock of a hyperbolic quadric corresponds to a spread with property (2).

In Bruen and Fisher ([5] Corollary to Theorem 3), it is proved that any subregular spread is a dual spread.

In Biliotti and Johnson [4], it is pointed out that any flock of an elliptic quadric in $PG(3, K)$ corresponds to a spread which may be obtained from a regular spread by multiple derivation. Hence, any such spread is a dual spread.

**Definition 5.9** *A semifield spread in $PG(3, K)$ is a spread $S$ that admits an automorphism group that fixes a line $L$ pointwise and acts regularly on the remaining lines of $S - \{L\}$. We call $L$ an axis of the spread $S$.*

*A nearfield spread in $PG(3, K)$ is a spread $S$ that admits an automorphism group that fixes two lines $L$ and $M$ of the spread, fixes $L$ pointwise and acts regularly on the remaining lines of $S - \{L, M\}$. The line $L$ is called an axis and the line $M$ is called a coaxis for the spread.*

**Corollary 5.10** *Any semifield spread $S$ in $PG(3, K)$ that contains a regulus $N$ sharing an axis $L$ of $S$ is a dual spread. Furthermore, $(S - N) \cup N^*$ is a spread which is a dual spread.*

*Any nearfield spread $S$ in $PG(3, K)$ that contains a regulus sharing an axis and coaxis for the spread is a dual spread. The derived structure is a spread which is a dual spread.*

*Proof.* Apply the groups to obtain the spread as a union of reguli that share $L$ or as a union of reguli that share $L$ and $M$. Now apply the above theorem.

**Theorem 5.11** *Let $S$ be a spread in $PG(3, K)$ which is a union of reguli sharing one line or a union of reguli sharing two lines. Let $X_o$ be any 1-dimensional $K$ subspace and let $L$ and $M$ be components of the corresponding translation plane such that $L$ contains $Y_o$. Let $(\infty)$ denote the parallel class containing $M$, and form $(\infty)\alpha$ for all $\alpha$ in $Y_o$.*

*Then $\{(\infty)\alpha$ for $\alpha$ in $Y_o\}$ is a derivation set for the associated dual translation plane.*

*Proof.* Previously, the net defined by the set in question was shown to be a derivable net. Since the spread is now also a dual spread, this forces the dual translation plane containing the net to be derivable as the subplanes in question now must be Baer in the translation plane and so are Baer in the dual translation plane.

# 6 Extensions of derivable nets

In a previous section, we have shown that any derivable net may be embedded in a dual translation plane. In this section, we show that any derivable net $N$ which may be properly embedded in a net with one additional parallel class not in $N$ can be naturally embedded in a dual translation plane constructed by an extension procedure.

Actually, all of this depends on the existence of a transversal to the net.

**Definition 6.1** *Let $N$ be an arbitrary net. A transversal $T$ is a set of points of cardinality the cardinality of a line which intersects each line in exactly one point and such that $T$ is contained in the set of such intersections.*

**Theorem 6.2** *Let $N$ be a derivable net.*

(1) *Then there exists a skewfield $K$ such that the points of $N$ may be identified with a 4-dimensional right and left vector space over $K$ and the lines may be represented in the form $x = (c_1, c_2), y = \delta x = (\delta x_1, \delta x_2) + (d_1, d_2)$ for all $\delta$ in $K$ and all $c_i, d_i$ in $K$ for $i = 1, 2$. The lines incident with $(0,0)$ are right 2-dimensional $K^{\mathrm{opp}}$-subspaces.*

*The Baer subplanes of the net incident with the zero vector are*

$$\{(\alpha a_1, \alpha a_2, \beta a_1, \beta a_2) \text{ for } \alpha, \beta \in K\}$$

*for fixed $a_1, a_2$ in $K$ not both zero. These Baer subplanes are 2-dimensional left $K$-subspaces.*

(2) *If $T$ is a transversal to $N$ then there is a function $f$ on $K \oplus K$ such that $T = \{(x, f(x))$ for all $x$ in $K \oplus K\}$ with the following properties:*

(*i*) *$f(a) - f(b) \neq (a - b)\delta$ for all $a \neq b$ and for all $\delta$ in $K$ and*

(*ii*) *for all $\delta$ in $K$ and $b$ in $K \oplus K$ there exists a $c$ in $K \oplus K$ such that $f(c) = \delta c + b$.*

(*iii*) *Conversely, any function $g$ with properties (i) and (ii) defines a transversal.*

(*iv*) *$\{(x, \alpha f(x) + \beta x + b)\}$ for fixed $\alpha, \beta$ in $K$ and $b$ in $K \oplus K$ is a transversal and*

(*v*) *there is a dual translation plane containing $N$ with lines $y = \alpha f(x) + \beta x + b$ for all $\alpha, \beta$ in $K^{\mathrm{opp}}$ and for all $b$ in $K \oplus K$.*

*Proof.* (1) follows from Johnson,[15],[18].

The proof to (2) is essentially in Theorems (1.4), (1.6), (1.7) and Lemma (1.5) of Johnson [13].

**Corollary 6.3** (1) *Any derivable net which admits a transversal may be embedded in a dual translation plane where the transversal is a line of the dual translation plane.*

(2) *Let $\pi$ be any dual translation plane which is constructed by a transversal of a derivable net $N$. Then the natural dual of $\pi$ defines a spread in $PG(3,K)$ for some skewfield $K$. Furthermore, if $T$ is any line of $\pi$ which is not in the derivable net $N$ then $T$ is a transversal to $N$ and the extension process described above produces a dual translation plane isomorphic to $\pi$.*

(3) *If $S$ is any spread in $PG(3,U)$, for $U$ an arbitrary skewfield, then any associated corresponding dual translation plane may be considered as constructable by the extension process.*

*Proof.* We may consider the extension process above and regard the set of points as either a right or left 4-dimensional $K$-vector space where $K$ is the skewfield $K$ arising from the derivable net.

For each $\alpha \neq 0$, $\beta$ of $K$, there is a transversal $g_{\alpha,\beta}(x)$ given by some combinattion of functions $\alpha f(x) + \beta x + b$ and which is zero when $x = 0$. Note that construction using $g_{\alpha,\beta}(x)$ in place of $f(x)$ does not change the set of lines.

We may regard $K \oplus K$ as either a left or right $K$-space and the original derivable net specifies a particular basis say $\{1, s\}$ where we may identify $0 \oplus K$ with $K$ so that $\delta x = s\delta x_1 + \delta x_2 = (\delta x_1, \delta x_2)$.

Since $g_{\alpha,\beta}(x)$ defines a transversal and $g(0) = 0$ then $g_{\alpha,\beta}(1)$ cannot be in $K$ as otherwise there would be two intersections with $y = \delta x$ for some $\delta$ in $K$. Since we obtain an affine plane, the set of lines incident with $(0,0) \equiv (0,0,0,0)$ cover the points of the plane. Hence, there exists a line incident with $(0,0)$ which contains $(1,s) \equiv ((0,1),(1,0))$. This cannot be a line $h(x)$ of the derivable net $N$ as noted above so there exists $\alpha_o \neq 0$ and $\beta_o$ in $K$ such that $g_{\alpha_o,\beta_o}(1) = s$. Again, we note that the extension process using $h(x)$ in place of $f(x)$ does not change the set of lines of the dual translation plane.

Hence, $h(x)$ has the property that $h(0) = 0$ and $h(1) = s$.

We define a multiplication "$\circ$" on $K \oplus K$ with right basis $\{s, 1\}$ as follows:

Let $x \circ (s\alpha + \beta) = \alpha h(x) + \beta x$ for all $\alpha, \beta$ in $K$.

Note that $x = sx_1 + x_2$ and that $\alpha x = x \circ \alpha$ for all $\alpha$ in $K$.

We assert that $a \circ (b \circ \alpha) = (a \circ b) \circ \alpha$ for all $a, b$ in $K \oplus K$ and for all $\alpha$ in $K$.

Let $a = sa_1 + a_2, b = sb_1 + b_2$. Then $a \circ (b \circ \alpha) = a \circ (\alpha b = s\alpha b_1 + \alpha b_2) = \alpha b_1 h(a) + \alpha b_2 a$.

Furthermore, $(a \circ b) \circ \alpha = (b_1 h(a) + b_2 a) \circ \alpha = \alpha(b_1 h(a) + b_2 a)$. Let $h(a) = sh_1 + h_2$ so that $b_1(sh_1 + h_2) + b_2(sa_1 + a_2) = s(b_1 h_1 + b_2 a_1) + (b_1 h_2 + b_2 a_2)$. Then $\alpha(s(b_1 h_1 + b_2 a_1) + (b_1 h_2 + b_2 a_2)) = s\alpha(b_1 h_1 + b_1 a_1) + \alpha(b_2 h_2 + b_2 a_2)$
$= (s\alpha b_1 h_1 + \alpha b_1 h_2) + (s\alpha b_2 a_1 + \alpha b_2 a_2) = \alpha b_1 h(a) + \alpha b_2 a$.

Now define a multiplication $*$ as follows: $a * b = b \circ a$.

It now follows that $(K \oplus K, +, *)$ is a right quasifield with (inner) kernel $(K, +, *)$. Note that $(K, +, *)$ is anti-isomorphic to $(K, +, \circ)$ and is isomorphic to $(K, +, \cdot)$ where $(K, +, \cdot)$ denotes the original skewfield.

Hence, each set $\{(x, x * m)\}$ is a 2-dimensional left $(K, +, *)$-subspace.

It now follows that a spread is obtained in $PG(3, K)$. This proves (2).

To prove (3), let $t$ be in $K \oplus K - 0 \oplus K$ and note that $y = x \circ (t \circ \alpha + \beta)) + b = (x \circ t) \circ \alpha + x \circ \beta + b$.

Define $f(x) = x \circ t$ then $f$ is a transversal function and lines are $y = \alpha f(x) + \beta x + b$.

This completes the proof of (3).

# References

[1] A.A. Albert. The finite planes of Ostrom. Lecture Notes, Univ. of Chicago.

[2] A. Barlotti. On the definition of Baer subplanes of infinite planes. J. Geom. 3(1973), 87–92.

[3] M. Bernardi. Esistenza di fibrazioni in uno spazio proiettivo infinito. Ins. Lombardo Accad. Sci. Lett. Rend. A 197 (1973), 528–542.

[4] M. Biliotti and N.L. Johnson. Bilinear flocks of quadratic cones. J. Geom. (to appear).

[5] A. Bruen and J.C. Fisher. Spreads which are not dual spreads. Canad. Math. Bull. 12 (1969), 801–803.

[6] P.J. Cameron. Projective and Polar Spaces. University of London, Queen Mary and Westfield College lecture notes. QMW Math. Notes 13.

[7] J. Cofman. Baer subplanes and Baer collineations of derivable projective planes. Abh. Math. Sem. Hamburg 44 (1975), 187–202.

[8] T. Gründhofer. Eine Charakterisierung ableitbarer Translationsebenen. Geom. Dedicata 11 (1981), 177–185.

[9] M. Hall, Jr. *The Theory of Groups.* Macmillan: New York, 1959,.

[10] D.R. Hughes. A class of non-Desarguesian projective planes. Canad. J. Math. 9 (1957),378–388.

[11] V. Jha and N.L. Johnson. Infinite flocks of quadratic cones. J. Geom. 57, no. 1/2 (1996), 123–150.

[12] V. Jha and N.L. Johnson. Conical, ruled and deficiency one translation planes. Preprint.

[13] N.L. Johnson A note on the construction of quasifields. Proc. Amer. Math. Soc., vol, 29 (1971), 138–142.

[14] N.L. Johnson. Derivation in infinite planes. Pacific J. Math. 42 (1972), 387–402.

[15] N.L. Johnson. Derivable nets and 3-dimensional projective spaces. Abh. Math. Sem. Hamburg, 58 (1988), 245–253.

[16] N.L. Johnson. Derivation is a polarity. J. Geom. vol. 35(1989), 97–102.

[17] N.L. Johnson. Derivation by coordinates. Note di Mat. 10 (1990), 89–96.

[18] N.L. Johnson. Derivable nets and 3-dimensional projective spaces II. The structure. Archiv Math. vol. 55 (1990), 84–104.

[19] N.L. Johnson. Derivation. Combinatorics '88. Research and Lecture Notes in Mathematics. Combinatorics '88, Volume 2, (1991), 97–114.

[20] N.L. Johnson. Flocks of infinite hyperbolic quadrics. J. Alg. Comb., 6, no. 1 (1997), 27–51.

[21] G. Lunardon. Piani di traslazione derivabili. Rend. Sem. Mat. Univ. Padova 61 (1979), 271–284.

[22] T.G. Ostrom. Translation planes and configurations in Desarguesian planes. Arch. Math. 11 91960), 457–464.

[23] T.G. Ostrom. Finite planes with a single $(p, L)$-transitivity. Arch. Math. 15 (1964), 378–384.

[24] T.G. Ostrom. Semi-translation planes. Trans. Amer. Math. Soc. 111 (1964), 1–18.

[25] O. Prohaska. Endliche ableitbare affine Ebenen. Geom. Ded. 1 (1972), 6–17.

[26] L.A. Rosati. Su una nuova classe di piani grafici. Ric. Mat. 13 (1964), 39–55.

Norman L. Johnson
Mathematics Dept.
University of Iowa
Iowa City, Iowa 52242, USA
e-mail: `njohnson@math.uiowa.edu`

Trends in Mathematics, © 1998 Birkhäuser Verlag Basel/Switzerland

# Regular Orbits and the $k(GV)$-Problem

Martin W. Liebeck

**Abstract**

We describe some recent results concerning regular orbits of quasisimple groups in coprime representations, and discuss an application to the $k(GV)$-problem in modular representation theory.

An old question of Richard Brauer from the 1950's asks whether it is true that if $B$ is a block of a finite group, with defect group $D$, then the number of ordinary irreducible characters in $B$ is at most $|D|$. By results of Fong [Fo] and Nagao [Na], for $p$-solvable groups Brauer's question reduces to the following:

$k(GV)$**-problem.** *Let $p$ be a prime, $G$ a finite $p'$-group, and $V$ a faithful finite-dimensional $\mathbb{F}_pG$-module. Denote by $k(GV)$ the number of conjugacy classes of the semidirect product $GV$. Show that $k(GV) \leq |V|$.*

Note that equality can be achieved, for any prime $p$ and any dimension $n$: for example, let $q = p^n$ and regard $V = \mathbb{F}_q$ as an $n$-dimensional space over $\mathbb{F}_p$. Let $G = \mathbb{F}_q^*$, with natural scalar action on $\mathbb{F}_q$. Then $k(GV) = k(G) + 1 = q = |V|$. Observe also that an affirmative answer to the problem would imply that any $p'$-subgroup of $GL_n(p)$ has at most $p^n - 1$ conjugacy classes, which seems far from obvious (at least to me).

A number of authors have worked on the $k(GV)$-problem, particular highlights being its solution in the following cases:

(1) if $G$ is supersolvable or $|G|$ is odd [Kn1, Gl];

(2) if $V \cong V^*$ as $\mathbb{F}_pG$-modules [Go];

(3) if there exists $v \in V$ such that $V \downarrow C_G(v)$ is a permutation module [Kn2].

Nevertheless the problem has proved rather stubborn, and it is only recently that it has been solved, for large primes $p$, by Robinson and Thompson [RT].

We discuss briefly the approach used in [RT]. First, character theory is used to solve the problem in the following case, extending (2) and (3) above:

(4) if there exists $v \in V$ such that $V \downarrow C_G(v)$ has a faithful self-dual submodule.

In particular, this will be satisfied if there exists $v \in V$ such that $C_G(v) = 1$, in which case the orbit $v^G$ is a *regular* orbit of $G$ on $V$.

After this, Robinson and Thompson reduce the proof of an affirmative solution of the $k(GV)$-problem to the following two situations:

(A) $G$ has a normal irreducible $q$-subgroup $Q$ of symplectic type ($q$ prime, $q \neq p$): that is, all characteristic abelian subgroups of $Q$ are central. (Such $q$-groups are determined by a well known result of Philip Hall.)

(B) $G$ has a normal irreducible quasisimple subgroup $H$.

In case (A) it is proved in [RT, Theorem 8] that, provided $p > 2^{27}$, such a group $G$ has a regular orbit on $V$, solving the problem by (4) above.

Now consider case (B). One might hope that here too, $G$ always has a regular orbit provided $p$ is sufficiently large. However, this is not the case, as the following example shows.

**Example** Let $H = A_c$, and let $p$ be a prime with $p > c$. Define

$$V = \{(\alpha_1, \ldots, \alpha_c) : \alpha_i \in \mathbb{F}_p, \sum \alpha_i = 0\}$$

and take $S_c$ to act on $V$ by permuting coordinates. As is well known, $V$ is an irreducible $\mathbb{F}_p S_c$-module; we call $V$ the *deleted permutation module* for $S_c$ over $\mathbb{F}_p$.

Now suppose that $p = c + 1$, and take $G = \langle -1 \rangle \times S_c$. We claim that $G$ has no regular orbit on $V$. To see this, let $v = (\alpha_1, \ldots, \alpha_c) \in V$. If $\alpha_i = \alpha_j$ for some $i \neq j$, then $(i\ j) \in C_G(v)$. Otherwise, $\{\alpha_1, \ldots, \alpha_c\} = \mathbb{F}_p \backslash \{\alpha\}$ for some $\alpha$, and so $\sum \alpha_i = (\sum_{\beta \in \mathbb{F}_p} \beta) - \alpha = -\alpha$, whence $\alpha = 0$. Reordering coordinates, we can therefore take

$$v = (1, -1, 2, -2, \ldots, \frac{p-1}{2}, \frac{p+1}{2}).$$

Defining $x = (1\ 2)(3\ 4)\ldots$, we then have $vx = -v$, so $-1.x \in C_G(v)$. Consequently $C_G(v) \neq 1$ for all $v \in V$, as claimed.

The following theorem, taken from [Li], shows that, at least for large primes, the above examples are the only obstacles to a regular orbit.

**Theorem** *Assume $p > 5^{30}$, and let $G$ be a $p'$-subgroup of $GL_n(p)$ containing a normal irreducible quasisimple subgroup $H$. Then one of the following holds:*

(i) *$G$ has a regular orbit on $V$;*

(ii) *$H = A_{n+1}$, $p > n + 1$, and $V_n(p)$ is the deleted permutation module for $H$ as described above.*

**Remark** Thompson (private communication) has shown that the only cases in conclusion (ii) in which the group $\mathbb{F}_p^* \times S_{n+1}$ has no regular orbit are those with $n + 2 \leq p \leq n + 4$.

It follows from the Theorem that for $p > 5^{30}$, the vector $v$ required for the application of (4) above exists in case (B) also. Thus the $k(GV)$-problem is solved for primes $p > 5^{30}$.

It should be pointed out that the proof of the Theorem relies on the classification of finite simple groups, and hence so does the solution of the $k(GV)$-problem.

Recently, Robinson [Ro] has taken further the work on case (A), and has shown that the required vector $v$ exists, assuming only that $p > 157$; and D. Goodwin, in his PhD work at Imperial College, London, has done the same for case (B). Consequently the $k(GV)$-problem is now in fact solved for primes $p > 157$.

We close with a few words on the proof of the above Theorem in [Li]. Suppose $G$ is as in the hypothesis, and $G$ has no regular orbit on $V = V_n(p)$. Then $C_G(v) \neq 1$ for all $v \in V$, whence $V = \bigcup_{g \in G^\sharp} C_V(g)$ (where $G^\sharp = G \backslash \{1\}$). Thus

(1) $p^n = |V| \leq \sum_{g \in G^\sharp} |C_V(g)|$.

To illustrate the argument for groups of Lie type, consider the case where $G = H$ is quasisimple, and $G/Z(G) \cong L_d(r)$. Observe

(2) $|G| < r^{d^2}$,

(3) $n \geq r^{d-1} - 1$ by [LS].

For $g \in G \backslash Z(G)$, let $r(g)$ be the minimum number of conjugates of $g$ required to generate $G$, and define $r(G)$ to be the maximal value of $r(g)$. By [HLS, Theorem 2], we have $r(G) \leq 8(2d-1)$. Therefore

(4) $\dim C_V(g) \leq n(1 - \frac{1}{8(2d-1)})$ for all $g \in G \backslash Z(G)$.

By (1) and (4), we have

$$p^n < |G| p^{n(1-1/(8(2d-1)))}.$$

Hence, using (2) and (3), we obtain

$$p < |G|^{8(2d-1)/n} < r^{d^2 . 8(2d-1)/(r^{d-1}-1)}.$$

It is readily checked that the right hand side of this inequality is less than $5^{30}$, contradicting the hypothesis that $p > 5^{30}$.

Similar arguments handle the other groups of Lie type (although in general matters are complicated by the fact that one only has $H \lhd G$ rather than $H = G$). The sporadic groups succumb easily, in view of the hypothesis $p > 5^{30}$. The alternating groups require a somewhat more delicate argument.

Finally, we offer some remarks on the above-mentioned improvement of the Theorem by D. Goodwin. For his proof, Goodwin obtains some new bounds for the numbers $r(G)$ defined above, when $G$ is sporadic; and for $G$ of Lie type he uses a recent improvement [GS] of the bounds for $r(G)$ given by [HLS].

## References


[Fo] P. Fong, "On the characters of $p$-solvable groups", Trans. Amer. Math. Soc. **98** (1961), 263–284.

[Gl] D. Gluck, "On the $k(GV)$-problem", *J. Algebra* **89** (1984), 46–55.

[Go] R. Gow, "On the number of characteris in a block and the $k(GV)$-problem for self-dual $V$", *J. London Math. Soc.* **48** (1993), 441–451.

[GS] R. Guralnick and J. Saxl, "Generation of classical groups by conjugates", to appear.

[HLS] J. Hall, M.W. Liebeck and G.M. Seitz, "Generators for finite simple groups, with applications to linear groups", *Quart. J. Math.* **43** (1992), 441–458.

[Kn1] R. Knorr, "On the number of characters in a block of a $p$-solvable group", *Illinois J. Math.* **28** (1984), 181-210.

[Kn2] R. Knorr, "On Brauer's $k(B)$-conjecture", *J. Algebra* **131** (1990), 444-450.

[LS] V. Landazuri and G.M. Seitz, "On the minimal degrees of projective representations of the finite Chevalley groups", *J. Algebra* **32** (1974), 418–443.

[Li] M.W. Liebeck, "Regular orbits of linear groups", *J. Algebra* **184** (1996), 1136–1142.

[Na] H. Nagao, "On a conjecture of Brauer for $p$-solvable groups", *J. Math. Osaka City Univ.* **13** (1962), 35–38.

[Ro] G.R. Robinson, "Further reductions for the $k(GV)$-problem", preprint, Univ. of Leicester.

[RT] G.R. Robinson and J.G. Thompson, "On Brauer's $k(B)$-problem", *J. Algebra* **184** (1996), 1143–1160.


Martin W. Liebeck,
Dept. of Mathematics,
Imperial College,
London SW7 2BZ, UK
e-mail: `m.liebeck@ic.ac.uk`

Trends in Mathematics, © 1998 Birkhäuser Verlag Basel/Switzerland

# Generating Minimally Transitive Groups

Andrea Lucchini

**Abstract**

It is described a general method to approach the study of the minimal number of generators of a finite group. In particular this method is applied to find a bound on the number of generators of minimally transitive permutation groups.

## 1 Introduction

A transitive permutation group $G \leq S_n$ is called minimally transitive if it has no proper transitive subgroups.

Let $n = \prod_{1\leq i\leq k} p_i^{\alpha_i}$ be the prime factorization of $n$, let $\lambda(n) = \sum_{1\leq i\leq k} \alpha_i$ and $\mu(n) = \max_{1\leq i\leq k} \alpha_i$.

In 1963 Shepperd and Wiegold [7] proved that, assuming that every finite simple group can be generated by 2 elements (now it is known from the classification of finite simple groups that this hypothesis is true, see the survey [3] by Di Martino and Tamburini) then any minimally transitive subgroup of $S_n$ can be generated by $\lambda(n)$ elements.

Neumann and Vaughan-Lee [5] gave an elementary proof of the fact that $\log_2(n)$ elements suffice.

Recently Pyber [6] has suggested to investigate whether $\mu(n) + 1$ elements could be always sufficient to generate a minimally transitive group of degree $n$ (as it occurs, for example, in the case of regular or nilpotent groups).

In this paper we deal with this question, proving

**Theorem** *Let $G$ be a minimally transitive subgroup of $S_n$. If $\mu(n)+1$ elements are not sufficient to generate $G$ then $\lambda(n) \geq 2$ and $G$ can be generated by $[\log_2(\lambda(n) - 1) + 3]$ elements.*

**Corollary** *Any solvable minimally transitive subgroup of $S_n$ can be generated by $\mu(n) + 1$ elements.*

Our approach to the problem relies on some recent results (most of which were proved using the classification of finite simple groups) which allow one to develop a method to compute the minimal number of generators of a finite group $G$ from the study of the factor groups of a particular prescribed type that $G$ admits. This method is described in section 2; we think that it could be of independent interest and useful in the study of other questions about generation of finite groups.

## 2 Preliminary results

In this section we describe an approach to the study of the minimal number of generators $d(G)$ of a finite group $G$, which relies on some results recently proved by Francesca Dalla Volta and the author.

Suppose $d(G) = d > 2$ and consider a factor group $G/M$ of $G$ which is minimal with respect to the property $d(G/M) = d$; that factor group has a very particular structure. To describe the structure we need to recall some definitions and results.

Let $H$ be a finite group with a unique minimal normal subgroup, $N$. If $N$ is abelian, assume also that $N$ has a complement in $H$.

For each positive integer $k$, let $H^k$ be the $k$-fold direct power of $H$ and define the subgroup $H_k$ by

$$H_k = \{ (h_1, \dots, h_k) \in H^k \mid h_1 \equiv \cdots \equiv h_k \bmod N \}.$$

Equivalently set $\mathrm{diag}(H^k) = \{(h, \dots, h) \in H^k | h \in H\}$ and $H_k = N^k \,\mathrm{diag}(H^k)$. If $m \geq d(H)$ then there is a (unique) $k$ such that $d(H_k) = m < d(H_{k+1})$: set $f(m) = k + 1$.

In [2] it is proved that if $m \geq 2$ and $K$ is a finite group such that every proper factor group of $K$ can be generated by $m$ elements, but $K$ itself cannot, then $K$ is isomorphic to one of the $H_{f(m)}$ constructed above.

In our particular case this implies that $G/M \cong H_{f(d-1)}$ for some choice of $H$.

In [2] it is described how the function $f$ (which of course depends on $H$) may be calculated. For any finite group $K$, let $\phi_K(m)$ denotes the number of $m$-basis of $K$ and let $\Gamma$ be the group of those automorphisms of $H$ that act trivially on $H/N$. If $m \geq d(H)$ then

$$f(m) = 1 + \begin{cases} \frac{\phi_H(m)}{|\Gamma|\phi_{H/N}(m)} & \text{if } N' = N, \\ \log_q \left(1 + (q-1)\frac{\phi_H(m)}{|\Gamma|\phi_{H/N}(m)}\right) & \text{if } N' = 1 \text{ and } q = |\operatorname{End}_{H/N}(N)|. \end{cases}$$

When $N$ is abelian, this formula can be written in the language of cohomology; if $N = H$ then of course $f(m) = m + 1$, otherwise $N$ is not centralized by $H$ and we have ([2], section 2)

$$f(m) = 1 + \log_q \left(|N|^{m-1} / |H^1(H/N, N)|\right).$$

Now $|H^1(H/N, N)| < |N|$ by the fact that the first cohomology group with coefficients in a faithful simple module is always smaller than the module itself (this was proved by Aschbacher and Guralnick [1] using the classification of finite simple groups). So $f(m) > 1 + \log_q(|N|^{m-2})$, which implies

(*) $f(m) \geq 2 + (m-2)\dim_{\mathrm{End}_{H/N}(N)} N \geq m$ *if* $N = \operatorname{soc} H$ *is abelian.*

When $N$ is nonabelian the number $f(m)$ is much larger than $m$. For example in [4] it is proved

(**) $f(m) \geq 2^{m-2} + 1$ *if* $N = \operatorname{soc} H$ *is nonabelian .*

From all this information we can deduce:

**Theorem 2.1** *If* $d(G) = d > 2$ *and* $G/M$ *is a factor group of* $G$ *which is minimal with respect to the property* $d(G/M) = d$ *then*

(i) $G/M \cong H_{f(d-1)}$ *where* $H$ *is one of the groups described above;*

(ii) $f(d-1) \geq 2 + (d-3)\dim_{\mathrm{End}_{H/N}(N)} N \geq d-1$ *if* $N = \operatorname{soc} H$ *is abelian;*

(iii) $f(d-1) \geq 2^{d-3} + 1$ *if* $N = \operatorname{soc} H$ *is nonabelian.*

This theorem can be used to deduce informations on the minimal number of generators of a finite group $G$ from the knowledge of the factor groups of kind $H_k$ that $G$ admits. In particular we have:

**Corollary 2.2** *Suppose that* $t_1, t_2$ *are two integers such that if some* $H_k$ *is a factor group of* $G$ *then* $k - 2 \leq t_1 \dim_{\mathrm{End}_{H/N}(N)} N$ *if* $N = \operatorname{soc} H$ *is abelian,* $k \leq t_2$ *if* $N = \operatorname{soc} H$ *is nonabelian. If* $d(G) > 2$ *then either* $d(G) \leq t_1 + 3$ *or* $t_2 \geq 2$ *and* $d(G) \leq [\log_2(t_2 - 1) + 3]$.

The previous results suggest that there is a sort of dichotomy according to whether or not there exists a group $H$ such that $\operatorname{soc} H = N$ is abelian and $H_{f(d(G)-1)}$ is an epimorphic image of $G$. In the first case the number of generators can be computed from the knowledge of the abelian chief factors of $G$ using cohomological methods and many of the results one can prove for number of generators of solvable groups remain true under this weaker hypothesis. In the latter case the study of $d(G)$ is more difficult and more strongly related to the properties of the finite non abelian simple groups.

# 3 The proof of the main theorem

We prove our main result:

**Theorem 3.1** *If* $G$ *is a minimally transitive subgroup of* $S_\Omega$ *with* $|\Omega| = n$ *and* $\lambda(n) \geq 2$ *then* $d(G) \leq \max(\mu(n) + 1, [\log_2(\lambda(n) - 1) + 3])$.

We note that there is no particular restriction in assuming $\lambda(n) \geq 2$: in fact a minimally transitive group of prime degree is cyclic.

*Proof.* Of course we may assume $d(G) = d > 2$. As we have seen in the previous section there exists a group $H$ such that $H_t$ is an epimormphic image of $G$, with

$t = f(d-1)$. Notice that, by Theorem 2.1, $d \geq 3$ implies $t \geq 2$. The socle of $H_t$ is $N_1 \times \ldots \times N_t$, the direct product of $t$ minimal normal subgroups, each isomorphic to $N = \operatorname{soc} H$.

For each subgroup $X$ of $H_t$ let denote by $\overline{X}$ the preimage of $X$ under the given epimorphism $\phi : G \to H_t$.

Now fix $\omega \in \Omega$, define $X_0 = 1$, $X_i = N_1 \times \ldots \times N_i$, $1 \leq i \leq t$, and consider the orbit $B_i = \omega^{\overline{X}_i}$ of $\omega$ under $\overline{X}_i$. We have

$$|B_i| = |\overline{X}_i : \overline{X}_i \cap G_\omega| = |\overline{X}_i G_\omega : G_\omega|$$

where $G_\omega$ denotes the stabilizer of $\omega$ in $G$. In particular, since $\overline{X}_i$ is a normal subgroup of $G$, $|B_i|$ divides $|\Omega| = n$. Moreover, for each $1 \leq i \leq t-1$,

$$\frac{|B_{i+1}|}{|B_i|} = \frac{|\overline{X}_{i+1} G_\omega|}{|\overline{X}_i G_\omega|} = \frac{|\overline{N}_{i+1} \overline{X}_i G_\omega|}{|\overline{X}_i G_\omega|} = \frac{|\overline{N}_{i+1}|}{|\overline{N}_{i+1} \cap \overline{X}_i G_\omega|} = \frac{|N_{i+1}|}{|\phi(\overline{N}_{i+1} \cap \overline{X}_i G_\omega)|}.$$

So we deduce:

(*) *For each* $0 \leq i \leq t-1$, $|B_{i+1}|/|B_i|$ *is an integer and divides* $|N|$.

Moreover we claim:

(**) $|B_i| < |B_{i+1}|$ *for each* $0 \leq i \leq t-2$ *and, if* $N = \operatorname{soc} H$ *is abelian, also for* $i = t-1$.

Indeed suppose by contradiction that $\omega^{\overline{X}_i} = B_i = B_{i+1} = \omega^{\overline{X}_{i+1}}$.
If $i \leq t-2$ define

$$Y_{i+1} = \operatorname{diag}(H^t) \prod_{\substack{1 \leq j \leq t, \\ j \neq i+1, i+2}} N_j.$$

If $N$ is abelian and $i = t-1$ recall that $N$ has a complement, say $K$, in $H$, and define

$$Y_t = \operatorname{diag}(K^t) \prod_{1 \leq j \leq t-1} N_j.$$

Since $N_{i+1} Y_{i+1} = H_t$ we have $G = \overline{H}_t = \overline{N}_{i+1} \overline{Y}_{i+1} = \overline{X}_{i+1} \overline{Y}_{i+1}$. Now $\overline{X}_i \leq \overline{Y}_{i+1}$ so $\omega^{\overline{X}_{i+1}} = \omega^{\overline{X}_i} \leq \omega^{\overline{Y}_{i+1}}$ and $\Omega = \omega^G = \omega^{\overline{X}_{i+1} \overline{Y}_{i+1}} = \omega^{\overline{Y}_{i+1}}$.

But then $\overline{Y}_{i+1}$ is a transitive subgroup of $G$; on the other hand $\overline{Y}_{i+1}$ is a proper subgroup of $G$ so we contradict the hypothesis that $G$ is minimally transitive.

Now we distinguish two cases:

a) $N = \operatorname{soc} H$ is an elementary abelian $p$-group.

By (*) and (**) $p$ divides $|B_{i+1}|/|B_i|$ for each $0 \leq i \leq t-1$ so $p^t$ divides $|B_t|$, hence divides $n$. So $t = f(d-1) \leq \mu(n)$. By Theorem 2.1 we conclude $d \leq t+1 \leq \mu(n)+1$.

b) $N = \operatorname{soc} H$ is non abelian.

By (*) and (**) $\lambda(|B_{i+1}|) > \lambda(|B_i|)$ for each $0 \leq i \leq t-2$. Moreover, since $\overline{X}_{t-1}$ is a proper normal subgroup of $G$, $|B_{t-1}|$ divides properly $|\Omega| = n$. So $\lambda(n) > \lambda(|B_{t-1}|) \geq t-1$. Again by Theorem 2.1 we deduce $d \leq \log_2(t-1)+3 \leq \log_2(\lambda(n)-1)+3$.

If $G$ is a solvable group and $H_k$ is an epimorphic image of $G$ then of course $N = \operatorname{soc} H$ must be abelian, so from the proof of the previous theorem we immediately deduce:

**Corollary 3.2** *If $G$ is a solvable minimally transitive subgroup of $S_\Omega$ with $|\Omega| = n$ then $d(G) \leq \mu(n) + 1$.*

# References

[1] M. Aschbacher and R. Guralnick, *Some Applications of the First Cohomology Group*, J. Algebra 90 (1984), 446–460

[2] F. Dalla Volta and A. Lucchini, *Finite groups that need more generators than any proper quotient* (to appear)

[3] L. Di Martino and M. C. Tamburini, *2-Generation of finite simple groups and some related topics*, in "Generators and relations in groups and geometries", proceedings of the NATO Advanced Study Institute, Castelvecchio Pascoli (Lucca), Italy, April 1–14, 1990, (eds. A. Barlotti et al.), Kluwer Academy Publishers, Dordrecht, Boston, 1991, 195–233

[4] A. Lucchini, *A bound on the presentation rank of a finite group* (to appear)

[5] P.M. Neumann and M.R. Vaughan-Lee, *An essay on BFC groups*, Proc. London Math. Soc. 35 (1977), 213–237

[6] L. Pyber, *Asymptotic results for permutation groups*, in "Groups and Computation" DIMACS: Series in Discrete Mathematics and Theoretical Computer Science 11 (eds. L. Finkelstein and W. M. Kantor), Amer. Math. Soc. Providence, 1993, 197–219

[7] J.A.M. Shepperd and J. Wiegold, *Transitive groups and groups with finite derived groups*, Math. Z. 81 (1963), 279–285

Andrea Lucchini
Dipartimento di Elettronica per l'Automazione
Università di Brescia
Via Branze, 25123 Brescia, Italy
e-mail: `lucchini@bsing.ing.unibs.it`

Trends in Mathematics, © 1998 Birkhäuser Verlag Basel/Switzerland

# Maximal Subgroups of Finite Exceptional Groups

Gary M. Seitz

**Abstract**

In this note we discuss some recent results on the subgroup structure of exceptional groups obtained jointly with Martin Liebeck and some related projects in progress.

Let $\bar{G}$ be a simple adjoint algebraic group of exceptional type over an algebraically closed field of positive characteristic $p$ and $\sigma$ a Frobenius morphism of $\bar{G}$. Then $G(q) = O^{p'}(\bar{G}_\sigma)$, with $q = p^a$, is a finite group of Lie type and we are interested in the maximal subgroups of groups $H$, where $G(q) \leq H \leq Aut(G(q))$.

To fix notation, let $M < H$ be a maximal subgroup. In [LS1] a theorem is established which reduces the study to the case where $F^*(M) = X$ is a simple group. This result is an analog for finite exceptional groups of the well-known result of Aschbacher on maximal subgroups of finite classical groups. In both situations the problem is reduced to the study of almost simple subgroups.

At this stage there is a natural subdivision of the problem according to whether or not $X$ is a group of Lie type defined over a finite field of characteristic p. The *generic case* is when $X = X(q_o)$, a group of Lie type with $q_o = p^b$.

We will first report on some contributions to the generic case. Later we will discuss some progress on the remaining cases. Our general approach to the generic case is based on the following key steps:

i). Lift the embedding $X < G(q)$ to an embedding $\bar{X} < \bar{G}$, where $\bar{X}$ is connected, semisimple and $X = O^{p'}(\bar{X}_\tau)$, for $\tau$ a Frobenius morphism of $\bar{X}$.

ii). Determine the possible subgroups $\bar{X} < \bar{G}$.

In other words, the idea is to lift the problem to one concerning algebraic groups and to then solve that problem. There already exist published results which address each of these issues:

1). It is shown in [ST] that a suitable group $\bar{X}$ exists, provided that $X$ is not contained in a parabolic subgroup of $G(q)$ and that $p$ is suitably large. The characteristic condition is a specific bound depending on the rank of $X(q_o)$ and the type of $\bar{G}$.

2). The maximal subgroups of positive dimension in exceptional algebraic groups were determined in [LS1] subject to certain characteristic restrictions, extending the work in [S1] which determined the maximal closed connected subgroups, subject to similar characteristic restrictions. The characteristic restrictions are required only in the case that the subgroup is simple and the restriction depends on the rank of the group. In particular, $p > 7$ is sufficient for all cases.

These results go a long way towards settling the generic case, as they provide complete information provided that $p$ is sufficiently large. It can be seen from representation theory that some field restrictions are necessary. For example, there are indecomposable representations of $SL_2(p)$ of dimension $p-1$ which do not extend to representations of $SL_2(K)$ for $K$ an algebraically closed field of characteristic $p$. So this gives an embedding $SL_2(p) < SL_{p-1}(p)$ which does not lift to a containment of algebraic groups.

The above examples indicate that some field restrictions are necessary for the lifting process. However, in the above examples the rank of the large group increases with $p$ and the indecomposable modules mentioned above occur only for the prime field. So for exceptional groups, where the rank is bounded, one could hope for lifting results which which require an hypothesis on only the size of the defining field for $X$. We will discuss some new results of this type. Proofs will appear in [LS2].

We require the following definition. Let $\Sigma = \Sigma(\bar{G})$ be the root system of $\bar{G}$. For $\alpha, \beta \in \Sigma$, call the element $\alpha - \beta$ a *root difference.* Define an integer $t(\Sigma)$ as follows.

**Definition 1** *For $L \leq Z\Sigma$, let $t(L)$ be the exponent of the torsion group of $Z\Sigma/L$. Set $t(\Sigma) = max\{t(L) : L$ is generated by root differences in $\Sigma\}$*

The integer $t(\Sigma)$ plays a role in our lifting result for rank 1 groups of Lie type. It has been computed by Lawther for exceptional groups $\bar{G} \neq E_8$.

Our first two theorems apply to more general situations than the analysis of maximal subgroups. After stating these results we give applications to the study of maximal subgroups. The first result is important for lifting embeddings.

**Theorem 1** *Assume $\bar{G}$ is a simple adjoint algebraic group of exceptional type over an algebraically closed field of characteristic $p$. Let $X = X(q_o) < \bar{G}$ be such that $q_o = p^b$ satisfies the following*

$q_o > t(\Sigma) \cdot (2, p-1)$ *if* $X = L_2(q_o), Sz(q_o), R(q_o)$.

$q_o > 9$, *otherwise (except for* $A_2^{\epsilon}(16)$*).*

*Then there exists a closed, connected, $\sigma$-invariant subgroup $\bar{X} \leq \bar{G}$ such that each $X$-invariant subspace of $L(\bar{G})$ is also $\bar{X}$-invariant.*

We make several observations about Theorem 1. First note that the conclusion holds for all but finitely many groups $X = X(q_o)$, as the rank of $X$ is bounded above. Moreover there is no maximality hypothesis.

On the negative side, the result gives no information if $X$ acts irreducibly on $L(\bar{G})$. We will address this is in the next theorem. Another issue is that the result makes no assertion about the structure of $\bar{X}$, nor the embedding of $X$ in $\bar{X}$. However, in Theorems 3 and 4 we will see that such information is available in the context of maximality. Finally, for the rank 1 cases mentioned in the theorem it should be possible to establish a much better bound than $t(\Sigma) \cdot (2, p-1)$.

The next result addresses the possibility of $X(q_o)$ being irreducible on $L(\bar{G})$. This can occur if the subgroup is the group of fixed points of $\bar{G}$ with respect to a Frobenius morphism. The theorem shows that for exceptional groups this is the only possibility. There is no characteristic restriction in the hypothesis and the result covers classical groups as well as exceptional groups.

**Theorem 2** *Let $X = X(q_o) < \bar{G}$, for $q_o = p^b$. Assume $X(q_o)$ is irreducible on each $\bar{G}$-composition factor of $L(\bar{G})$. Then one of the following holds:*

*(i). $X(q_o) = O^{p'}(\bar{G}_\delta)$ for some Frobenius morphism $\delta$ of $\bar{G}$.*

*(ii). $\bar{G} = C_n$ or $B_n$, $p = 2$, and $X = D_n^\epsilon(q_o)$ or $G_2(q_o)$ $(n = 3)$.*

The upshot of Theorem 2 is that the subgroup $\bar{X}$ in Theorem 1 is proper in $\bar{G}$ unless $X$ is of the same type as $\bar{G}$.

The following results are applications of Theorems 1 and 2 to the analysis of maximal subgroups of finite exceptional groups. We maintain the notation $G(q) = O^{p'}(\bar{G}_\sigma)$, for $\bar{G}$ an adjoint group of exceptional type and $q = p^a$, and $X = X(q_o)$ simple of Lie type with $q_o = p^b$. Also, recall that $G(q) \leq H \leq Aut(G(q))$.

**Theorem 3** *Let $M = N_H(X(q_o))$ be maximal in $H$ and assume $q_o$ satisfies the hypothesis of Theorem 1. Then one of the following holds.*

*(i). $X(q_o) = O^{p'}(\bar{G}_\delta)$, for $\delta$ a Frobenius morphism of $\bar{G}$.*

*(ii). $X(q_o) = O^{p'}(\bar{X}_\sigma)$, for $\bar{X}$ maximal among closed, connected, reductive, $M\langle\sigma\rangle$-invariant subgroups of $\bar{G}$.*

It is possible to refine part (ii) of the result to show that either $\bar{X}$ contains a maximal torus of $\bar{G}$ (and hence is known precisely) or $q_o = q$. Moreover, if $p$ satisfies the conditions of [LS1] (e.g. $p > 7$ suffices for all cases), then the possibilities for $\bar{X}$ are determined explicitly by the results of [LS1]. In view of Theorem 3 it is now of high priority to improve the results in [LS1] to remove all characteristic restrictions. Efforts in this direction will be discussed later.

Our last result from [LS2] is the following.

**Theorem 4** *There is a constant c such that if $M$ is a maximal subgroup of $H$ and $|M| > c$, then one of the following holds*

*(i). $M = N_H(\bar{G}_\tau)$, for some Frobenius morphism $\tau$ of $\bar{G}$.*

*(ii). $M = N_H(\bar{X}_\sigma)$, where $\bar{X}$ is maximal among closed, connected, reductive, $M\langle\sigma\rangle$-invariant subgroups of $\bar{G}$.*

In (i) of Theorem 4, $\bar{G}_\tau$ is a subgroup of $G(q)$ of the same type (or twisted version) over a smaller field. For example, if $G(q) = E_6(q)$, then $\bar{G}_\tau$ could be $E_6(q_o)$ or $^2E_6(q_o)$.

In the following we indicate a rough sketch of our approach to the proof of Theorem 1. The proof is long and difficult, the difficulty arising from showing that $q_o > 9$ suffices for all but certain rank 1 groups and $A_2^\epsilon(16)$. If one only wants the existence of some bound, then this can be obtained from the following general result, whose proof is short and which also applies to classical groups.

**Proposition 1** *Let $\bar{G}$ be a simple adjoint algebraic group and let $x \in \bar{G}$ be a semisimple element of finite order such that $|x| > t(\Sigma)$. Then there is a closed subgroup $J \leq \bar{G}$ of positive dimension such that $J$ leaves invariant each $x$-invariant subspace of $L(\bar{G})$.*

For the rank 1 groups mentioned in Theorem 1 we simply apply the above proposition. So now assume that $X$ is of (BN)-rank at least 2 or $PSU_3(q_o)$. Then $X(q_o)$ contains a subgroup $A = A_2^\epsilon(q_o)$ or $B_2(q_o)$ and we work with subgroups of $A$. In particular, $A \geq \langle a\rangle \cdot S$, where $S \cong SL_2(q_o)$ and $a$ is an element of order dividing $q_o \pm 1$, commuting with $S$.

We first establish a result showing that it suffices to work within the group $E_8$. Of course this case must be handled in any case and it has the advantage that centralizers of semisimple elements are always connected.

Now $a$ is a semisimple element of $\bar{G} = E_8$, so it follows that $C_{\bar{G}}(a) = T_kD$, where $T_k$ is a $k$-dimensional torus and $D$ is a semisimple group, in fact a subsystem subgroup of $\bar{G}$.

If $k$ is small, say $k = 1, 2$, then in many cases it is possible to determine the precise connection between $a$ and $T_k$ and to show that there is a 1-dimensional torus of $T_k$ which stabilizes each $a$-invariant subspace of $L(\bar{G})$. So here we let $\bar{X}$ be the group generated by this torus and $X$.

On the other hand, when $k$ is larger, $D$ is usually a product of classical groups. Here we analyze the embedding $S < D$ and attempt to show that $S$ is contained in a connected subgroup $\bar{S}$ of $D$ of type $A_1$. In some cases there are infinitely many possible overgroups of type $A_1$ and the goal is to produce one which stabilizes all $S$-invariant subspaces of $L(\bar{G})$. For example, we establish a result showing that this holds if all composition factors of $\bar{S}$ on $L(\bar{G})$ have high weight $c\lambda_1$, with $c < q_o$. Once a suitable $\bar{S}$ is found we set $\bar{X} = \langle X, \bar{S}\rangle$.

The above ideas are quite easy in principle, but there are many obstacles that occur along the way and this accounts for the level of difficulty of the actual proof.

**Future Directions.** In this section we discuss several problems concerning the subgroup structure of exceptional groups as well as classical groups. Several of these are projects currently underway where reasonable results appear to be in reach. Others, such as the problems in **3** deserve serious consideration, but are not currently being worked on.

**Problem 1.** Extend the work on maximal subgroups of exceptional algebraic groups, removing all characteristic restrictions. In view of the above results this problem is now fundamental for the further understanding of the maximal subgroups of finite exceptional groups.

Prior to the results described in this note, the lifting results all had characteristic restrictions which were stronger than those assumed in the papers determining the subgroup structure of exceptional algebraic groups. But now we have lifting results independent of characteristic and so the results on exceptional algebraic groups must be improved.

Liebeck and I have made considerable progress on this problem. In particular, for subgroups of type $A_1$ there is nearly a complete result. Interestingly enough the small characteristic configuration allows for some very nice Lie algebra methods to come into play, which were not available for larger characteristics. The point is that in large characteristic the relevant representation theory for the subgroup and its Lie algebra are roughly the same, while in small characteristic there are important differences that can be exploited.

We are hopeful that a complete result will be forthcoming in the not too distant future.

**Problem 2.** Carry out a complete analysis of the nongeneric case. This is another project fundamental to the study of the subgroup structure of finite exceptional groups. Liebeck and I began an analysis of this problem in summer 1996, prior to the Siena conference. The first step is to determine embeddings $X < \bar{G}$, where $X$ is simple , but not of Lie type in characteristic $p$. This problem can be regarded as a modular version of the problem of determining the finite simple subgroups of complex exceptional groups.

We were able to obtain a list of such embeddings, with very few question marks. Most of our work was aimed at reducing the number of potential subgroups. Our list closely resembles the list over the complex numbers, although there are some differences. For example, $L_4(5) < E_8 = \bar{G}$ only when $p = 2$. Most existence issues are either easily resolved or follow from existing results for complex exceptional groups. In particular, the recent work of Serre is very helpful here.

This is a good start, but there remain a number of issues to be settled before we would have satisfactory results for the finite groups. In particular, there are conjugacy class questions, field of definition problems, and maximality issues.

**Problem 3.** Establish better lifting results. It is undoubtedly possible to improve on the lifting results mentioned above. The bound in Theorem 1 for rank 1 groups is probably much too large and it would be an important contribution to improve this bound. It should also be possible to improve the bounds for large rank subgroups. As indicated above, our approach was independent of the rank of $X$, except for the rank 1 cases mentioned. It should be possible to establish a result taking the ranks of $X$ and $\bar{G}$ into account. For example, we mention the result of [LST] showing that $q_o > 2$ suffices for lifting, provided the rank of $X$ is greater than half that of $\bar{G}$.

**Problem 4.** Establish lifting results for applications to representation theory. The use of lifting techniques to study subgroups of exceptional type have been successful, and there are indications that lifting techniques can also be useful in the analysis of subgroups of classical groups and more generally to problems in representation theory.

For example, given a finite group of Lie type it is usually very difficult to analyze possible extensions between simple modules in the defining characteristic. On the other hand, there are more techniques available to analyze similar questions at the level of algebraic groups. In particular, one knows that extensions between simple modules must occur as images of Weyl modules. Therefore, lifting results for indecomposable modules could be an important tool in analyzing extensions of simple modules for finite groups of Lie type. Liebeck and I have started work on this problem using some ideas from [S2], although this is still in the preliminary stages.

**Problem 5.** Determine all subgroups of groups of Lie type which are irreducible on the adjoint module. This is a problem that may have important applications. Theorem 2 establishes such a result for subgroups which are simple and of Lie type in the same characteristic. But in future lifting results a more general result will be needed.

Strong information on exceptional groups will follow from Theorem 2 and the work mentioned in Problem 2. In particular, this should yield a complete result for simple subgroups. For classical groups, the problem is open. Guralnick and I have some results on the problem provided the defining field of the subgroup is reasonably large and Maagard [M] has a result in this direction. But no general result is yet available.

Let me give one example of how such a result might be used. For purposes of illustration I will go back to my paper on groups containing maximal tori [S3]. I use this example, since it is now clear that lifting techniques can be used to strengthen the results of this paper. Say $G(q)$ is a group of Lie type and $X$ is a subgroup containing a maximal torus, $T$, of $G(q)$. So for example, $X$ might contain a Singer cycle of a classical group or large cyclic subgroup of an exceptional group.

There is a result in [S3] showing that lifting is possible except for small values of $q$ ($q > 4$ is sufficient in almost all cases and for cyclic tori much better results should be possible). This means that there is a maximal torus, $\bar{T}$ of $\bar{G}$ stabilizing the same subspaces of $L(G)$ as $T$. Hence the subgroup $\bar{X} = \langle X, \bar{T} \rangle$ is $\sigma$-stable of positive dimension and leaves invariant all $X$- invariant subspaces of $L(G)$. The possibilities for such subgroups are determined by the root system, giving strong information on $X$, unless $X$ happens to be irreducible on $L(G)$ (or irreducible on all $G$-composition factors of $L(G)$ in those situations where $G$ acts reducibly on its Lie algebra). For in this case we would have $\bar{X} = G$.

So the problem of determining subgroups of algebraic groups irreducible on the adjoint module is one which may play a significant role in the analysis of subgroups of groups of Lie type.

## References

[LS1] Liebeck, M. and Seitz, G., Maximal subgroups of exceptional groups of Lie type, finite and algebraic, Geom. Ded, 36, (1990), 353–387.

[LS2] Liebeck, M. and Seitz, G., Finite subgroups of exceptional groups of Lie type, (to appear).

[LST] Liebeck, M., Saxl, J., and Testerman, D., Simple subgroups of large rank in groups of Lie type, Proc. LMS, 65, (1996),425–457.

[M] Maagard, K., On the irreducibility of alternating powers and symmetric squares, Archiv. der Mathematik, 63, (1994), 211–215.

[ST] Seitz, G., and Testerman, D., Extending morphisms from finite to algebraic groups, J. Alg., 131, (1990), 559–574.

[S1] Seitz, G., The maximal subgroups of exceptional algebraic groups, Memoirs AMS, 441, (1991), 1–197.

[S2] Seitz, G., Abstract homomorphisms of algebraic groups,(to appear Proc. LMS).

[S3] Seitz, G., The root groups for maximal tori in finite groups of Lie type, Pacific J. Math., 106, (1983), 153–244.

Gary M. Seitz
Department of Mathematics
University of Oregon
Eugene, OR 97403-1222, USA
e-mail: `seitz@math.uoregon.edu`

Trends in Mathematics, © 1998 Birkhäuser Verlag Basel/Switzerland

# Subgroup Structure, Fractal Dimension, and Kac-Moody Algebras

Aner Shalev *

**Abstract**

This paper describes recent joint work with Barnea, and with Barnea and Zelmanov, where ideas from fractal geometry and Kac–Moody algebras are applied in studying the subgroup structure of profinite groups. We shall be interested in the spectrum of a finitely generated profinite group $G$, which is the set of Hausdorff dimensions of closed subgroups of $G$. This leads to questions about the subalgebra structure of affine Kac–Moody algebras. We determine the maximal graded subalgebras of affine Kac–Moody algebras, and derive applications to the spectrum of certain groups (such as matrix groups over local rings), whose subgroup structure is far from clear. We also examine the spectrum of $p$–adic analytic groups, and formulate several problems and conjectures.

## 1 Subgroup Structure

The main motivation for the work described below is the study of the subgroup structure of matrix groups over local fields and over their valuation rings. The maximal subgroups of $SL_d(K)$ have been described successfully when $K$ is an algebraically closed field (see Seitz [S]) and when $K$ is a finite field (see Aschbacher [A], Kleidman-Liebeck [KL] and the references therein). On the other hand it is known that there cannot be such good describtions in the case $K = \mathbb{Q}$ (the rational numbers), since $SL_d(\mathbb{Q})$ has uncountably many conjugacy classes of maximal subgroups (see Margulis and Soifer [MS]). This paper provides some indication that the maximal subgroups of $SL_d(K)$ when $K$ is a local field could possibly be described in the long run. The two fundamental cases are $K = \mathbb{Q}_p$ and $K = \mathbb{F}_p((t))$, but we shall be mostly interested here in $SL_d(R)$, where $R$ is the valuation ring of $K$. The more challenging case is that of $S = SL_d(\mathbb{F}_p[[t]])$, which will be our main object of study. Note that $S$ is a profinite group which has an open pro-$p$ subgroup (i.e. $S$ is virtually pro-$p$); some attention will therefore be devoted to pro-$p$ groups in general. I will survey recent joint work with Yiftach Barnea [BSh]

*I am grateful to All Souls College and the University of Oxford for their hospitality during the preparation of this paper.

(in Section 2) and with Yiftach Barnea and Efim Zelmanov [BShZ] (in Section 3) on this subject. Occasionally, some arguments which are not included in [BSh] and [BShZ] will be presented, and things will be viewed from a somewhat different angle. I will also provide a list of related open problems, which I hope will stimulate further research.

First, some clarifications. Let $G$ be a (topologically) finitely generated infinite profinite group. By a subgroup of $G$ we shall always mean a *closed* subgroup (closed in the profinite topology). The maximal subgroups of $G$ correspond to the maximal subgroups of the Frattini quotient $G/\Phi(G)$, where $\Phi(G)$ denotes the Frattini subgroup of $G$. If $G$ is a pro-$p$ group then $G/\Phi(G)$ is a finite (elementary abelian) $p$-group, and so the maximal subgroups of $G$ have index $p$ and are not of great interest. Similarly, if $G$ is virtually pro-$p$, then the Frattini quotient $G/\Phi(G)$ is finite, and so finding the maximal subgroups of $G$ is a finitary problem. We therefore have to slightly modify the notion of a maximal subgroup in order to pinpoint the real objects we have in mind.

**Definition 1.1** A (closed) subgroup $H$ of $G$ is said to be *weakly maximal* if
(i) $H$ has infinite index, and
(ii) $H$ is maximal with respect to this property.

In other words, the maximal non-open subgroups of $G$ are termed weakly maximal. This seems to be the right concept, since in our context it is natural to study subgroups up to commensurability (so open subgroups are not considered proper).

Do weakly maximal subgroups exist? Let $G$ be the cartesian product of all cyclic groups of prime order (each occurring once). Then $G$ is a finitely generated profinite group, in fact $G$ is generated (topologically) by one element. However, it is easy to see that every non-open subgroup of $G$ is contained in a larger non-open subgroup, and so $G$ does not have weakly maximal subgroups. However, for the profinite groups we have in mind, one can show that weakly maximal subgroups do exist. In fact we have

**Lemma 1.2** Let $G$ be a profinite group which is virtually pro-$p$. Then every non-open subgroup of $G$ can be extended to a weakly maximal subgroup.

*Proof.* Let $\mathcal{F}$ denote the set of non-open subgroups of $G$. Using Zorn's lemma it suffices to show that every ascending chain of subgroups in $\mathcal{F}$ has an upper bound in $\mathcal{F}$. Let $\{H_\alpha\}$ be such a chain. Let $H$ be the closure of the union $\cup_\alpha H_\alpha$. If $H$ is not open then $H \in \mathcal{F}$ and we are done. So suppose $H$ is open in $G$. Since $G$ is virtually pro-$p$, so is $H$. Hence $H/\Phi(H)$ is finite. Let $K_\alpha$ be the image of $H_\alpha$ in $H/\Phi(H)$. Since $\cup_\alpha H_\alpha$ is dense in $H$, we have $\cup_\alpha K_\alpha = H/\Phi(H)$. It follows that $K_\alpha = H/\Phi(H)$ for some $\alpha$, so $H_\alpha$ is dense in $H$. But $H_\alpha$ is closed. Hence $H_\alpha = H$, so $H_\alpha$ is open in $G$, a contradiction. □

It would be interesting to find the largest class of profinite groups for which the conclusion of Lemma 1.2 holds.

We remark that the weakly maximal subgroups of $SL_d(\mathbb{F}_p[[t]])$ include various geometric subgroups, e.g. stabilizers of subspaces, stabilizers of direct sum decompositions, groups defined over (weakly maximal) subrings such as $SL_d(\mathbb{F}_p[[t^r]])$ ($r$ prime), etc. Indeed, it is our hope that an Aschbacher type theorem could eventually be estabished in this context.

**Problem 1.** Describe the weakly maximal subgroups of the groups $SL_d(\mathbb{Z}_p)$ and $SL_d(\mathbb{F}_p[[t]])$.

In particular, it would be interesting to find out whether these groups have only countably many weakly maximal subgroups up to conjugacy. The case of $SL_d(\mathbb{Z}_p)$ seems more approachable. During the Siena meeting Nikolai Vavilov expressed the view that Problem 1 in the $p$-adic case could in principle be solved with present methods, though this may require a considerable amount of work.

Let me also mention some other questions concerning the subgroup structure of $SL_d(\mathbb{F}_p[[t]])$.

**Problem 2.** Does $SL_d(\mathbb{F}_p[[t]])$ have a free nonabelian pro-$p$ subgroup?

It is conjectured in [LSh] that the answer is negative. This is known only in the case $d = 2, p > 2$ (see Zubkov [Zu]). Note that the $p$-adic version of this question is easily solved using the theory of $p$-adic analytic groups (see [DDMS]). Indeed, a closed subgroup of a $p$-adic analytic group must itself be $p$-adic analytic, and since free nonabelian pro-$p$ groups are not $p$-adic analytic, they cannot be embedded in $SL_d(\mathbb{Z}_p)$. A negative solution to Problem 2 is likely to pave the way for an interesting theory of topological identities in linear groups over profinite rings.

A question of a somewhat different flavour is the following.

**Problem 3.** Can $SL_m(\mathbb{Z}_p)$ ($m > 1$) be embedded in $SL_d(\mathbb{F}_p[[t]])$?

Note that some $p$-adic analytic groups (such as $\mathbb{Z}_p^{d-1}$) can be embedded in $SL_d(\mathbb{F}_p[[t]])$, but embedding non-soluble $p$-adic analytic groups seems harder (and possibly impossible).

Some of the questions on the subgroup structure of $SL_d(\mathbb{F}_p[[t]])$ arise in probabilistic contexts. Recall that profinite groups are equipped with a canonical normalized Haar measure, and can thus be viewed as a probability space. In recent years the topic of random generation of finite classical groups and finite simple groups in general has received considerable attention; see [Di], [B], [KaL], [GKS], [LiSh1], [LiSh2], [LiSh3]. It is now known that a randomly chosen pair of elements of a finite simple group $T$ generates $T$ with probability $\to 1$ as $|T| \to \infty$. Similar questions have been studied for profinite groups by Mann and others (see [KaL], [M], [MSh], [BPSh]). When the ambient profinite group $G$ is virtually pro-$p$, questions about random generation of $G$ are really questions about the finite Frattini quotient $G/\Phi(G)$, and so generation probabilities can in principle be computed using finitary methods; note also that these probabilities are always strictly less than 1. Following the approach of studying subgroups up to commensurability, we shall focus instead on the probability of generating an open subgroup.

**Problem 4.** Does a randomly chosen pair (or $k$-tuple) of elements of $SL_d(\mathbb{F}_p[[t]])$ generate an open subgroup with probability 1?

I would guess that the answer is positive. It is known that $p$-adic analytic pro-$p$ (or virtually pro-$p$) groups have the property that, for some $k$, almost all $k$-tuples generate an open subgroup; and it was suggested that this property might in fact characterize $p$-adic analytic groups among virtually pro-$p$ groups (see [M, pp. 435-436]). An affirmative answer to Problem 4 would provide a counter-example to this conjecture. It could be regarded as a local field analogue of results of Kantor and Lubotzky [KaL] for $SL_d(\mathbb{F}_q)$ (and other finite classical groups).

While Aschbacher's theorem plays a key role in [KaL], solving Problem 4 seems to require information on weakly maximal subgroups of $S = SL_d(\mathbb{F}_p[[t]])$, namely, a solution of Problem 1. Indeed, if $x, y \in S$ do not generate an open subgroup, then they both lie in some weakly maximal subgroup $M$ of $S$ (by Lemma 1.2), so information on the various possibilities for $M$ should be helpful. A solution to Problem 1 is also likely to yield solutions to Problems 2 and 3, using some kind of induction on the dimension. However, Problem 1 (in the characteristic $p$ case) seems unapproachable with present methods; it is even open for $d = 2$.

In these circumstances it seems reasonable to start with less ambitious tasks. In the case of finite Chevalley groups, various results on the *sizes* of maximal subgroups were proven before the Classification of finite simple groups and Aschbacher's theorem were available; see Patton [P], Landazuri and Seitz [LS], Cooperstein [C] and Kantor [Ka]. Proceeding in a similar direction, we shall try to shed light on the possible sizes of subgroups of $SL_d(\mathbb{F}_p[[t]])$. In particular, we shall address the following.

**Problem 5.** Find the maximal size of a weakly maximal subgroup of $SL_d(\mathbb{F}_p[[t]])$.

Of course, we first need a precise (and satisfactory) definition of size. This is the content of the next section.

## 2 Hausdorff Dimension

The study of fractals requires some nonstandard notions of dimension (see Falconer [F]), among which the concept of Hausdorff dimension seems to play the most important role. This concept was originally defined over the reals, but can be defined in exactly the same manner over any metric space.

**Definition 2.1**
Let $(X, d)$ be a metric space, let $Y \subseteq X$ and let $\alpha, \epsilon$ be positive numbers. Define

$$\mu_\epsilon^\alpha(Y) = \inf \sum_i (\text{diam } S_i)^\alpha,$$

where $\{S_i\}_{i=0}^\infty$ is a cover of $Y$ by balls of diameter at most $\epsilon$, and the infimum is taken over all such covers. Note that $\mu_\epsilon^\alpha(Y)$ is non-increasing with $\epsilon$, and so the limit

$$\mu^\alpha(Y) = \lim_{\epsilon \to 0} \mu^\alpha_\epsilon(Y)$$

exists (though it may be infinite). The function $\mu^\alpha$ is usually referred to as the *$\alpha$-dimensional Hausdorff measure on $X$*. It can be shown that, if $\mu^\alpha(Y) < \infty$ and $\alpha < \alpha'$, then $\mu^{\alpha'}(Y) = 0$.

We can now define the *Hausdorff dimension* of a set $Y \subseteq X$, as follows.

$$\mathrm{Dim}(Y) = \sup\{\alpha | \mu^\alpha(Y) = \infty\} = \inf\{\alpha | \mu^\alpha(Y) = 0\}.$$

Clearly, if $Y \subseteq Y'$ then $\mathrm{Dim}(Y) \leq \mathrm{Dim}(Y')$. It can also be shown that $\mathrm{Dim}(\bigcup_{n=0}^\infty Y_n) = \sup \mathrm{Dim}(Y_n)$ for subsets $Y_n \subseteq X$. The Hausdorff dimension of a subset $Y$ may exceed its topological dimension, and it need not be an integer; for instance, the Hausdorff dimension of the standard Cantor set in $[0, 1]$ is $\frac{\log 2}{\log 3}$.

In order to apply the notion of Hausdorff dimension to profinite groups $G$, we need to regard them as metric spaces. This can be done once we fix a *filtration* $\{G_n\}_{n \geq 0}$ of $G$, namely a descending chain of open normal subgroups (with $G_0 = G$) which form a base for the neighbourhoods of $\{1\}$.

**Definition 2.2** Given a filtration $\{G_n\}$, define an invariant metric $d$ on $G$ by

$$d(x, y) = \inf\left\{ |G : G_n|^{-1} \;\middle|\; xy^{-1} \in G_n \right\}.$$

The balls in $G$, with respect to this metric, are the cosets of $G_n$, and the diameter of such a ball is $|G : G_n|^{-1}$. In particular, $G$ has diameter 1. It turns out that, in the metric space $(G, d)$, the Hausdorff dimension of a subgroup $H$ is easily determined from the orders of the projections of $H$ into the finite quotients $G/G_n$.

**Theorem 2.3** *Let $G$ be a profinite group with a filtration $\{G_n\}_{n \geq 0}$, and let $H \leq G$ be a (closed) subgroup. Then*

$$\mathrm{Dim}(H) = \liminf_{n \to \infty} \frac{\log |HG_n/G_n|}{\log |G/G_n|},$$

*where the Hausdorff dimension is computed with respect to the metric associated with the filtration $\{G_n\}$.*

The main part of the theorem, namely the inequality $\geq$, has been established by Abercrombie in [Ab, 2.6], who – as far as I am aware – was the first to apply the Hausdorff dimension concept to profinite groups. Theorem 2.3 can be used as an algebraic definition of Hausdorff dimension, for those who find the analytic definition too awkward. It also serves as a basic tool in the results below.

It follows from the theorem that $\mathrm{Dim}(H) = 1$ for all open subgroups $H$ of $G$ (recall that $G$ is infinite). Similarly, finite subgroups of $G$ have Hausdorff dimension zero, and commensurable subgroups have the same Hausdorff dimension (a desirable property). There are several indications that the Hausdorff dimension is the right tool in studying intermediate (i.e. not finite or of finite index) subgroups of profinite groups. One setback though is that $\mathrm{Dim}(H)$ may depend on the filtration $\{G_n\}$ which defines the underlying metric (see [BSh, 2.5]). The question of finding a canonical filtration therefore arises. While I do not know a natural candidate for a canonical filtration in arbitrary profinite groups, it seems natural to let $G_n$ be the subgroup generated by all $p^n$th powers if $G$ is a pro-$p$ group. Experience shows that other natural choices (such as the lower $p$-series, the dimension subgroups in characteristic $p$, the congruence subgroups when available, etc) usually give rise to the same Hausdorff dimension function. A general result guaranteeing this would be helpful.

We need another definition.

**Definition 2.4** Let $G, d$ de as above. Define the *spectrum* of $G$ by

$$\mathrm{Spec}(G) = \{\mathrm{Dim}(H) : H \text{ is a closed subgroup of } G\}.$$

Then we have

$$\{0, 1\} \subseteq \mathrm{Spec}(G) \subseteq [0, 1],$$

and so $\mathrm{Spec}(G)$ can be viewed as a picture reflecting the subgroup structure of $G$.

Before embarking on our main object of study, namely $SL_d(\mathbb{F}_p[[t]])$, let me discuss briefly the situation for $p$-adic analytic pro-$p$ groups. If $G$ is $p$-adic analytic then its dimension $\dim G$ (as a $p$-adic manifold) is well defined; moreover, if $H \leq G$ then $\dim H$ is also well defined, since $H$ too is $p$-adic analytic. The following result shows that the Hausdorff dimension of $H$ is proportional to its dimension as a manifold.

**Theorem 2.5** *Let $G$ be a $p$-adic analytic pro-$p$ group. Then*

$$\mathrm{Dim}(H) = \frac{\dim H}{\dim G}$$

*for all subgroups $H \leq G$.*

We see that the notion of Hausdorff dimension extends the notion of dimension in Lie groups; of course it is also applicable when no Lie group structure is available.

**Corollary 2.6** *The spectrum of a $p$-adic analytic pro-$p$ group is finite and consists of rational numbers.*

It is natural to ask whether the converse holds.

**Problem 6.** Let $G$ be a finitely generated pro-$p$ group such that $\mathrm{Spec}(G)$ is finite. Does it follow that $G$ is $p$-adic analytic?

Though we are not able to solve this problem, we do have some sort of a characterization of $p$-adic analytic groups in terms of Hausdorff dimension. Note that it follows from Theorem 2.5 that infinite subgroups of a $p$-adic analytic pro-$p$ group have positive Hausdorff dimension (since they have positive dimension as Lie groups). It turns out that this property characterizes $p$-adic analytic groups.

**Theorem 2.7** *The following are equivalent for a finitely generated pro-p group $G$.*
*(i) The Hausdorff dimension of every infinite subgroup of $G$ is positive.*
*(ii) $G$ is p-adic analytic.*

The proofs of Theorem 2.5 and 2.7 rely on the theory of $p$-adic analytic groups developed by Lazard, Lubotzky and Mann, and others (see [DDMS]). It is interesting that, in proving Theorem 2.7, we also apply Zelmanov's solution to the Restricted Burnside Problem in a strengthened form: torsion pro-$p$ groups are locally finite [Z]. This implies that, if $G$ is not $p$-adic analytic, then it has an infinite procyclic subgroup $H$; it can then be shown, combining Theorem 2.3 and results of Lazard, that $\mathrm{Dim}(H) = 0$ (so (i) implies (ii)).

Let us now consider the spectrum of $S = SL_d(\mathbb{F}_p[[t]])$. We regard $S$ as a metric space, where the metric is induced from the congruence filtration $\{S(n)\}_{n\geq 0}$, given by $S(0) = S$ and

$$S(n) = Ker(S \to SL_d(\mathbb{F}_p[[t]]/(t^n))) \quad (n \geq 1).$$

It is easy to see that, in contrast with the $p$-adic analytic case, the spectrum of $S$ is not discrete. For example, let $H$ be a standard root subgroup (so $H$ is isomorphic to the additive group of $\mathbb{F}_p[[t]]$). Then one easily verifies that $\mathrm{Dim}(H) = \frac{1}{d^2-1}$, and that for each $\alpha \in [0, \frac{1}{d^2-1}]$ there is a closed subgroup $H_\alpha$ of $H$ such that $\mathrm{Dim}(H_\alpha) = \alpha$. Therefore

$$\mathrm{Spec}(S) \supseteq [0, \frac{1}{d^2-1}],$$

so the spectrum of $S$ contains intervals. Roughly speaking we can say that root subgroups are 'soft' in that they can be deformed continuously to the trivial subgroup. In fact the Borel subgroup $B$ of upper triangular matrices (with arbitrary diagonal) is also soft in the same sense; this gives rise to the following.

**Proposition 2.8**

$$\mathrm{Spec}(SL_d(\mathbb{F}_p[[t]])) \supseteq [0, \frac{d(d+1)-2}{2(d^2-1)}] \supset [0, \frac{1}{2}].$$

It will turn out that the spectrum of $SL_d(\mathbb{F}_p[[t]])$ contains isolated points; in fact 1 is such a point. To study this phenomenon we need the following.

**Definition 2.9** Let $G$ be a finitely generated virtually pro-$p$ group (euipped with a filtration defining the metric). Define

$$\lambda(G) = \sup\{\mathrm{Dim}(H) : H < G \text{ is weakly maximal}\}.$$

Thus $\lambda(G)$ can be regarded as the supremal size of a weakly maximal (equivalently, of a non-open) subgroup of $G$. Since the open subgroups of $G$ have Hausdorff dimension 1, we see that

$$\mathrm{Spec}(G) \cap (\lambda(G), 1) = \emptyset.$$

Therefore, if $\lambda(G) < 1$, then 1 is an isolated point in $\mathrm{Spec}(G)$.

We can now restate Problem 5 in precise terms.

**Problem 5'.** Compute $\lambda(SL_d(\mathbb{F}_p[[t]]))$.

Taking $H$ to be the stabilizer in $S$ of a line, or of a subspace of codimension 1, one easily finds that

$$\mathrm{Dim}(H) = \frac{d^2-d}{d^2-1} = 1 - \frac{1}{d+1}.$$

This yields

$$\lambda(SL_d(\mathbb{F}_p[[t]])) \geq 1 - \frac{1}{d+1}. \tag{1}$$

Do we have equality?

Our solution to this problem starts with reductions to Lie-theoretic questions. Suppose, from now on, that $p > 2$. Let $S(n)$ $(n \geq 1)$ be as above, and set

$$L = S(1)/S(2) \oplus S(2)/S(3) \oplus S(3)/S(4) \oplus \dots$$

Multiplication in $S$ induces on $L$ the structure of an infinite dimensional linear space over $\mathbb{F}_p$, and commutation in $S$ induces on $L$ a binary operation which turns it into an $\mathbb{N}$-graded Lie algebra (see Chpater 8 of [HB] for an extensive discussion of such constructions and their applications). Let $L_n = S(n)/S(n+1)$ be the $n$th homogeneous component of $L$. Then $\dim_{\mathbb{F}_p} L_n = d^2 - 1$ for all $n$ and $L = \oplus_{n\geq 1} L_n$.

**Definition 2.10** The *density* of a graded subalgebra $K = \oplus_{n\geq 1} K_n$ of $L$ is defined by

$$D(K) = \liminf_{m\to\infty} \frac{\sum_{n\leq m} \dim K_n}{\sum_{n\leq m} \dim L_n}.$$

Thus, roughly speaking, the density of $K$ is the average dimension of a homogenous component of $K$ divided by $d^2 - 1$.

It turns out that the density function on graded Lie subalgebras of $L$ is strongly related to the Hausdorff dimension function on the subgroups of $S$. To see this, note that for each subgroup $H \leq S$ we can associate a graded subalgebra $L(H)$ of $L$, given by

$$L(H) = \oplus_{n\geq 1} (H \cap S(n))S(n+1)/S(n+1).$$

It can then be verified, using Theorem 2.3, that

$$\mathrm{Dim}(H) = D(L(H)). \tag{2}$$

Now, if $H$ is not open in $G$, then $L(H)$ has infinite codimension in $L$ (as $\mathbb{F}_p$-algebras). In order to solve Problem 5' we therefore need to find how large the density of graded subalgebras of infinite codimension can be.

Let us say that a subalgebra of $L$ is *weakly maximal* if it is graded of infinite codimension, and it is maximal with respect to these properties. It can be shown that every graded subalgebra of infinite codimension in $L$ can be extended to a weakly maximal one. By a slight abuse of notation, set

$$\lambda(L) = \sup\{D(K) : K < L \text{ is weakly maximal}\}.$$

Then equality (2) yields

$$\lambda(S) \leq \lambda(L), \tag{3}$$

so it may be useful to compute $\lambda(L)$.

So far the Lie-theoretic reduction has been fairly general, and can actually be applied for other types of virtually pro-$p$ groups (where $L$ is the Lie algebra induced by the prescribed filtration). However, in our case the Lie algebra $L$ can be easily identified. Let $\mathcal{G}$ be the finite Lie algebra $sl_d(\mathbb{F}_p)$ consisting of the $d \times d$ matrices of trace zero over $\mathbb{F}_p$. Then by identifying $L_n$ with $\mathcal{G} \otimes t^n$ we obtain a Lie algebra isomorphism

$$L \cong \oplus_{n \geq 1} \mathcal{G} \otimes t^n = \mathcal{G} \otimes t\mathbb{F}_p[t].$$

Combining this isomorphism with (1) and (3), we obtain:

**Corollary 2.11**

$$1 - \frac{1}{d^2 - 1} \leq \lambda(SL_d(\mathbb{F}_p[[t]])) \leq \lambda(sl_d(\mathbb{F}_p) \otimes t\mathbb{F}_p[t]).$$

We are led to the following.

**Problem 7.** Compute $\lambda(sl_d(\mathbb{F}_p) \otimes t\mathbb{F}_p[t])$.

Naturally, we will have to address a somewhat more general question.

**Problem 8.** Describe the weakly maximal subalgebras of $sl_d(\mathbb{F}_p) \otimes t\mathbb{F}_p[t]$.

Recall that $p$ is odd. Hence the finite Lie algebra $sl_d(\mathbb{F}_p)$ is simple when $p$ does not divide $d$, and simple modulo its 1-dimensional center otherwise. For simplicity I will assume below that $p$ does not divide $d$, but the arguments can be adapted to the general case (see [BSh]). Since $sl_d(\mathbb{F}_p)$ is simple, the infinite-dimensional Lie algebra $sl_d(\mathbb{F}_p) \otimes t\mathbb{F}_p[t]$ can be regarded as the positive part of an affine Kac-Moody algebra, namely $sl_d(\mathbb{F}_p) \otimes \mathbb{F}_p[t, t^{-1}]$. In the next section we shall study the subalgebra structure of affine Kac-Moody algebras in general, and then explore the implications to Problems 7 and 8.

# 3 Kac-Moody Algebras

The maximal subalgebras of the simple finite-dimensional Lie algebras over $\mathbb{C}$ were determined in classical works of Dynkin [D1],[D2]. Since the late sixties there has been considerable interest in some classes of infinite dimensional simple Lie algebras, most notably affine Kac-Moody algebras, and various classical results were extended to the infinite-dimensional case (see Kac [K]). However, while the representation theory of affine Kac-Moody algebras is now fairly developed, it seems that virtually nothing is known on the subalgebra structure of these objects (or on the subgroup structure of Kac-Moody groups). In [BShZ] we begin the study of the maximal subalgebras of affine Kac-Moody algebras and of related objects, such as loop algebras. Our results are fairly general, in that they hold over arbitrary fields $F$, though they take a particularly simple form in the important case $F = \mathbb{C}$. The case of finite fields leads to the solution of some of the problems mentioned in previous sections.

Given a finite-dimensional Lie algebra $\mathcal{G}$ over a field $F$, consider the infinite-dimensional $F$-algebra

$$L(\mathcal{G}) = \mathcal{G} \otimes_F F[t, t^{-1}],$$

where $F[t, t^{-1}]$ is the ring of Laurent polynomials. Then (derived) affine Kac-Moody algebras (corresponding to an indecomposable extended Cartan matrix) $\tilde{L}$ can be realized as central extensions of $L(\mathcal{G})$ by a 1-dimensional center $Z$, where $\mathcal{G}$ is a simple finite-dimensional Lie algebra over $F$. The study of the maximal subalgebras of $\tilde{L}$ is therefore reduced to the study of the maximal subalgebras of $L(\mathcal{G})$. The latter algebra is $\mathbb{Z}$-graded, and so it is natural to focus first on its maximal *graded* subalgebras.

Recall that a simple Lie algebra $\mathcal{G}$ over $F$ is said to be *central simple* if $F$ coincides with the centroid of $\mathcal{G}$, which consists of all elements $T \in \mathrm{End}_F\mathcal{G}$ satisfying $[T(x), y] = T([x, y])$ $(x, y \in \mathcal{G})$. Clearly, if $p$ does not divide $d$, then $sl_d(\mathbb{F}_p)$ is central simple over $\mathbb{F}_p$.

We need to define the important notion of a *loop algebra*.

**Definition 3.1** Let $k$ be a positive integer, and let $\mathcal{G} = \oplus_{i=0}^{k-1}\mathcal{G}_i$ be a $C_k$-grading of $\mathcal{G}$. Let $\alpha$ be the $k$-tuple $(\mathcal{G}_0, \ldots, \mathcal{G}_{k-1})$. Define

$$L(\mathcal{G}, k, \alpha) = \oplus_{n\in\mathbb{Z}}\mathcal{G}_{n \bmod k} \otimes t^n.$$

Then $L(\mathcal{G}, k, \alpha)$ is said to be a *loop algebra* on $\mathcal{G}$ (corresponding to $\alpha$). Loop algebras are often regarded as twisted Kac-Moody algebras.

We allow the trivial $C_k$-grading $\alpha = (\mathcal{G}, 0, \ldots, 0)$, which gives rise to the loop algebra $L(\mathcal{G}, k, \alpha) = \mathcal{G} \otimes F[t^k, t^{-k}]$.

We can now state the main result of this section.

**Theorem 3.2** *Let $F$ be any field and let $\mathcal{G}$ be a central simple finite-dimensional Lie algebra over $F$. Let $M$ be a maximal graded $F$-subalgebra of $L(\mathcal{G})$. Then one of the following holds:*

*(i)* $M = L(\mathcal{G}, r, \alpha)$ *for some prime* $r$ *and a* $C_r$*-grading* $\alpha$ *of* $\mathcal{G}$.
*(ii)* $M = L(\mathcal{H}) = \mathcal{H} \otimes F[t, t^{-1}]$, *where* $\mathcal{H}$ *is a maximal subalgebra of* $\mathcal{G}$.
*(iii)* $M$ *is commensurable with* $\mathcal{G} \otimes F[t]$ *or with* $\mathcal{G} \otimes F[t^{-1}]$.

Some remarks are in order. First, note that the subalgebras of types (ii) and (iii) (as well as $\mathcal{G} \otimes F[t]$ and $\mathcal{G} \otimes F[t^{-1}]$) are all maximal. Secondly, the main result of [BShZ] is slightly more general, in that simple algebras which are not central simple are also dealt with; this is done by allowing a certain twisted version of case (ii). Thirdly, let us examine the situation in the important case $F = \mathbb{C}$. In this case $\mathcal{G}$ is one of the known (classical or exceptional) simple complex Lie algebras. According to the theorem, in order to determine all the possibilities for $M$ (up to commensurability), one needs to know

(a) the maximal subalgebras of $\mathcal{G}$, and

(b) the elements of prime order in Aut$\mathcal{G}$ (these elements determine all cyclic gradings of prime period of $\mathcal{G}$).

This data is available in the literature (see Dynkin [D1, D2] and Kac [K, p. 96]), and so Theorem 3.2 is rather satisfactory.

For the group-theoretic applications we have in mind, we need to define the positive part of Kac-Moody algebras and of loop algebras. For $\mathcal{G}$ and $F$ as above, set

$$L^+(\mathcal{G}) = \mathcal{G} \otimes_F tF[t].$$

Define the positive part of the loop algebra $L(\mathcal{G}, k, \alpha)$ by

$$L^+(\mathcal{G}, k, \alpha) = L(\mathcal{G}, k, \alpha) \cap L^+(\mathcal{G}).$$

Instead of tackling Problem 8 directly, we address a more general problem of describing the weakly maximal subalgebras of Lie algebras of type $L^+(\mathcal{G})$. This can be done using our methods, and the description is even simpler than the one obtained in Theorem 3.2 (as subalgebras of type (iii) do not occur).

**Theorem 3.3** *Let* $\mathcal{G}$ *be a central simple finite-dimensional Lie algebra over a field* $F$, *and let* $M$ *be a weakly maximal subalgebra of* $L^+(\mathcal{G})$. *Then one of the following holds.*

*(i)* $M = L^+(\mathcal{G}, r, \alpha)$ *for some prime* $r$ *and a* $C_r$*-grading* $\alpha$ *of* $\mathcal{G}$.
*(ii)* $M = L^+(\mathcal{H}) = \mathcal{H} \otimes tF[t]$, *where* $\mathcal{H}$ *is a maximal subalgebra of* $\mathcal{G}$.

We can now return, better equipped, to Problem 7. Let $\mathcal{G}, L^+(\mathcal{G}), M$ be as in Theorem 3.3. What can be said about the density of $M$? If $M$ is as in part (i) (i.e. the positive part of a loop algebra of period $r$), then we have

$$D(M) = \frac{1}{r}, \quad \text{while if } M = L^+(\mathcal{H}) \text{ as in part (ii), then} \quad D(M) = \frac{\dim \mathcal{H}}{\dim \mathcal{G}}.$$

Defining

$$m(\mathcal{G}) = \max\{\dim \mathcal{H} : \mathcal{H} \text{ is a proper subalgebra of } \mathcal{G}\},$$

we obtain the following.

**Corollary 3.4** *With the above notation we have*

$$\lambda(L^+(\mathcal{G})) = \max\{\frac{1}{2}, \frac{m(\mathcal{G})}{\dim \mathcal{G}}\}.$$

Note that, if $\mathcal{G}$ is a classical Lie algebra (e.g. $\mathcal{G} = sl_d(\mathbb{F}_p)$), then $m(\mathcal{G}) \geq \frac{1}{2}\dim \mathcal{G}$, and so

$$\lambda(L^+(\mathcal{G})) = \frac{m(\mathcal{G})}{\dim \mathcal{G}} \tag{4}$$

in this case. Problem 7 is now reduced to the following question on finite Lie algebras.

**Problem 9.** Compute $m(sl_d(\mathbb{F}_p))$.

If $F$ has characteristic zero, then it follows from the results of Dynkin that $m(sl_d(F)) = d^2 - d$ (the maximum being attained by the obvious parabolic subalgebra). Though Dynkin's results are not applicable for $sl_d(\mathbb{F}_p)$, we are able to show the following (in [BSh]).

**Proposition 3.5** *Suppose* $p > 2$. *Then* $m(sl_d(\mathbb{F}_p)) = d^2 - d$.

Combining (4) with Proposition 3.5 we can now provide an answer to Problem 7 as follows.

**Proposition 3.6** *Suppose* $p > 2$. *Then*

$$\lambda(sl_d(\mathbb{F}_p) \otimes t\mathbb{F}_p[t]) = \frac{d^2 - d}{d^2 - 1} = 1 - \frac{1}{d+1}.$$

Applying Corollary 2.11, we deduce the main result of this paper, namely, the solution of Problem 5 (5').

**Theorem 3.7** *Suppose* $p > 2$. *Then*

$$\lambda(SL_d(\mathbb{F}_p[[t]])) = 1 - \frac{1}{d+1}.$$

In other words, *among all weakly maximal subgroups of* $SL_d(\mathbb{F}_p[[t]])$, *the parabolic subgroups stabilizing a line or a subspace of codimension 1 are of largest size.* This is of course a local field analog of the result of Patton [P] from 1972, determining the maximal size of a proper subgroup of $SL_d(q)$ (it is usually attained by similar parabolic subgroups). Note, however, that it does *not* follow from our arguments that all weakly maximal subgroups of $S$ whose Hausdorff dimension is $1 - \frac{1}{d+1}$ are the above mentioned parabolic subgroups. A result of this type would require a deeper understanding of the subgroups $H < S$ corresponding to a given graded Lie subalgebra $K < sl_d(\mathbb{F}_p) \otimes t\mathbb{F}_p[t]$.

We conclude this article with three general problems. First, it would be interesting to try to extend Dynkin's work [D1], [D2] to the modular case.

**Problem 10.** Study the maximal subalgebras of (the known) finite-dimensional simple modular Lie algebras.

Secondly, our results on maximal subalgebras of affine Kac-Moody algebras only covered graded subalgebras.

**Problem 11.** Determine the non-graded maximal subalgebras of affine Kac-Moody algebras.

This should be useful in tackling the following problem (of which our present discussion of $SL_d(\mathbb{F}_p[[t]])$ could be regarded as a special case).

**Problem 12.** Study the maximal subgroups of Kac-Moody groups.

# References


[Ab] J.L. Abercrombie, Subgroups and subrings of profinite rings, *Math. Proc. Cambr. Phil. Soc.* **116** (1994), 209–222.

[A] M. Aschbacher, On the maximal subgroups of the finite classical groups, *Invent. Math.* **76** (1984), 469–514.

[B] L. Babai, The probability of generating the symmetric group, *J. Comb. Th. Ser. A* **52** (1989), 148–153.

[BSh] Y. Barnea and A. Shalev, Hausdorff dimension, Pro-$p$ groups, and Kac-Moody algebras, Preprint, 1996.

[BShZ] Y. Barnea, A. Shalev and E.I. Zelmanov, Graded subalgebras of affine Kac-Moody algebras, Preprint, 1996.

[BPSh] A. Borovik, L. Pyber and A. Shalev, Maximal subgroups of profinite groups, *Trans. Amer. Math. Soc.*, to appear.

[C] B.N. Cooperstein, Minimal degrees for a permutation representation of a classical group, *Israel J. Math.* **30** (1978), 213–235.

[DDMS] J. Dixon, M.P.F. du Sautoy, A. Mann and D. Segal, *Analytic Pro-p Groups*, London Math. Soc. Lecture Note Series **157**, Cambridge University Press, Cambridge, 1991.

[Di] J.D. Dixon, The probability of generating the symmetric group, *Math. Z.* **110** (1969), 199–205.

[D1] E.B. Dynkin, Semisimple subalgebras of semisimple Lie algebras, *Amer. Math. Soc. Transl (2)* **6** (1957), 111–244.

[D2] E.B. Dynkin, Maximal subgroups of the classical groups, *Amer. Math. Soc. Transl (2)* **6** (1957), 245–378.

[F] K. Falconer, *Fractal Geometry: mathematical foundations and applications*, John Wiley & Sons, New York, 1990.

[GKS] R.M. Guralnick, W.M. Kantor and J. Saxl, The probablity of generating a classical group, *Comm. in Alg.* **22** (1994), 1395–1402.

[HB] B. Huppert and N. Blackburn, *Finite Groups, II*, Springer, Berlin, 1982.

[K] V.G. Kac, *Infinite Dimensional Lie Algebras*, Progress in Mathematics **44**, Birkhäuser, Boston, 1983.

[Ka] W.M. Kantor, Permutation representations of the finite classical groups of small degree or rank, *J. Algebra* **60** (1979), 158–168.

[KaL] W.M. Kantor and A. Lubotzky, The probability of generating a finite classical group, *Geom. Ded.* **36** (1990), 67–87.

[KL] P.B. Kleidman and M.W. Liebeck, *The Subgroup Structure of the Finite Classical Groups*, London Math. Soc. Lecture Note Series **129**, Cambridge University Press, Cambridge, 1990.

[LS] V. Landazuri and G.M. Seitz, On the minimal degrees of projective representations of the finite Chevalley groups, *J. Algebra* **32** (1974), 418–443.

[LiSh1] M.W. Liebeck and A. Shalev, The probability of generating a finite simple group, *Geom. Ded.* **56** (1995), 103–113.

[LiSh2] M.W. Liebeck and A. Shalev, Classical groups, probabilistic methods, and the $(2,3)$-generation problem, *Annals of Math.* **144** (1996), 77–125.

[LiSh3] M.W. Liebeck and A. Shalev, Simple groups, probabilistic methods, and a conjecture of Kantor and Lubotzky, *J. Algebra* **184** (1996), 31–57.

[LSh] A. Lubotzky and A. Shalev, On some $\Lambda$-analytic pro-$p$ groups, *Israel J. Math.* **85** (1994), 307–337.

[M] A. Mann, Positively finitely generated groups, *Forum Math.* **8** (1996), 429–459.

[MSh] A. Mann and A. Shalev, Simple groups, maximal subgroups, and probabilistic aspects of profinite groups, *Israel J. Math.*, to appear.

[MS] G.A. Margulis and G.A.Soifer, Maximal subgroups of infinite index in finitely generated linear groups, *J. Algebra* **69** (1981), 1–23.

[P] W.H. Patton, *The Minimum Index for Subgroups in some Classical Groups: a generalization of a theorem of Galois*, Ph.D. Thesis, University of Illinois at Chicago Circle, Chicago, 1972.

[S] G.M. Seitz, The maximal subgroups of classical algebraic groups, *Mem. Amer. Math. Soc.* **67** No. 36 (1987).

[Z] E.I. Zelmanov, On periodic compact groups, *Israel J. Math.* **77** (1992), 83–95.

[Zu] A. Zubkov, Non-abelian free pro-$p$ groups cannot be represented by 2-by-2 matrices, *Sib. Math. J.* **28** (1987), 742–747.

Aner Shalev
Institute of Mathematics
The Hebrew University
Jerusalem 91904, Israel
e-mail: `shalev@math.huji.ac.il`

Trends in Mathematics, © 1998 Birkhäuser Verlag Basel/Switzerland

# Aspects of Buildings

Ernest Shult *

**Abstract**

A precise characterization of buildings as chamber systems holds for all ranks. It has long been known that the corresponding idea of a building as a geometry, gets a bit strange in infinite rank. Some precise description of the departure of these two concepts in infinite rank is given. Some speculations are offered about which "classical" infinite-rank geometries may still possess characterizations by point-line axioms.

## 1 Basics on graphs and subgraphs

We consider here only simple graphs – that is, edges are not directed; there are no multiple edges nor loops. Thus edges can be regarded as 2-subets of the set of vertices. Let $(V, E)$ be a simple graph. A **path of length** $n$ is a sequence $(v_0, \ldots, v_n)$ of vertices with $\{v_i, v_{i+1}\}$ an edge and $v_i$ not equal to $v_{i+2}$, for appropriate $i$.

The **distance** $d(x, y)$ **between vertex** $x$ **and vertex** $y$ is the length of a shortest path connecting them. Such shortest paths are called **geodesics**. Distance is a metric

$$d : V \times V \to \mathbf{N}.$$

For any subset $X$ of the vertex set $V$ let $E_X$ be the set of all edges both of whose vertices lie in $X$. A **subgraph** of $(V, E)$ is a pair $(X, E')$ where $X \subseteq V$ and $E' \subseteq E_X$. The **intersection** over a family $(X_i, E_i)$ of subgraphs, is the subgraph $(\cap_i X_i, \cap_i E_i)$.

A subgraph is said to be an **induced subgraph** if and only if it has the form $(X, E_X)$. It is **convex** if any geodesic path connecting two of its vertices has all its intermediate vertices in the subgraph. (Warning! This does not mean that the *edges* of the connecting geodesic belong to the convex subgraph.) The class of induced subgraphs and the class of convex subgraphs are both closed under taking intersections.

A subgraph $(X, E')$ of $(V, E)$ inherits two metrics:

1. the restriction of the global metric $d$ from $(V, E)$, and
2. the internal metric $d_{X,E'}$ defined by those geodesics of the graph $(X, E')$.

*The author gratefully acknowledges the support of the US. National Science foundation of the research leading to these remarks.

The subgraph $(X, E')$ is said to be **isometrically embedded** if the two metrics agree.

Let us display the logical relations between these properties of subgraphs.

1. "Isometrically embedded" implies "induced".
2. "Induced and convex" implies "isometrically embedded".

But of the two concepts "isometrically embedded" and "convex", neither alone implies the other.

Let $(V, \mu)$ be a set with a metric $\mu : V \times V \to \mathbf{R}$, with values in the real number system. A subset $X$ of $V$ is said to be **gated with respect to element** $v$ if and only if there exists an element $g_v$ in $X$ (called the "gate with respect to $v$) such that for every element $x$ in $X$, distance from $v$ "can be measured through the gate" – i.e.

$$\mu(v, x) = \mu(v, g_v) + \mu(g_v, x).$$

Clearly, the gate $g_v$ is uniquely determined by $X$ and $v$. Note that if $v$ is in $X$, then $v$ is its own gate, i.e. $g_v = v$.

$X$ is said to **be gated** if and only if $X$ is gated with respect to every element $v$ of $V$ (Dress and Scharlau [DS]). In that case one has

1. If $X$ is gated in $V$, then $X$ is convex.

For subgraphs there is a stronger condition. A subgraph $(X, E')$ is said to be **strongly gated with respect to vertex** $v$ if and only if there exists a gate $g_v$ in $X$ such that for every $x$ in $X$,

$$d(v, x) = d(v, g_v) + d_{X,E'}(g_v, x),$$

where $d$ is the natural global metric and $d_{X,E'}$ is the internal metric of the subgraph $(X, E')$. The subgraph $(X, E')$ is **strongly gated** if and only if it is strongly gated with respect to every vertex. One now has:

1. "Strong gatedness" implies "gatedness".
2. Strong gatedness of a subgraph implies its "convexity" and its being an induced subgraph.

# 2 Chamber systems

## 2.1 Chamber systems

Let $\Gamma = (V, E)$ be a simple graph. A **labelling** of the graph $\Gamma$ is a mapping

$$\lambda : \text{unordered pairs of distinct vertices} \to 2^I$$

into subsets of a **set of labels** $I$ such that $\lambda(\{u, v\})$ is non-empty *precisely when* $\{u, v\}$ *is an edge.* In general we shall write $\lambda(u, v)$ for $\lambda(\{u, v\})$.

A **chamber system over** $I$ is a quartette $C = (V, E, \lambda, I)$ where $(V, E)$ is a simple graph with a labelling $\lambda$ by subsets of $I$ such that if $\{x, y\}$ and $\{y, z\}$ are edges then

$$\lambda(x, y) \cap \lambda(y, z) \subseteq \lambda(x, z).$$

The cardinal number $|\cup_E \lambda(e)|$, the number of different labels which make an appearance, is called the **rank** of the chamber system.

Two vertices $x$ and $y$ are **$i$-adjacent** if and only if $i$ is a member of $\lambda(x, y)$. (Note that this forces $\{x, y\}$ to be an edge.) The chamber system is **connected** if and only if $(V, E)$ is a connected graph.

Let $J$ be a subset of $I$. Set $E_J := \{e \in E | J \cap \lambda \neq \emptyset\}$ and for any 2-set of vertices $\{u, v\}$, set $\lambda_J(u, v) := \lambda(u, v) \cap J$. Note that since $\lambda_J$ has a non-empty-set value only at the edges $E_J$, we see that $(V, E_J, \lambda_J, J)$ is a chamber system. The connected components $R$ of the graph $(V, E_J)$, each yield a chamber system

$$R = (R, E_J \cap E_R, \lambda_J, J),$$

called a **residue of type** $J$. The cardinal number $|\cup_{E_R \cap E_J} \lambda_J(e)|$ is the **rank of the residue** $R$. If $J = I - \{i\}$, then $R$ is called a residue of **cotype** $i$ and **corank one.**

A chamber system is said to **separate chambers** if and only if $|\lambda(e)| = 1$ for each edge $e$. In this case, attached to each path $p = (x_0, \ldots, x_n)$ of length $n$ is a word

$$\lambda(x_0, x_1)\lambda(x_1, x_2) \cdots \lambda(x_{n-1}, x_n),$$

in the free monoid $I^*$ with alphabet $I$, called the **type** of the path.

## 2.2 Generalized polygons

A generalized $m$-gon is a chamber system $C$ with these properties:

**(GP0)** (Rank 2 version of residual connectedness:) *$C$ is a connected rank 2 chamber system in which distinct rank 1 residues meet in at most one vertex.*

**(GP1)** *For each vertex $v$ and label $i$, there is a vertex $i$-adjacent to $v$.*

**(GP2)$_m$** *$C$ contains no proper circuit of length less than $2m$.*

**(GP3)$_m$** (Elementary $M$-homotopy of paths) *If chambers $c$ and $c'$ can be joined by a path of type $ijij\cdots$ ($m$ factors), then it can also be joined by a path of type $jiji\cdots$ ($m$ factors).*

Let us clear up a mysterious word here. A **proper circuit** is a circular path for which consecutive edges have *different* labels. If $m = \infty$ then axiom (GP2)$_m$ is understood to mean that no proper circuits exist, and so (GP3)$_m$ is vacuous. ( A little confusion is introduced in the literature by a tendency to call these "trees" even though, no meaningful graph in sight is a tree. The confusion is not remedied by omitting axiom (GP1).)

### 2.3 Chamber systems of type $M$

A **Coxeter matrix** is a function $M : I \times I \to \mathbf{N}$ such that $M(i,i) = 1$ and for $i \neq j$, $M(i,j)$ is a positive integer. For each such matrix $M$ there is a **Coxeter group** $W(M)$ defined as a presented group in the usual way. Then there is a monoid epimorphism $f : I^* \to W(M)$ . A word $w$ in $I^*$ is **reduced** if there is no shorter word $w'$ such that $f(w') = f(w)$.

A **chamber system of type** $M$, where $M$ is a Coxeter matrix, is a chamber system over $I$ such that for any 2-subset $\{i,j\}$, any residue of type $\{i,j\}$ is a generalized $m_{ij}$-gon, where $m_{ij}$ is the $(i,j)$-th entry of the symmetric Coxeter matrix $M$. This is a very strong global hypothesis determined by the matrix $M$, called a "diagram".

## 3 Buildings

### 3.1 The main characterizations

Consider the following conditions for a chamber system of type $M$.

**($\mathbf{G}_c$)** Every path of reduced type beginning at chamber $c$ is a geodesic path.

**($\mathbf{P}_c$)** Any two paths of reduced type beginning at $c$ and ending at the same vertex $v$ are $M$-**homotopic** – that means one path can be transformed into the other by a sequence of elementary $M$-homotopies.

**(G)** Condition ($\mathrm{C}_c$) for all vertices (chambers) $c$.

**(P)** Condition ($\mathrm{P}_c$) for all chambers $c$.

**($\mathbf{RG}_2$)** Every residue of rank 1 or 2 is strongly gated in $C$.

**(RG)** Every residue of $C$, regardless of rank, is strongly gated in $C$.

The main theorem about buildings is this:

**Theorem 1** *If $C$ is a chamber system of type $M$, then all of the conditions just listed above are equivalent.*

Please note that in the above theorem (whose proof is recounted in [Sh]), the rank $|\tau(I)|$ is arbitrary. We call any chamber system of type $M$ satisfying any of the above axioms a **building**. On a pedagogical level this is a very simple way for unsophisticated persons such as myself to reach the concept of a building with little intellectual baggage: there are no simplicial complexes, no foldings nor the rest of it; just graphs.

One of the most beautiful theorems illustrating the "naturalness" of the concept of "building" is that of Rudolf Scharlau, where the so-called "type-matrix" $M$

is re-interpreted as a matrix of type-driven diameters of rank-two residues [S]. You might recognize special cases of his theorem in the above where conditions ($RG_2$) and the infinite condition ($RG$) above are made equivalent to the rest. This wonderful conception of Scharlau is too important to remain tainted with a minor flaw. It's correct version involves only substituting the phrase "strongly gated" for "gated". With this modification, we have

**Theorem 2** (Scharlau's Theorem) *Let $C$ be a chamber system over the typeset $I$. Suppose*

1. *$C$ separates chambers.*
2. *For every vertex $c$ and label $i$, there is a vertex $i$-adjacent to $c$.*
3. *Every residue of rank at most 2 is strongly gated in $C$.*
4. *For every 2-subset $\{i, j\}$ of $J$, there is a number $m_{ij}$ such that every residue of type $\{i, j\}$ has diameter (as a simple graph) $m_{ij}$, an entry of Coxeter matrix $M$.*

*Then $C$ is a building of type $M$.*

REMARK: A. Kasikova has an example in which the strongly gated hypothesis (RG) holds but is not type $M$. This means Scharlau's hypothesis on uniform diameters of rank two residues of a given type is indeed necessary.

That "strongly gated"cannot be replaced by "gated' is revealed by these examples:

EXAMPLE NO. 1: Suppose $n$ is an odd positive integer. There are $2n$ vertices $\{x_0, x_1, \ldots, x_{2n-2}, x_{2n-1}\}$. Pairs $(x_{2k}, x_{2k+1})$, $k = 0, \ldots, n-1$, are edges labeled "1"; similarly, pairs $(x_{2k+1}, x_{2k+2})$ are edges labeled "2" thus producing a circuit of length $2n$ with alternating labels. Finally, pairs $(x_k, x_{2k})$ are edges labeled "3". Clearly the residue of type $\{1, 2\}$ is the $2n$-circuit encompassing all vertices. Similarly, as $n$ is odd, the residues of types $\{1, 3\}$ and $\{2, 3\}$ are also circuits of length $2n$. Thus, as they encompass all vertices, all rank 2 residues are gated with each vertex acting as its own gate. Also, since $n$ is odd, this trivalent graph is bipartite. That makes each rank 1 residue gated. But it is not a building.

EXAMPLE NO. 2: Take any rank 3 building $\Delta$ over $I = \{1, 2, 3\}$ and choose two types, say $\{1, 2\}$. We are going to take a certain collection $R$ of the residues of type $\{1, 2\}$ in $\Delta$ and transpose their labels – every edge in such a residue which was labeled "1" is now labelled "2" and *vice versa.* It is normally possible to choose the set $R$ of "doctored" residues so that now all vertices are encompassed in a single (connected) residue of type $\{1, 3\}$, and in one residue of type $\{2, 3\}$. It then follows that all residues of rank one or two are gated. Nevertheless, this chamber system is not a building.

One can add a modest contribution to the characterization of buildings of arbitrary rank:

**Theorem 3** *Suppose $C$ is a connected chamber system which seperates chambers. Suppose $C$ satisfies property:*

**($\mathbf{RG}^1$)** *Every residue of corank 1 is strongly gated.*

*Then (RG) holds – that is, every residue of $C$ is strongly gated.*

*Proof:* Assume every residue of corank 1 in chamber system $C = (V, E)$ is strongly gated. Then each such residue is an induced convex graph, and also, so is any intersection of them. (Note that we are using intersections of subgraphs as defined at the beginning of Section 1.) Now let $R$ be any non-empty residue of $C$ – say of type $J$. Let $K$ be the intersection of all corank one residues which contain $R$. Then $K$ itself is a union of connected components of the subgraph $(V, E_J)$ in the notation of Section 1. Choose vertex $c \in K - R$, let $r$ be any vertex in $R$ and let $(r = x_0, x_1, \ldots, x_n = c)$ be a geodesic path in $C$ from $r$ to $c$ (exists since $C$ is connected). Let $j$ be minimal such that $e = (x_j, x_{j+1})$ is an edge all of whose set of labels $\lambda(e)$ meet $J$ trivially. (Some such $j$ exists, otherwise $c$ would be in $R$.) Since $K$ is an intersection of convex subgraphs, it is convex, and so $x_{j+1}$ lies in $K - R$. But also, $K$ is an intersection of induced subgraphs, so for each type $i \in I - J$, there is a label in $\lambda(e)$ distinct from $i$. But since $C$ separates chambers, $\lambda(e)$ contains a unique label distinct from all $i \in I - J$ – i.e. it is in $J$, a contradiction to the choice of subscript $j$. Thus we have proved $K = R$.

It remains to show that $R$ is strongly gated. Suppose false. Then there exists vertices $y$ such that $R$ is not gated with respect to $y$. Among these choose $y$ so that $d = d(y, R)$, the length of a shortest geodesic in $C$ from $y$ to an element of $R$, is as small as possible. Since $C$ is connected, this distance from $y$ to a nearest point of $R$ is finite.

Let us first show that $d$ is not zero, i.e. $y$ is not in $R$. Since $K = R$, $R$ is convex and at the same time, an induced graph. From the remarks of the first section, $R$ is isometrically embedded in $C$. This means that for any vertex $r$ of $R$, $d(y, r) = d_R(y, r)$ where $d_R$ is the internal metric of $R$ – so $R$ is strongly gated with respect to $y$, using $y$ itself as the gate.

So $y$ is not in $R$, and $d > 0$. Then since $K = R$, there is a corank one residue $M$ containing $R$ but not $y$. Since $M$ is strongly gated, there exists a vertex $g$ in $M$ such that for any vertex $r$ in $R$,

$$d(y, r) = d(y, g) + d_M(g, r). \tag{1}$$

It follows from this statement (universally quantified on $r$), and the fact that $d(y, g) > 1$ that

$$d(g, R) < d(y, R) = d.$$

By the minimal choice of $d$, $R$ is strongly gated with respect to $g$. So, again, there is a gate $h \in R$ such that for any vertex $s$ of $R$,

$$d(g,s) = d(g,h) + d_R(h,s), \text{ for all } s \in R. \tag{2}$$

But as a special case of (1):

$$d(y,h) = d(y,g) + d_M(g,h). \tag{3}$$

Now as $M$ is isometric $d(g,s) = d_M(g,s)$, for all $s \in R$. By the substitution of bound variables we may replace $s$ by $r$ to get

$$\begin{aligned} d(y,r) &= d(y,g) + d_M(g,r) \\ &= d(y,g) + d(g,r) \\ &= d(y,g) + d(g,h) + d_R(h,r) \\ &= (d(y,g) + d_M(g,h)) + d_R(h,r) \\ &= d(y,h) + d_R(h,r) \end{aligned}$$

for all $r \in R$. Thus $R$ is gated with respect to $y$ after all.

**Corollary 4** *A chamber system of type $M$ is a building if and only if it satisfies condition* ($\mathrm{RG}^1$).

# 4 Chamber systems and geometries in infinite rank

## 4.1 The functors connecting geometries and chamber systems

It is time to define geometries. As we know, it is possible for a bipartite graph to be bipartite in several ways. In a geometry this ambiguity is artificially removed. A **geometry** is a multipartite graph $(V, E)$ whose defining parts have been labeled by a set $I$. We thus have a **type mapping** $\tau : V \to I$, whose fibres are induced subgraphs with no edges. In any graph, a **clique** is a subgraph of mutually adjacent vertices. In a geometry, such a clique $F$ is called a **flag**, and if it involves one vertex of each type, – so that $\tau$ resticted to $F$ is a bijection – then $F$ is called a **flag chamber**.

Under label- or type-preserving graph morphisms, the chamber systems over $I$ form a category $C(I)$. Similarly, the geometries over $I$, form a category $\Gamma(I)$ whose morphisms are the graph homomorphisms which preserve types of each vertex. These two categories are connected by functors. For each geometry $\Gamma = (V, E, \tau, I)$ where $(V, E)$ is multipartite with each part the fibre of a type map $\tau : V \to I$, one can define a chamber system $\mathbf{C}(\Gamma)$ whose vertices (chambers) are the flag-chambers of $\Gamma$, two being declared $i$-adjacent if and only if they differ exactly at their vertices of type $i$.

*The chamber systems* $\mathbf{C}(\Gamma)$ *always separate chambers.*

Similarly, if $C = (V, E, \lambda, I)$ is a chamber system over $I$, we may form a multipartite graph $\mathbf{\Gamma}(C)$, whose vertices are the residues of corank 1, two of them being adjacent if and only they possesses a non-empty intersection – that is some chamber in common. The type $\tau(R)$ of each such vertex is defined to be the cotype of the rank one residue $R$ which it represents. (Be aware that residues of different cotypes may represent the same set of chambers. They are nonetheless distinct objects of $\Gamma(C)$ because of the distinct types attached to them.) Since residues of the same cotype are defined as connected components of $(V, E_{I-\{i\}})$, the fibres of $\tau$ are cocliques as required.

We remark

*Both* **C** *and* $\mathbf{\Gamma}$ *are functors, the former from the category* $C(I)$ *of chamber systems over* $I$ *to the category* $\Gamma(I)$ *of geometries over* $I$*, the latter in the opposite direction.*

For each chamber system $C$, composition of the functors induces a chamber system morphism

$$\mu : C \to C(\Gamma(C)),$$

which takes each chamber (vertex of graph $C$) to the collection of all corank one residues of $C$ which contain it. The latter is clearly a flag chamber of $\Gamma(C)$.

Of course it is always possible to invent pathological chamber systems $C$ for which only one object of each type is represented in $\Gamma(C)$. Similar pathologies exist in the converse direction. This is why, geometries and chamber systems over $I$ need not realize all labels: that way $\mathbf{\Gamma}$ and **C** are still functors. ("Rank", on the other hand, is defined in terms of the number of realized types.)

The furthest departure from these anomalies that one could imagine is when the morphism $\mu : C \to \mathbf{C}(\mathbf{\Gamma}(C))$ is an isomorphism of chamber systems. In this case, we say $C$ is **geometric**.

## 4.2 Chamber systems of infinite rank

Buildings of finite rank are geometric, and because of this we are accustomed to passing back and forth, defining buildings either as chamber systems or as (certain simply connected) geometries. But the two notions completely diverge in infinite rank. This divergence extends far beyond buildings as the next result reveals:

**Theorem 5** (Kasikova and S. [KS]) *Let $C$ be a chamber system over an infinite set $I$. Suppose*

1. *$C$ seperates chambers.*

2. *Every chamber realizes every $i$-adjacency, $i \in I$.*

3. *Every corank one residue is an an induced subgraph of $C$.*

*Then*

**(i)** *The canonical mapping $C \to \mathbf{C}(\mathbf{\Gamma}(C))$, is not surjective on chambers.*

**(ii)** *$C$ is not residually connected.*

**(iii)** *Moreover, if all corank one residues are gated (in the weaker original sense) then $\mathbf{C}(\mathbf{\Gamma}(C))$ is not even a connected chamber system.*

**Corollary 6** *If $C$ is a building of infinite rank, then all three conclusions (i)-(iii) of Theorem 5 hold.*

So in infinite rank we cannot pass back and forth between buildings as geometries and buildings as chamber systems. As Theorem 1 shows, the notion of building as a chamber system is in perfectly good shape. So what is a "building geometry" of infinite rank?

## 4.3 Classical geometries of infinite rank

There are certainly geometries of infinite rank that we would like to think of as "classical". But even the notion of "projective geometry" in infinite rank holds some ambiguities. In all cases listed below, the objects are certain proper subspaces of a vector space $V$ and incidence is inclusion among these subspaces. There is also a uniform type mapping, which we now describe.

There is an equivalence relation induced on all subspaces of $V$: we say that a subspace $A$ is **commensurate-equivalent** (*ce*) to a subspace $B$ if and only if

$$\dim(A/(A \cap B)) = \dim(B/(A \cap B)) < \infty.$$

So if we let $T$ denote the collection of all commensurate-equivalence classes, there is a natural mapping

$$\tau : \text{proper subspaces of } V \ \to T.$$

which takes each subspace $W$ to its *ce*-class $[W]$. $T$ is our type set and $\tau$ is our type mapping. The geometries are then completey described by naming the set of objects to which the type map is to be restricted.

1. Objects are all proper finite dimensional subspaces of $V$. The type mapping here simply records the dimension of each object. (We could also take all proper subspaces of finite codimension, but when $V$ is a dual subspace of some other space, this is the same thing.)

2. Objects are all proper subspaces of $V$.

3. Let $W$ be any infinite-dimensional subspace of $V$. The objects are all proper subspaces $A$ of $V$ such that
   (a) $\dim(W/(A \cap W))$ is finite,
   (b) $\dim(A/(A \cap W))$ is finite, or
   (c) both of the previous.

Note that all five of these geometries are connected. Which of them is the "projective geometry" associated with the infinite-dimensional vector space $V$?

Now again suppose $V$ is a right vector space over $D$ of infinite dimension. Suppose

$$B : V \times V \to D$$

is a non-degenerate $(\sigma, \epsilon)$-Hermitian form. There is then a collection $\mathcal{I}$ of totally isotropic subspaces of $V$. We can then restrict the type mapping $\tau$ to $\mathcal{I}$ to obtain classical geometries which will have infinite rank if $\mathcal{I}$ contains a subspace of infinite dimension.

Similarly, we may consider a pseudo-quadratic form $Q : V \to D/[D,D]$ (in characteristic 2), and replace $\mathcal{I}$ above by the collection $\mathcal{S}$ of all proper totally singular subspaces to obtain at least four further geometries, when some member of $\mathcal{S}$ has infinite dimension.

These are "good" geometries in that the algebraic situation which gives rise to them are locally characterizable. For example, the famous von Staudt-Hilbert-Veblen-Young characterization of classical projective spaces extends perfectly well to infinite rank. The projective planes determine the division ring $D$, while the dimension of the matroid of finite dimensional subspaces determines $\dim V$.

Polar spaces is another success story. The characterization of finite rank thick polar spaces (Veldkamp-Tits-Buekenhout-Shult [V], [T], [BS]) was extended to infinite rank by Peter Johnson [J], using ideas of Buekenhout- LeFevre [BL] and Cuypers-Johnson-Pasini [CJP]). Again, the algebraic data consisting of a division ring $D$, a vector space $V$ over $D$, and forms $B$ or $Q$ as above, are determined.

But which geometry is being characterized ?

## 4.4 Point-line geometries in infinite rank

The very modest point being made here is this: a characterization in terms of points and lines already indicates which geometry is relevant. The points and lines are two types of objects of the classical geometry, and as the point-line geometry is connected, the points must have finite distance from a given point, a fact that has to be realized in the appropriate rank two truncation of the classical geometry. All other subspaces derived from the point-line axioms which are to be realized by objects of the classical geometry are constructible by a finite process from the point-line axioms, and this, too, has to be reflected in the classical geometry.

Thus the Veblen-Young point-line axioms are characterizing the geometry of finite-dimensional subspaces of a vector space $V$, not the other geometries. Similarly, the point-line axioms for polar spaces are characterizing the geometry of finite dimensional isotropic/singular subspaces of a vector space with a Hermitian/pseudoquadratic form.

On the other hand there are several sets of axioms on points and lines characterizing Grassmann spaces in finite singular rank. If these were to be extended to a characterization of one of the four or more infinite projective geometries, it is clear that it must be the geometry 3(c) above whose points are the *ce*-class $[W]$, for some infinite dimensional subspace $W$ of $V$. Lines are then pairs of subspaces $(A, B)$ such that for some subspace $U \in [W]$, $A < U < B$ is an unrefineable chain of subspaces of $V$ – i.e. $\dim(A/B) = 2$. All other constructible subspaces (singular or otherwise) correspond to flags of subspaces $S$ of $V$ with both

$$\dim(S/(W \cap S)) < \infty \text{ and } \dim(W/(W \cap S)) < \infty. \tag{4}$$

Similarly, one may imagine a point-line characterization, say, of a spin geometry (for example that of Cameron [C]) extended to infinite rank. Again the target geometry is clearly indicated. Let $B$ (resp. $Q$) be a non-degenerate Hermitan (resp. pseudoquadratic) form on vector space $V$ over division ring $D$. By the Zorn-hypothesis, there is a maximal totally isotropic (singular) subspace $W$ of $V$. Points are the set $P = \mathcal{I} \cap [W]$ $(P = \mathcal{S} \cap [W])$, and lines are maximal subspaces of these subspaces. Again all constructible objects are subspaces of finite codimension in some member of $P$.

It is similarly easy to see which classical geometry would be targeted by an extension to infinite singular rank of one of the half-spin characterizations.

At any rate, while the world of chamber system buildings of infinite rank sails serenely onward, secure in its many-fold characterizations, the world of classical "building-like" geometries of infinite rank seems to march to the drum of points and lines, a fact which pleases me.

# References

[BL] F. BUEKENHOUT and C. LEFEVRE, Semiquadratic sets in projective spaces, J. Geom. **7** (1976), 17–42.

[BS] F. BUEKENHOUT and E. SHULT, On the foundations of polar geometry, Geom. Dedicata **3** (1974), 155–170.

[C] P. CAMERON, Dual polar spaces, Geom. Dedicata **12** (1982), 75–85.

[CC] A. COHEN and B. COOPERSTEIN, A characterization of some geometries of exceptional Lie type, Geom. Dedicata **15** (1983), 73–105.

[CJP] H. CUYPERS, P. JOHNSON and A. PASINI, On the embeddability of polar spaces, Geom. Dedicata **44** (1992), 56–62.

[DS] A. DRESS and R. SCHARLAU, Gated sets in metric spaces, Aequ. Math. **34** (1987), 112–120.

[J] P. JOHNSON, Polar spaces of arbitrary rank, Geom. Dedicata **34** (1990), 229–250.

[KS] A. KASIKOVA and E. SHULT, Chamber systems which are not geometric, Comm. Algebra **24** (1996), 3471–3481.

[S] R. SCHARLAU, A characterization of Tits buildings by metrical properties, J. London Math. Soc. **32** (1985), 317–327.

[Sh] E. SHULT, Freiburg Lecture Notes, 1989. Unpublished.

[T] J. TITS, Buildings of Spherical Type and Finite $BN$-pairs, Lecture Notes in Mathematics, Springer Verlag, New-York-Berlin, 1974.

[V] F. VELDKAMP, Polar geometry I–IV, Proc. Kon. Ned. Akad. Wet. **A62**,**A63** (1959), 512–551, 207–212.

Ernest Shult
Kansas State University
Manhattan, KS 66502, USA
e-mail: `shult@math.ksu.edu`

Trends in Mathematics, © 1998 Birkhäuser Verlag Basel/Switzerland

# Generalized Quadrangles Arising from Groups Generated by Abstract Transvection Groups

Anja Ingrid Steinbach

**Abstract**

We show that generalized quadrangles $\Gamma$ arising from groups generated by abstract transvection groups satisfy the Moufang condition (and vice versa). Furthermore, if the abstract transvections of $\Gamma$ act as linear transvections on some finite-dimensional vector space over a commutative field of characteristic $\neq 2$, we construct a weak embedding of $\Gamma$ in a projective space, which is full over a subfield. This yields that $\Gamma$ is a symplectic or a hermitian quadrangle in this case.

## 1 Introduction and statement of the Theorems

To be able to state the theorems, we first give the definition of so-called abstract transvection groups and of the associated point-line space, according to TIMMESFELD [Ti2]. Let $G$ be a group. A set $\Sigma$ of abelian subgroups of $G$ is called a set of abstract transvection groups of $G$, if the following holds: $G = \langle\Sigma\rangle$ with $\Sigma^g \subseteq \Sigma$ for $g \in G$ and for different elements $A, B$ in $\Sigma$, we have either $[A, B] = 1$ or for $1 \neq a \in A$ there exists some $1 \neq b \in B$ such that $A^b = B^a$ and vice versa.

Let $G$ be a quasi-simple group generated by the class $\Sigma$ of abstract transvection groups such that there exist different commuting elements in $\Sigma$. By $\wp(\Sigma)$ we denote the following point-line space associated to $G$ and $\Sigma$: The points are the elements of $\Sigma$ and the lines are the sets $\ell_{A,C} := \mathrm{C}_\Sigma(\mathrm{C}_\Sigma(\mathrm{A}, \mathrm{C}))$, where $A$ and $C$ are different commuting elements in $\Sigma$. (Here $\mathrm{C}_\Sigma(\Lambda) = \{\mathrm{A} \in \Sigma \mid [\mathrm{A}, \mathrm{B}] = 1 \text{ for all } B \in \Lambda\}$ for $\Lambda \subseteq \Sigma$. Since $G$ is quasi-simple, the definition of $\wp(\Sigma)$ agrees with the one given in [Ti2, § 1] by [Ti2, (5.11)].)

We assume that $G$ and $\Sigma$ satisfy the following additional hypothesis (H) of [Ti2, § 1], which ensures a certain richness (similarly as in the situation where $X := \langle A, B\rangle \simeq (\mathrm{P})\mathrm{SL}_2(\mathrm{L})$, $L$ a division ring with at least four elements, for non-commuting elements $A, B \in \Sigma$).

(H) If $A, B \in \Sigma$ satisfy $[A, B] \neq 1$ and $o(a) \leq 3$ for each $a \in A$, then there exist $A_0 \leq A$, $B_0 \leq B$ such that $\langle A_0, B_0\rangle \simeq (\mathrm{P})\mathrm{SL}_2(\ell)$ with $\ell$ a field, $|\ell| \geq 4$, and $A_0$, $B_0$ full unipotent subgroups of $\langle A_0, B_0\rangle$.

Then $\wp(\Sigma)$ is a non-degenerate polar space of rank at least 2 by [Ti2, Th. 3]. Hypothesis (H) yields that lines are thick (see [Ti2, (6.2)]). Further, the intersection $A \cap X'$, $X$ as above, is non-trivial, and there exists $A \neq C \in \Sigma$ with $[A, C] = 1$, $[B, C] \neq 1$ (see [Ti2, (2.10), (4.15)]). Hypothesis (H) is used in the description of the so-called radical R(G) of $G$ (see [Ti2, (5.11)]).

Using the classification of polar spaces, TIMMESFELD achieves a list of possibilities for $\overline{G} := G/\mathrm{Z(G)}$ and $\overline{\Sigma}$, if the rank of $\wp(\Sigma)$ is at least 3; see [Ti2, Th. 4]. In this paper we deal with the case, where the polar space $\wp(\Sigma)$ has rank 2, that is, where $\wp(\Sigma)$ is a generalized quadrangle. In Section 3 we prove:

**1.1 Theorem.** *Let $G$ be a quasi-simple group generated by the class $\Sigma$ of abstract transvection groups satisfying Hypothesis* (H). *We assume that there exist different commuting elements in $\Sigma$ and that the point-line space $\wp(\Sigma)$ is a generalized quadrangle. Then $\wp(\Sigma)$ satisfies the Moufang condition. Conversely, let $\Gamma$ be a Moufang quadrangle admitting central elations. Then the class $\Sigma$ of central elation groups is a class of abstract transvection groups of $\langle\Sigma\rangle \leq \mathrm{Aut}(\Gamma)$.*

In Section 2, we determine the possibilities for $G$ and $\Sigma$ in the case that $\wp(\Sigma)$ is a generalized quadrangle, under the assumption that $G$ is a subgroup (generated by transvections) of some finite-dimensional linear group over a commutative field of characteristic not 2. This is accomplished by the following result:

**1.2 Theorem.** *Let $K$ be a commutative field with* char $\mathrm{K} \neq 2$ *and $V$ be a finite-dimensional vector space over $K$,* $\dim\mathrm{V} \geq 3$. *We set $Y = \mathrm{SL(V)} = \langle\Sigma\rangle$, where $\Sigma$ is the class of linear transvection groups. Let $G$ be a quasi-simple subgroup of $Y$ such that $\Sigma^0 := \{A^0 \mid A \in \Sigma, A^0 \neq 1\}$, where $A^0 := A \cap G$ for $A \in \Sigma$, is a class of abstract transvection groups of $G$ satisfying $|A^0| \geq 4$ for $A^0 \in \Sigma^0$. We assume that the point-line space $\wp(\Sigma^0)$ is a generalized quadrangle. Then $\wp(\Sigma^0)$ is fully embedded in a subspace of $V$ defined over a subfield of $K$.*

Now results of DIENST [Di] show that $\wp(\Sigma^0)$ is associated to a $(\sigma, \epsilon)$-hermitian form or to a quadratic form. Combining this with the results of [St], we obtain the following corollary:

**1.3 Corollary.** *For $Y$ and $G$ as in Theorem* (1.2), *one of the following holds:*

(a) *$G \simeq \mathrm{Sp(W)}$, where $W$ is a 4-dimensional non-degenerate symplectic space over some field $L$ with* char $\mathrm{L} \neq 2$ *and* $|L| > 4$.

(b) *$G \simeq \mathrm{SU(W,f)}$, where $W$ is a finite-dimensional vector space over some field $L$ with* char $\mathrm{L} \neq 2$, *and $f : W \times W \to L$ is a non-degenerate anti-hermitian form of Witt index 2 (with respect to an involutory automorphism $\sigma$ of $L$ such that $\{c \in L \mid c^\sigma = c\}$ contains more than four elements).*

*In both cases the elements of $\Sigma^0$ are the symplectic or unitary transvection groups. Further, $V = [V, G] \oplus \mathrm{C_V(G)}$ and $\dim_{\mathrm{K}}[\mathrm{V, G}] = \dim_{\mathrm{L}}\mathrm{W}$. Denote by $\rho$ the above* *isomorphism from $G$ to the symplectic or unitary group. Then there exists an embedding $\alpha : L \to K$ and an injective semi-linear mapping $\varphi : W \to [V, G]$ (with respect to $\alpha$) such that $[V, G] = \langle W\varphi \rangle_K$ and $(w(g\rho))\varphi = (w\varphi)g$ for $w \in W$, $g \in G$ (that is, we may regard the commutator space $[V, G]$ as a natural module for $G$, tensored with $K$).*

In the following, we use the notion of generalized quadrangles weakly embedded in projective space, which has been introduced by LEFÈVRE-PERCSY [LP]. A weak embedding (see [SVM], [TVM]) of some generalized quadrangle $\Gamma$ in a projective space $P$ is an injective mapping from the set of points of $\Gamma$ into the set of points of $P$ such that

(*i*) for each line $L$ of $\Gamma$ the subspace spanned by $\{\pi(x) \mid x \text{ point on } L\}$ is a line of $P$,

(*ii*) the set $\{\pi(x) \mid x \text{ point of } \Gamma\}$ generates $P$,

(*iii*) if $x, y$ are points such that $\pi(y)$ is contained in the subspace of $P$ generated by the set $\{\pi(z) \mid z \text{ collinear with } x\}$, then $y$ is collinear with $x$.

In the proof of Theorem (1.2), we show first that the generalized quadrangle $\wp(\Sigma^0)$ is weakly embedded in the projective space $\mathbf{P}([\mathbf{V}, \mathbf{G}])$. Then we construct a subset $F$ of $K$ such that the lines of $\wp(\Sigma^0)$ are coordinatized by $F \cup \{\infty\}$. Under the assumption that $K$ is commutative and char $\mathrm{K} \neq 2$, we obtain that $F$ is a subfield of $K$. Without these assumptions this conclusion is not valid. For example, if $\wp(\Sigma^0)$ is a so-called mixed quadrangle, then lines are not necessarily coordinatized over a field. For quaternion division rings $K$ (over $L$), there exists a further example with lines parametrized by $L + Lx + Ly$, where $K = L + Lx + Ly + Lxy$ is written in a standard way, see [SVM, Section 3].

Theorem (1.2) can be used in the classification of generalized quadrangles weakly embedded in projective space. Let $\Gamma$ be a generalized quadrangle weakly embedded in the projective space $\mathbf{P}(\mathbf{V})$, such that secant lines contain at least three points of $\Gamma$. Then $\Gamma$ admits non-trivial central elations (see Section 3), by a result of LEFÈFVRE-PERCSY [LP, Th.1]. If the underlying field is commutative and of characteristic $\neq 2$, one can show that each central elation of $\Gamma$ with center $p$ is induced by a linear transvection corresponding to the hyperplane spanned by all $\pi(x)$, $x$ collinear with $p$, and to the point $\pi(p)$. In this case $\Gamma$ is determined by Theorem (1.2) and Corollary (1.3). This approach is independent from the classification of Moufang quadrangles, compare [SVM].

We close this section with some notation for generalized quadrangles. If $p$ is a point and $\ell$ is a line of some (thick) generalized quadrangle with $p$ not on $\ell$, then $p$ is collinear to a unique point on $\ell$, called the neighbour of $p$ on $\ell$.

## 2 Proof of Theorem (1.2)

For $Y$ and $G$ as in Theorem (1.2), we construct a weak embedding of the generalized quadrangle $\wp(\Sigma^0)$ in the projective space $\mathbf{P}(\mathbf{V_0})$, where $V_0 := [V,G] = \langle [V,T] \mid T^0 \in \Sigma^0 \rangle$. Then we show that this embedding is full over a subfield of $K$.

**2.1** By $A^0 \mapsto [V,A]$ we obtain an injective mapping from the set of points of $\Sigma^0$ into the set of points of $\mathbf{P}(\mathbf{V_0})$. Let $\ell_{A^0,C^0}$ be a line of $\wp(\Sigma^0)$. We set $M_{A^0} := \mathrm{R}(\mathrm{C_{A^0}})\mathrm{A}^0$, where $\mathrm{R}(\mathrm{C_{A^0}})$ is the so-called radical of $\langle \mathrm{C}_{\Sigma^0}(\mathrm{A}^0) - \{\mathrm{A}^0\} \rangle$ defined in [Ti2, (6.1)]. By [Ti2, (6.2)] we have

$$\ell_{A^0,C^0} = \{A^0\} \cup (C^0)^{M_{A^0}} = \{C^0\} \cup (A^0)^{M_{C^0}}.$$

For $A \in \Sigma$, let $M_A \leq Y$ be the unipotent radical of $C(A)$. Then $M_{A^0} \leq M_A$ by [St, (3.4.2)] and $[\mathrm{C_V(A)}, \mathrm{M_A}] \leq [\mathrm{V}, \mathrm{A}]$. Hence $\langle [V,T] \mid T^0 \in \ell_{A^0,C^0} \rangle = [V,A] + [V,C]$ is a line in $V_0$.

Let $A^0, C^0 \in \Sigma^0$ with $[V,C] \subseteq \langle [V,T] \mid T^0 \in \mathrm{C}_{\Sigma^0}(\mathrm{A}^0) \rangle$. Then $[V,C] \subseteq \mathrm{C_V(A)}$ and $C^0 \in \mathrm{C}_{\Sigma^0}(\mathrm{A})$. Hence we have a weak embedding of the generalized quadrangle $\wp(\Sigma^0)$ in $\mathbf{P}(\mathbf{V_0})$.

**2.2** *Let $\ell_{A^0,C^0}$ be a line of $\wp(\Sigma^0)$. Then $X := \langle M_{A^0}, M_{C^0} \rangle$ acts on the 2-dimensional subspace $[V,A]+[V,C]$ of $V$. Let $[V,C] = \langle v_2 \rangle$. There exists some $m_0 \in M_{A^0}$ with $(C^0)^{m_0} \neq C^0$. Further, $[v_2, m_0] \neq 0$.*

*Proof.* As in (2.1), we have $M_{A^0} \leq M_A$, $[[V,A], M_A] = 0$ and $[\mathrm{C_V(A)}, \mathrm{M_A}] \subseteq [\mathrm{V}, \mathrm{A}]$, which yields the first claim. The existence of $m_0$ follows from the thickness of lines of $\wp(\Sigma^0)$. Then $\langle v_2 \rangle = [V,C] \neq [V, C^{m_0}] = \langle v_2 m_0 \rangle$ and $[v_2, m_0] \neq 0$. □

**2.3 Definition:** *Let $\mathcal{B} = \{\sqsubseteq_\infty, \sqsubseteq_\in\}$ be a basis of $[V,A]+[V,C]$ with $v_1 := [v_2, m_0]$ as in (2.2). For $x \in X = \langle M_{A^0}, M_{C^0} \rangle$, we write $x \sim M$ if $M_{\mathcal{B}}^{\mathcal{B}}(x) = M$. Then*

$$m_0 \sim \begin{pmatrix} 1 & \\ 1 & 1 \end{pmatrix}, \quad m_a \sim \begin{pmatrix} 1 & \\ \lambda & 1 \end{pmatrix}, \quad m_c \sim \begin{pmatrix} 1 & \lambda \\ & 1 \end{pmatrix}$$

*for $m_a \in M_{A^0}$, $m_c \in M_{C^0}$. Empty entries should be read as 0. We define $F := \{\lambda \in K \mid$ there exists some $m \in M_{A^0}$ with $m \sim \begin{pmatrix} 1 & \\ \lambda & 1 \end{pmatrix}\}$. Then $0, 1 \in F$. Our aim is to show that $F$ is a subfield of $K$, using $K$ commutative and* $\mathrm{char}\, \mathrm{K} \neq 2$.

**2.4** *If $\lambda, \mu \in F$, then $\lambda + \mu \in F$ and $-\lambda \in F$.*

*Proof.* Let $m_1 \sim \begin{pmatrix} 1 & \\ \lambda & 1 \end{pmatrix}$ and $m_2 \sim \begin{pmatrix} 1 & \\ \mu & 1 \end{pmatrix}$. Then $m_1 m_2 \sim \begin{pmatrix} 1 & \\ \lambda+\mu & 1 \end{pmatrix}$ and $m_1{}^{-1} \sim \begin{pmatrix} 1 & \\ -\lambda & 1 \end{pmatrix}$ with $m_1 m_2, m_1{}^{-1} \in M_{A^0}$. Hence $\lambda + \mu, -\lambda \in F$. □

**2.5** *If $0 \neq \lambda \in F$, then $\lambda^{-1} \in F$.*

*Proof.* Let $m_a \in M_{A^0}$ with $m_a \sim \left(\begin{smallmatrix} 1 & \\ \lambda & 1 \end{smallmatrix}\right)$. Since $C^0 \neq (C^0)^{m_a} \in \ell_{A^0,C^0} = \{C^0\} \cup (A^0)^{M_{C^0}}$, there exists some $m_c \in M_{C^0}$ with $(C^0)^{m_a} = (A^0)^{m_c}$. Let $m_c \sim \left(\begin{smallmatrix} 1 & \mu \\ & 1 \end{smallmatrix}\right)$. Then $\langle \lambda v_1 + v_2 \rangle = [V,C]m_a = [V,A]m_c = \langle v_1 + \mu v_2 \rangle$ and $\mu = \lambda^{-1}$.

For $m_0 \in M_{A^0}$ as in (2.2), we choose $m_1 \in M_{C^0}$ with $(C^0)^{m_0} = (A^0)^{m_1}$. Then $m_1 \sim \left(\begin{smallmatrix} 1 & 1 \\ & 1 \end{smallmatrix}\right)$ and $\omega := m_0 {m_1}^{-1} m_0 \sim \left(\begin{smallmatrix} & -1 \\ 1 & \end{smallmatrix}\right)$. Further, $[V,C]{m_0}^{-1} = \langle -v_1 + v_2 \rangle = \langle v_1 - v_2 \rangle = [V,A]{m_1}^{-1}$, hence $(C^0)^{{m_0}^{-1}} = (A^0)^{{m_1}^{-1}}$. This yields $(A^0)^{\omega} = C^0$ and $(M_{A^0})^{\omega} = M_{C^0}$. Finally, $(m_c^{-1})^{\omega^{-1}} \sim \left(\begin{smallmatrix} 1 & \\ \lambda^{-1} & 1 \end{smallmatrix}\right)$ with $(m_c^{-1})^{\omega^{-1}} \in M_{A^0}$. Hence $\lambda^{-1} \in F$. □

**2.6** We define $M := \{t \in K^* \mid$ there exists some $x \in X = \langle M_{A^0}, M_{C^0} \rangle$ with $x \sim \begin{pmatrix} t^{-1} & \\ & t \end{pmatrix}\}$. Then $M$ is closed under multiplication and multiplicative inverses, since $K$ is commutative. Further, we have $F \subseteq M \cup \{0\}$. Namely, let $0 \neq \lambda \in F$ and $m_a \sim \left(\begin{smallmatrix} 1 & \\ \lambda & 1 \end{smallmatrix}\right)$. For $m_c$ as in the proof of (2.5), we have $x := m_a {m_c}^{-1} m_a \omega^{-1} \in X$ and $x \sim \begin{pmatrix} \lambda^{-1} & \\ & \lambda \end{pmatrix}$.

**2.7** *If $t \in M$ and $\lambda \in F$, then $t\lambda t \in F$.*

*Proof.* We may assume $\lambda \neq 0$. Then $\lambda^{-1} \in F$ by (2.5). Let $m_a \sim \left(\begin{smallmatrix} 1 & \\ \lambda^{-1} & 1 \end{smallmatrix}\right)$ and $x \sim \begin{pmatrix} t^{-1} & \\ & t \end{pmatrix}$. Then $x^{-1} m_a x \sim \left(\begin{smallmatrix} 1 & \\ t^{-1}\lambda^{-1}t^{-1} & 1 \end{smallmatrix}\right)$ with $x^{-1} m_a x \in (M_{A^0})^x = M_{A^0}$. Hence $t^{-1}\lambda^{-1}t^{-1} \in F$, and $t\lambda t \in F$ by (2.5). □

**2.8** *The set $F$ defined in (2.3) is a subfield of $K$. Further, $\{M_{\mathcal{B}}^{\mathcal{B}}(m) \mid m \in M_{A^0}\} = \{\left(\begin{smallmatrix} 1 & \\ \lambda & 1 \end{smallmatrix}\right) \mid \lambda \in F\}$ and $\{[V,T] \mid T^0 \in \ell_{A^0,C^0}\} = \{\langle v_1 \rangle, \langle \lambda v_1 + v_2 \rangle \mid \lambda \in F\}$.*

*Proof.* We have to show that $F$ is closed under multiplication. Let $c, d \in F$. Since $\mathrm{char}\, \mathrm{K} \neq 2$, we have $\frac{1}{2}c = \frac{1}{2}(2c)\frac{1}{2} \in F$ by (2.7). Hence $cd = (\frac{1}{2}c + d)^2 - (\frac{1}{2}c)^2 - d^2 \in F$ by (2.7). The second statement is clear by the definition of $F$ in (2.3). □

**2.9** *If $\ell_{E^0,D^0}$ is an arbitrary line in $\Sigma^0$, then there exists some $g \in G$ with $A^{0g} = E^0$, $C^{0g} = D^0$.*

*Proof.* Let $x \in G$ with $A^{0x} = E^0$. Then $C^{0x} \neq E^0$. If $C^{0x} \in \ell_{E^0,D^0}$, then $C^{0x} = D^{0m}$ with $m \in M_{E^0} \leq \mathrm{C_G(E^0)}$, and we may choose $g := xm^{-1}$. If $C^{0x} \notin \ell_{E^0,D^0}$, then $[D^0, C^{0x}] \neq 1$, since $\wp(\Sigma^0)$ is of rank 2. Hence $D^0$ and $C^{0x}$ are conjugated in $\langle D^0, C^{0x} \rangle \leq \mathrm{C_G(E^0)}$. This proves the claim. □

**2.10** *If $\ell_{E^0,D^0}$ is an arbitrary line in $\Sigma^0$, then there exist $v_e \in [V,E]$, $v_d \in [V,D]$ with $\{[V,T] \mid T^0 \in \ell_{E^0,D^0}\} = \{\langle v_e \rangle, \langle \lambda v_e + v_d \rangle \mid \lambda \in F\}$ where $F$ is the subfield constructed in (2.8). Further, $\{v_d m_e \mid m_e \in M_{E^0}\} = \{\lambda v_e + v_d \mid \lambda \in F\}$.*

*Proof.* We choose $g$ as in (2.9) and $v_1, v_2$ as in (2.2). Then the first claim follows from (2.8) with $v_e := v_1 g$, $v_d := v_2 g$. Let $m_e \in M_{E^0}$ and $\lambda \in F$ with $\langle v_d m_e \rangle = \langle \lambda v_e + v_d \rangle$. Then there exists some $k \in K$ such that $k(\lambda v_e + v_d) = v_d m_e = [v_d, m_e] + v_d$. Since $[v_d, m_e] \in [\mathrm{C_V(E)}, \mathrm{M_E}] \subseteq [\mathrm{V}, \mathrm{E}] = \langle \mathrm{v_e} \rangle$, comparing coefficients yields $k = 1$. Hence $v_d m_e = \lambda v_e + v_d$. This yields (2.10). □

**2.11** Let $A^0 \in \Sigma^0$ be arbitrary and $\ell_{A^0,C^0}$, $\ell_{A^0,B^0}$ be different lines through $A^0$. By (2.10) there are $v_1 \in [V,A]$, $v_2 \in [V,C]$ such that

$$\{[V,T] \mid T^0 \in \ell_{A^0,C^0}\} = \{\langle v_1\rangle, \langle \lambda v_1 + v_2\rangle \mid \lambda \in F\}.$$

Because $\wp(\Sigma^0)$ is of rank 2, we have $[B^0, C^0] \neq 1$. Let $c \in C^0$, $b \in B^0$ with ${C^0}^b = {B^0}^c$. With $v_1 bc^{-1} = v_1$ and $v_2 bc^{-1} := v_4$ we obtain

$$\{[V,T] \mid T^0 \in \ell_{A^0,B^0}\} = \{\langle v_1\rangle, \langle \lambda v_1 + v_4\rangle \mid \lambda \in F\},$$

since ${\ell_{A^0,C^0}}^{bc^{-1}} = \ell_{A^0,B^0}$. Next, let $\ell$ be an arbitrary line through $B^0$ different from $\ell_{A^0,B^0}$. By $D^0$ we denote the neighbour of $C^0$ on $\ell$. Then $D^0 \neq B^0$ and $[A^0, D^0] \neq 1$. Let $a \in A^0$, $d \in D^0$ with ${A^0}^d = {D^0}^a$. Then ${\ell_{C^0,A^0}}^{da^{-1}} = \ell_{C^0,D^0}$ and ${\ell_{B^0,A^0}}^{da^{-1}} = \ell_{B^0,D^0}$. With $v_1 da^{-1} =: v_3$ and $v_2 da^{-1} = v_2$, $v_4 da^{-1} = v_4$ we obtain

$$\begin{aligned} \{[V,T] \mid T^0 \in \ell_{C^0,D^0}\} &= \{\langle v_3\rangle, \langle \lambda v_3 + v_2\rangle \mid \lambda \in F\}, \\ \{[V,T] \mid T^0 \in \ell_{B^0,D^0}\} &= \{\langle v_3\rangle, \langle \lambda v_3 + v_4\rangle \mid \lambda \in F\}. \end{aligned}$$

**2.12** *We consider the configuration*

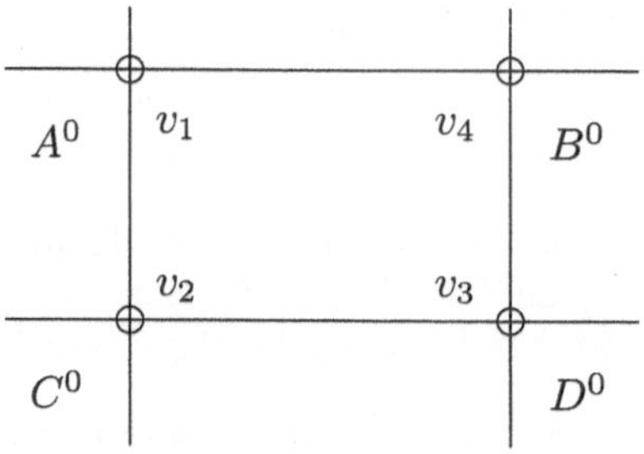


*as in* (2.11). *If* $d \in D^0$, *then* $[v_1, d] = f v_3$ *with* $f \in F$.

*Proof.* Let $d \in D^0$. Since $[v_1, d] \in [V, D] = \langle v_3\rangle$, we have $[v_1, d] = k v_3$ for some $k \in K$. By (2.10) there exists some $m_c \in M^{C^0}$ with $v_3 m_c = v_2 + v_3$, since $1 \in F$. Applying $m_c d^{-1}$ yields

$$\begin{aligned} & [v_1, d] m_c d^{-1} = k v_3 m_c d^{-1} = k(v_2 + v_3) d^{-1} = k(v_2 + v_3) \\ = \; & v_1 d m_c d^{-1} - v_1 m_c d^{-1} \quad \text{with } d m_c d^{-1} \in M_{C^0} \\ = \; & (\lambda v_2 + v_1) - (\mu v_2 + v_1) d^{-1} \quad \text{with } \lambda, \mu \in F \text{ by (2.10)} \\ = \; & \lambda v_2 + v_1 - \mu v_2 - [v_1, d^{-1}] - v_1 \\ = \; & \lambda v_2 - \mu v_2 + k v_3 \end{aligned}$$

since $[v_1, d^{-1}] = -k v_3$. Comparing coefficients yields $k = \lambda - \mu \in F$. □

**2.13** *We consider the configuration of* (2.12). *Let* $A^0 \neq (C^0)^m$ *be a point on* $\ell_{A^0,C^0}$ *and* $E^0$ *be the neighbour of* $(C^0)^m$ *on* $\ell_{B^0,D^0}$. *If* $[V,(C^0)^m] = \langle v_c \rangle$ *with* $v_c = \alpha v_1 + v_2$, $\alpha \in F$, *and* $[V,E] = \langle v_e \rangle$ *with* $v_e = \mu v_4 + v_3$, $\mu \in F$, *then we have* $\{[V,T] \mid T^0 \in \ell_{(C^0)^m,E^0}\} = \{\langle v_c \rangle, \langle \lambda v_c + v_e \rangle \mid \lambda \in F\}$.

*Proof.* Because of $[V,(C^0)^m] \neq [V,A]$ and $[V,E] \neq [V,B]$ we may choose $v_c, v_e$ as indicated. We have $[A^0,E^0] \neq 1$. If $a \in A^0$, $e \in E^0$ with $(A^0)^e = (E^0)^a$ then $ea^{-1}$ maps $\ell_{(C^0)^m,A^0}$ to $\ell_{(C^0)^m,E^0}$. Applying $ea^{-1}$ to $\{[V,T] \mid T^0 \in \ell_{(C^0)^m,A^0}\} = \{\langle v_1 \rangle, \langle \lambda v_1 + v_c \rangle \mid \lambda \in F\}$, we are done. Namely, $v_c ea^{-1} = v_c$ and $v_1 ea^{-1} \in [V,E] = \langle v_e \rangle$. Hence $v_1 ea^{-1} = kv_e$ with $k \in K$. We apply $a$ and obtain $v_1 e = k([v_e,a] + v_e) = k(fv_1 + v_e)$ with $f \in F$ by (2.12). Thus $kfv_1 + kv_e - v_1 = [v_1,e] \in [V,E] = \langle v_e \rangle$. Comparing coefficients yields $kf = 1$ and $k = f^{-1} \in F$. This yields the claim. □

**2.14** *Let* $\ell_{A^0,C^0}$ *and* $\ell_{A^0,B^0}$ *be different lines through* $A^0$ *with* $[V,A] = \langle v_1 \rangle$, $[V,C] = \langle v_2 \rangle$, $[V,B] = \langle v_4 \rangle$ *as in* (2.12). *If* $A^0 \neq (C^0)^m$ *is an arbitrary point on* $\ell_{A^0,C^0}$, *then* $[v_4,c^m] \in \langle v_1,v_2 \rangle_F$ *for* $c \in C^0$.

*Proof.* We choose a line $\ell$ through $B^0$ different from $\ell_{A^0,B^0}$ and we denote by $D^0$ the neighbour of $C^0$ on $\ell$ and by $E^0$ the neighbour of $(C^0)^m$ on $\ell$. Then we obtain

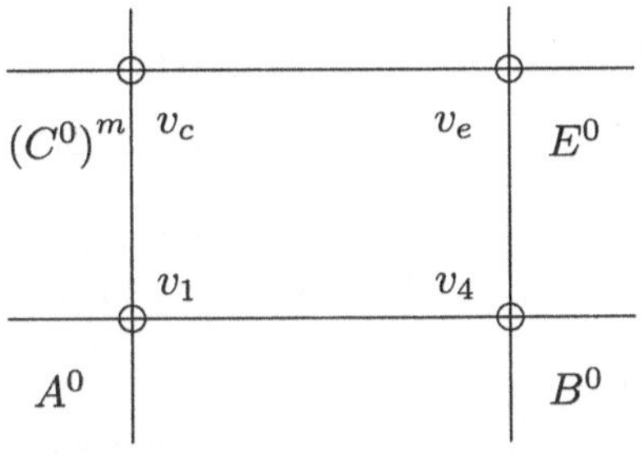


with $v_c, v_e$ as in (2.13). Now (2.12) yields $[v_4,c^m] = fv_c$ with $f \in F$. Since $v_c = \alpha v_1 + v_2$, this shows $[v_4,c^m] \in \langle v_1,v_2 \rangle_F$. □

**2.15** *We consider a configuration as in* (2.12). *For each line* $\ell_i$ *through* $A^0$ *different from* $\ell_{A^0,C^0}$, *we denote by* ${E_i}^0$ *the neighbour of* $D^0$ *on* $\ell_i$. *We choose* $e_i \in {E_i}^0$, $c \in C^0$ *with* $(C^0)^{e_i} = ({E_i}^0)^c$ *and set* $v_i := v_2 e_i c^{-1}$. *Then* $V_1 := \langle v_1,v_2,v_3,v_i | i \in I \rangle$ *is invariant under* $G$.

*Proof.* We have $\{[V,T] \mid T^0 \in \ell_{A^0,{E_i}^0}\} = \{\langle v_1 \rangle, \langle \lambda v_1 + v_i \rangle \mid \lambda \in F\}$, since $v_1 e_i c^{-1} = v_1$. Let $A^0 \neq T^0 \in C_{\Sigma^0}(A)$. Then $T^0 \in \ell_{A^0,C^0}$ or $T^0 \in \ell_{A^0,{E_j}^0}$ for some $j \in I$. In the first case we set $C^0 =: {E_j}^0$, $v_2 =: v_j$. Let $t \in T^0$. Then $v_1 t = v_1$, $v_j t = v_j$. For $j \neq i \in I$ we have $[v_i,t] \in \langle v_1,v_j \rangle_F$ by (2.14). Hence $v_i t \in V_1$. So we are left with $v_3 t$. If $T^0 = {E_j}^0$, then $v_3 t = v_3$, since $[{E_j}^0,D^0] = 1$. If $T^0 \neq {E_j}^0$, then $[v_3,t] \in \langle v_j,v_1 \rangle_F \subseteq V_1$ by (2.14).

For $d \in D^0$, we have $v_2 d = v_2$, $v_3 d = v_3$ and $v_i d = v_i$ for $i \in I$, since $[E_i{}^0, D^0] = 1$. By (2.12) we have $[v_1, d] = f v_3$ with $f \in F$, hence $v_1 d \in V_1$. Thus $V_1$ is invariant under $\langle \mathrm{C}_{\Sigma^0}(\mathrm{A}), \mathrm{D}^0 \rangle$, which is $G$ by [Ti2, (5.5)]. □

**2.16 Proof of Theorem (1.2):** Let $V_1$ be as constructed in (2.15). The mapping $T^0 \mapsto [V_1, T^0]$ $(T^0 \in \Sigma^0)$, $\ell \mapsto \langle [V_1, T^0] \mid T^0 \in \ell \rangle$ ($\ell$ a line of $\wp(\Sigma^0)$) maps points to points and lines to lines, because of $A^0 \mapsto \langle v_1 \rangle_F$, $\ell_{A^0, C^0} \mapsto \langle v_1, v_2 \rangle_F$. This yields a weak embedding of $\wp(\Sigma^0)$ in $\mathbf{P}(\mathbf{V_1})$ because of (2.1). Every point in $\langle v_1, v_2 \rangle_F$ has an inverse image $T^0 \in \ell_{A^0, C^0}$. Hence we obtain a full embedding of the generalized quadrangle $\wp(\Sigma^0)$ in the projective space $\mathbf{P}(\mathbf{V_1})$ over $F$.

**2.17 Proof of Corollary (1.3):** By the result of DIENST [Di] on generalized quadrangles fully embedded in a projective space over a commutative field, we know that $\wp(\Sigma^0)$ is associated to a non-degenerate $(\sigma, \epsilon)$-hermitian or quadratic form $f$ on the vector space $V_1$ over $F$. Since $\mathrm{char}\ \mathrm{F} \neq 2$, we may assume that $f$ is a symplectic or an anti-hermitian form and $\overline{G} = G/\mathrm{Z(G)} \simeq \mathrm{PSp(V_1)}$ resp. $\mathrm{PSU(V_1, f)}$, compare [Ti2, (7.12)]. Further, the elements of $\overline{\Sigma^0}$ are contained in the projective symplectic resp. unitary transvection groups. Since $\mathrm{char}\ \mathrm{F} \neq 2$, [St, (2.4.5)] yields equality. Hence we have achieved the situation handled in [St] and Corollary (1.3) follows from [St, (5.3.2), (9.4.2)].

# 3 Generalized quadrangles arising from abstract transvection groups and the Moufang condition

In this section, we prove Theorem (1.1). For each point $p$ (each path $\gamma$ of length 2) of some (thick) generalized quadrangle $\Gamma$, a central elation with center $p$ (a $\gamma$-elation) is an automorphism of $\Gamma$ fixing all points collinear with $p$ (all elements incident with at least one element of $\gamma$).

Let $G$ be as in the first part of Theorem (1.1). For the converse see (3.6). Proofs of the Moufang condition for $\wp(\Sigma)$ have been given by TIMMESFELD [Ti1, (7.5)] and CUYPERS [Cu, (5.7)], under the additional assumption that $\langle A, B \rangle \simeq \mathrm{(P)SL_2(K)}$ for some commutative field $K$ whenever $A, B \in \Sigma$ with $[A, B] \neq 1$. The proof of TIMMESFELD relies on the so-called maximality condition.

For $A \in \Sigma$, we set $\Omega_A := \{B \in \Sigma \mid [A, B] \neq 1\}$. Further, $M_A := \mathrm{R(C_A)} \cdot \mathrm{A}$, where $\mathrm{R(C_A)}$ is the so-called radical of $C_A := \langle \mathrm{C}_\Sigma(\mathrm{A}) - \{\mathrm{A}\} \rangle$, see [Ti2, (6.1)]. We will use that $M_A$ has the following properties: We have $A \leq M_A \trianglelefteq \langle \mathrm{C}_\Sigma(\mathrm{A}) \rangle$ and $M_A{}' \subseteq \mathrm{Z}(\langle \mathrm{C}_\Sigma(\mathrm{A}) \rangle)$. Further, for each line $\ell$ we have $\ell = \{A\} \cup B^{M_A}$, whenever $A$ and $B$ are different points on $\ell$, see [Ti2, (6.2)].

**3.1** *For $B \in \Omega_A$, we have $\langle \mathrm{C}_\Sigma(\mathrm{A}) \rangle = M_A \langle \mathrm{C}_\Sigma(\mathrm{A}) \cap \mathrm{C}_\Sigma(\mathrm{B}) \rangle$. Further, $M_A$ acts transitively on $\Omega_A$.*

*Proof.* Let $C \in \mathrm{C}_\Sigma(\mathrm{A}) - \{\mathrm{A}\}$. Then there exists $D \in \ell_{A,C} - \{A\}$ with $[B, D] = 1$. Since $\ell_{A,C} = \ell_{A,D} = \{A\} \cup D^{M_A}$, we obtain $C \leq M_A D$ and the first claim holds. To prove (3.1), it suffices to show that $\langle \mathrm{C}_\Sigma(\mathrm{A})\rangle$ acts transitively on $\Omega_A$. Let $B, C \in \Omega_A$. If $[B, C] = 1$, then $B$ and $C$ are conjugated in $\langle \mathrm{C}_\Sigma(\mathrm{A})\rangle$ by [Ti2, (4.1)]. So we may assume $[B, C] \neq 1$. If there exists some $D \in \mathrm{C}_\Sigma(\mathrm{B}) \cap \mathrm{C}_\Sigma(\mathrm{C})$ with $D \notin \mathrm{C}_\Sigma(\mathrm{A})$, then we use the previous argument twice. Finally, if $\mathrm{C}_\Sigma(\mathrm{B}) \cap \mathrm{C}_\Sigma(\mathrm{C}) \subseteq \mathrm{C}_\Sigma(\mathrm{A})$, then we choose $D, E \in \mathrm{C}_\Sigma(\mathrm{B}) \cap \mathrm{C}_\Sigma(\mathrm{C})$ with $D \neq E$. Since lines have at least 3 points by [Ti2, (6.2)], there exists $D' \in \ell_{D,C} - \{D, C\}$. Let $E'$ be the neighbour of $D'$ on $\ell_{E,B}$. Then $E' \neq E, B$ and $D', E' \in \Omega_A$. Thus we may conclude as above. □

**3.2** *Moufang condition for points: Let $A$ be a point and $\ell, s$ be different lines through $A$. Let $B$ be a point on $\ell$ with $B \neq A$, and $m, n$ be lines through $B$ with $m, n \neq \ell$. Then there exists some $(\ell, A, s)$-elation in $G$ mapping $m$ to $n$.*

*Proof.* Let $C$ be a point on $s$ with $C \neq A$, and let $X$ and $Y$ be the neighbour of $C$ on $m$ and $n$, respectively. Then $X, Y \in \Omega_A$ and there exists $g \in M_A$ with $X^g = Y$ by (3.1). Since $B = \ell \cap \mathrm{C}_\Sigma(\mathrm{X})$ and $C = s \cap \mathrm{C}_\Sigma(\mathrm{X})$, we have $B^g = B$, $C^g = C$ and $m^g = n$. All lines through $A$ are of the form $\{A\} \cup F^{M_A}$ with $F \in \mathrm{C}_\Sigma(\mathrm{A})$ and hence fixed by $g$. The points on $\ell - \{A\}$ and $s - \{A\}$ are of the form $B^m$ and $C^m$ with $m \in M_A$, respectively, and hence fixed by $g$. □

**3.3** *Moufang condition for lines: Let $\ell$ be a line and $A, B$ be different points on $\ell$. Let $s$ be a line through $A$ with $s \neq \ell$, and $C, D$ be points on $s$ with $C, D \neq A$. Then there exists some $(A, \ell, B)$-elation in $G$ mapping $C$ to $D$.*

*Proof.* We have $C, D \in \Omega_B$. Let $1 \neq b \in B$. By definition there exist $1 \neq c \in C$ and $1 \neq d \in D$ such that $C^b = B^c$ and $B^d = D^b$. This yields $C^y = D$ for $y := (c^{-1})^{b^{-1}} d^{b^{-1}}$. By [Ti2, (4.9)] we have $b^c = c^{-b}$, hence $b^{c^{-1}} = (c^{-1})^{b^{-1}}$ and $B^{c^{-1}} = C^{b^{-1}}$. For $E := D^{b^{-1}c}$, this yields $[B, E] = 1$. Since also $[A, E] = 1$, we obtain $E \in \ell - \{A\}$. Hence $[E, C] \neq 1$. For $e := (d^{-1})^{b^{-1}c}$, we calculate $C^e = E^c$. Further, $\tau := y^c = be^{-1} \in \langle \ell \rangle$ and $C^\tau = D$.

Hence $\tau = be^{-1} \in \langle E^{M_A}\rangle$ and $\tau$ normalizes $C^{M_A}$ where $[E, C] \neq 1$. Now the definition of $\mathrm{R}(\mathrm{C_A})$ yields $\tau \in \mathrm{R}(\mathrm{C_A}) \leq \mathrm{M_A}$. Since $\ell = \{B\} \cup A^{M_B}$, we obtain $\langle \ell \rangle = (\langle \ell \rangle \cap M_B)A$ and $\langle \ell \rangle \cap M_A = (\langle \ell \rangle \cap M_A \cap M_B)A$. Let $g \in \langle \ell \rangle \cap M_A \cap M_B$ and $a \in A$ such that $\tau = ga$. Then the element $g$ has all the desired properties. □

In the remainder of this section, let $\Gamma$ be a (thick) Moufang quadrangle admitting (non-trivial) central elations. We show that the class $\Sigma$ of central elation groups is a class of abstract transvection groups of $\langle \Sigma \rangle \leq \mathrm{Aut}(\Gamma)$.

Let $p, q$ be opposite points of $\Gamma$ and $x, y$ be different points collinear with $p$ and $q$. We denote by $A$, $B$ and $C$ the group of all $(py, p, px)$-, $(qx, q, qy)$- and $(px, x, xq)$-elations, respectively. Further, let $A^0 \leq A$, $B^0 \leq B$ and $C^0 \leq C$ be the central elation group with centers $p$, $q$ and $x$, respectively.

**3.4** *For $1 \neq a \in A$, there exists $1 \neq b \in B$ such that $B^a = A^b$. If $a$ is central, then also $b$ is central.*

*Proof.* Let $1 \neq a \in A$. By the Moufang condition for points there exists some $b \in B$ mapping $px$ to $q^a x$. Since $p$ is the neighbour of $y$ on $px$, we obtain that $p^b$ is the neighbour of $y$ on $q^a x$. Hence $p^b = q^a$ and $B^a = A^b$.

Next, we assume that $a$ is central. Let $\ell$ be a line through $q$ different from $qx$. By the Moufang condition for points, there exists some $(qx, q, \ell)$-elation $\theta$ mapping $p^b x$ to $px$. Denote by $r$ the neighbour of $p$ on $\ell$. Then $r$ is collinear with $q$ and hence with $q^a = p^b$, since $a$ is central. This yields $(p^b)^\theta = p$ and $r^b = r$ as above. The automorphism $b\theta$ of $\Gamma$ fixes all points of $qx$, all lines through $q$ and the points $p$ and $r$. Thus $b\theta = \mathrm{id}$ by [VM, (4.3.2)(v)]. Hence $b = \theta^{-1}$ fixes all points on $\ell$. Since $\ell$ was arbitrary, $b$ is central. □

**3.5** *The element $a \in A$ is a central elation, if and only if $[a, C] = 1$. We have $A^0 \leq \mathrm{Z(A)}$. In particular, the central elation groups are abelian.*

*Proof.* The set of lines through $p$ different from $px$ is $\{(py)^c \mid c \in C\}$ by the Moufang condition for points. If $a \in A$ is central and $c \in C$, then $y^{c^{-1}a} = y^{c^{-1}}$ and $y^{[a,c]} = y$. Since $[a, c]$ is a $(p, px, x)$-elation and the group of $(p, px, x)$-elations acts semi-regularly on the set of points different from $p$ on $py$, we obtain $[a, c] = 1$; see [Tt, p. 456] or [VM, (5.2.3)(ii), (4.3.3)]. We next assume $[a, C] = 1$. Let $c \in C$ and $z$ be a point on $(py)^c$. Then $z^{c^{-1}}$ is a point on $py$ and $z^{c^{-1}a} = z^{c^{-1}}$. This yields $z^a = z^{c^{-1}ac} = z$ and $a$ is central.

Let $1 \neq u$ be a $(y, py, p)$-elation. We have $[C^0, A] = 1$ as above and $[u, A] = 1$, again by [VM, (5.2.3)(ii)]. Hence the 3-subgroup-lemma yields $[[u, C^0], A] = 1$. Let $1 \neq c \in C^0$. We have $[u, c] = at$ for a unique $a \in A$ and some $(p, px, x)$-elation $t$ (with $[t, A] = 1$), again by [VM, (5.2.3)(ii)]. By [Tt, (2.1)] we have $a = \mu(u)c\mu(u)^{-1}$, where the element $\mu(u)$ satisfies $\mu(u)C\mu(u)^{-1} = A$. Hence $A^0 = \mu(u)C^0\mu(u)^{-1} \leq \mathrm{Z(A)}$. □

**3.6** *The class $\Sigma$ of central elation groups of $\Gamma$ is a class of abstract transvection groups of $\langle\Sigma\rangle \leq \mathrm{Aut}(\Gamma)$.*

*Proof.* By (3.5) the central elation groups are abelian. Let $S, T$ be central elation groups with centers $v$ and $w$, respectively. If $v$ and $w$ are collinear, then $[S, T] = 1$ by (3.5). If $v$ and $w$ are opposite, we use (3.4). □

# References

[Cu] Cuypers, H.: *The geometry of k-transvection groups.* Preprint TU Eindhoven (1994).

[Di] Dienst, K. J.: *Verallgemeinerte Vierecke in pappusschen projektiven Räumen.* Geom. Ded. **9** (1980), 199-206.

[LP] Lefèvre-Percsy, C.: *Projectivités conservant un espace polaire faiblement plongé*, Acad. Roy. Belg. Bull. Cl. Sci. **67** (1981), 45 – 50.

[St] Steinbach, A. I.: *Subgroups of classical groups generated by transvections or Siegel transvections I, II.* To appear in Geom. Ded.

[SVM] Steinbach, A. I., Van Maldeghem, H.: *Generalized quadrangles weakly embedded of degree $> 2$ in projective space.* Submitted.

[TVM] Thas, J., Van Maldeghem, H.: *Orthogonal, symplectic and unitary polar spaces sub-weakly embedded in projective space.* Comp. Math. **103** (1996), 75-93.

[Ti1] Timmesfeld: F. G.: *Groups generated by $k$-transvections.* Invent. Math. **100** (1990), 167-206.

[Ti2] Timmesfeld, F. G.: *Abstract root subgroups and quadratic action.* To appear in Adv. Math.

[Tt] Tits, J.: *Moufang Polygons, I. Root data.* Bull. Belg. Math. Soc. **3** (1994), 455-468.

[VM] Van Maldeghem, H.: *Generalized polygons, A geometric approach.* Birkhäuser, to appear.

Anja Ingrid Steinbach,
Mathematisches Institut,
Justus-Liebig-Universität Gießen,
Arndtstraße 2, D 35392 Gießen, Germany
e-mail: `anja.steinbach@math.uni-giessen.de`

Trends in Mathematics, © 1998 Birkhäuser Verlag Basel/Switzerland

# Embeddings of Geometries in Finite Projective Spaces: a Survey

Josef A. Thas

**Abstract**

A survey is given on embeddings in finite projective spaces of generalized polygons, polar spaces, partial quadrangles, partial geometries, semipartial geometries, dual semipartialgeometries and $(0, \alpha)$-geometries.

## 1 Definitions

A *lax embedding* of a point-line geometry $\mathcal{S}$ with point set $P$ in a finite projective space $\mathrm{PG}(d, K), d \geq 2$ and $K$ a commutative field, is a monomorphism $\theta$ of $\mathcal{S}$ into the geometry of points and lines of $\mathrm{PG}(d, K)$ satisfying

**(WE1) the set $P^\theta$ generates $\mathrm{PG}(d, K)$.**

In such a case we say that the image $\mathcal{S}^\theta$ of $\mathcal{S}$ is *laxly embedded* in $\mathrm{PG}(d, K)$.

A *weak embedding* in $\mathrm{PG}(d, K)$ is a lax embedding which also satisfies

**(WE2) for any point $x$ of $\mathcal{S}$, the subspace generated by the set $X = \{y^\theta \,\|\, y \in P$, is collinear with $x\}$ meets $P^\theta$ precisely in $X$.**

In such a case we say that the image $\mathcal{S}^\theta$ of $\mathcal{S}$ is *weakly embedded* in $\mathrm{PG}(d, K)$.

A *full embedding* in $\mathrm{PG}(d, K)$ is a lax embedding with the additional property that for every line $L$ of $\mathcal{S}$, all points of $\mathrm{PG}(d, K)$ on the line $L^\theta$ have an inverse image under $\theta$. In such a case we say that the image $\mathcal{S}^\theta$ of $\mathcal{S}$ is *fully embedded* in $\mathrm{PG}(d, K)$.

Usually, we simply say that $\mathcal{S}$ is laxly, or weakly, or fully embedded in $\mathrm{PG}(d, K)$ without referring to $\theta$, that is, we identify the points and lines of $\mathcal{S}$ with their images in $\mathrm{PG}(d, K)$.

In this talk we will restrict ourselves to $K = \mathrm{GF}(q)$. However several of the problems are solved also for any commutative field $K$ (in particular for polar spaces, see Thas and Van Maldeghem[1996],[19**a]).

The (finite) geometries we will consider are generalized polygons, polar spaces, partial quadrangles, partial geometries, semipartial geometries, dual semipartial geometries and $(0, \alpha)$-geometries.

# 2 Some important finite point-line geometries

## 2.1 Generalized polygons

A *generalized n-gon*, $n \geq 2$, or a *generalized polygon*, is a point-line geometry the incidence graph of which has diameter $n$ (i.e. two elements are at most at distance $n$) and girth $2n$ (i.e. the length of any shortest circuit is $2n$). A *thick* generalized polygon is a generalized polygon for which each element is incident with a least three elements. In this case, the number of points on a line is a constant, say $s+1$, and the number of lines through a point is also a constant, say $t+1$. The pair $(s,t)$ is called the *order* of the polygon; if $s=t$ we say the polygon has *order s*.

If $\mathcal{S}$ is a finite thick generalized $n$-gon, then by the Theorem of Feit and Higman[1964] we have $n \in \{2,3,4,6,8\}$. The digons ($n=2$) are trivial incidence structures, the thick generalized 3-gons are the projective planes (here $s=t$), and the generalized 4-gons, 6-gons, 8-gons are also called *generalized quadrangles, generalized hexagons, generalized octagons*, respectively.

## 2.2 Polar spaces

A (finite) point-line geometry is called a *polar space* if it is a (finite) generalized quadrangle, or if it is isomorphic to the geometry formed by the points and lines of a quadric of rank at least three in $\mathrm{PG}(d,q)$, or if it is isomorphic to the geometry formed by the points and lines of a hermitian variety of rank at least three in $\mathrm{PG}(d,q^2)$, or if it is isomorphic to the geometry formed by the points of $\mathrm{PG}(d,q), d \geq 2$, together with the totally isotropic lines with respect to some symplectic polarity of $\mathrm{PG}(d,q)$; note that we also include the singular (or degenerate) cases.

## 2.3 Partial geometries

A (finite) *partial geometry* is a point-line geometry $\mathcal{S} = (P,B,\mathrm{I})$ satisfying the following axioms

(i) any two distinct points are incident with at most one line and any point is incident with a constant number $t+1$ ($t \geq 1$) of lines;

(ii) any two distinct points are incident with at most one point and any line is incident with a constant number $s+1$ ($s \geq 1$) of points;

(iii) if $x$ is a point and $L$ is a line not incident with $x$, then there are exactly $\alpha$ point-line pairs $(y_i, M_i), \alpha \geq 1$, for which

$$x \,\mathrm{I}\, M_i \,\mathrm{I}\, y_i \,\mathrm{I}\, L, i = 1,2,\ldots,\alpha.$$

The integers $s,t,\alpha$ are the *parameters* of the partial geometry.

The partial geometries can be divided into four (nondisjoint) classes.

(a) The partial geometries with $\alpha = 1$. That are the generalized quadrangles having a constant number of points on any line and a constant number of lines through every point.

(b) The partial geometries with $\alpha = s+1$ or dually $\alpha = t+1$, i.e. the $2-(v, s+1, 1)$ designs and their duals.

(c) The partial geometries with $\alpha = s$ or dually $\alpha = t$. The partial geometries with $\alpha = t$ are the *Bruck nets of order* $s + 1$ *and degree* $t + 1$.

(d) Finally, the so-called *proper* partial geometries having $1 < \alpha < \min(s, t)$.

## 2.4 $(0, \alpha)$-geometries, semipartial geometries and partial quadrangles

A (finite) $(0, \alpha)$-*geometry* with *parameters* $s, t, \alpha$ is a connected point-line geometry $\mathcal{S} = (P, B, \mathrm{I})$ satisfying (i) and (ii) in the definition of partial geometry, together with

(iii)$'$ if $x$ is a point and $L$ is a line not incident with $x$, then there are exactly 0 or $\alpha$, $\alpha \geq 1$, point-line pairs $(y, M)$ for which

$$x \,\mathrm{I}\, M \,\mathrm{I}\, y \,\mathrm{I}\, L.$$

A (finite) *semipartial geometry* with *parameters* $s, t, \alpha, \mu$ is a $(0, \alpha)$-geometry with parameters $s, t, \alpha$, also satisfying

(iv) any two noncollinear points are collinear with $\mu, \mu > 0$, common points.

The partial geometries are the semipartial geometries with $(t + 1)\alpha = \mu$. If we want to exclude the partial geometries we will speak about *proper* semipartial geometries.

The semipartial geometries with $\alpha = 1$ are the so-called *partial quadrangles*; a partial quadrangle is *thick* if $s \geq 2$ and $t \geq 2$. The thick generalized quadrangles are the thick partial quadrangles with $t + 1 = \mu$. A partial quadrangle which is not a generalized quadrangle will be called *proper*.

### 2.5 Scheme

If "⟶" means "generalizes to", then, restricting ourselves to thick geometries, we have the following scheme

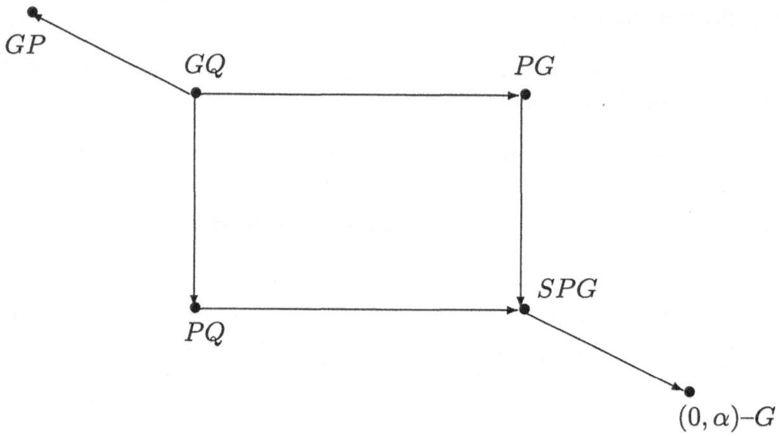


$GQ$ : generalized quadrangle
$PG$ : partial geometry
$PQ$ : partial quadrangle
$SPG$ : semipartial geometry
$GP$ : generalized polygon
$(0,\alpha)$-$G$ : $(0,\alpha)$–geometry

## 3 Full embeddings

### 3.1 Generalized polygons

The following beautiful theorem is due to Buekenhout and Lefèvre[1974].

**Theorem 1**. *Let $S$ be a finite nondegenerate (that is, in $S$ there is no point collinear with all points of $S$) generalized quadrangle fully embedded in* $\mathrm{PG}(d,q)$. *Then one of the following holds :*

*(a) $S$ is formed by the points and lines of a nonsingular quadric of rank two in* $\mathrm{PG}(d,q), d=3,4,5$;

*(b) $q$ is a square and $S$ is formed by the points and lines of a nonsingular hermitian variety of rank two in* $\mathrm{PG}(d,q), d=3,4$;

*(c) $S$ is formed by the points of* $\mathrm{PG}(3,q)$ *together with the totally isotropic lines for some nonsingular symplectic polarity of* $\mathrm{PG}(3,q)$.

Tits[1959] defines two classes of thick generalized hexagons arising from trialities on the nonsingular hyperbolic quadric in the projective 7-dimensional space over a commutative field. In the finite case the two classes are related to Dickson's simple group $G_2(q)$ respectively the triality group ${}^3D_4(q)$; we call them the *classical generalized hexagons.* In the first case one obtains a hexagon of order $(q, q)$ which lies in a hyperplane of $\mathrm{PG}(7, q)$ (which is nontangent to the hyperbolic quadric), and denoted by $H(q)$, in the second case one obtains a hexagon of order $(q, \sqrt[3]{q})$, here denoted by $H(q, \sqrt[3]{q})$. So the former can be represented on a nonsingular quadric $Q(6, q)$ in 6-dimensional space (its points are all the points of $Q(6, q)$ while its lines are some lines of $Q(6, q)$; see Tits[1959] for more details). In the even characteristic case, the polar space $Q(6, q)$ is isomorphic to the nonsingular symplectic polar space $W(5, q)$ and hence in this case one obtains a representation of $H(q)$ in 5-dimensional space. We call these three representations of the classical generalized hexagons the *natural embeddings.*

Any representation $\mathcal{H}$ of any thick finite generalized hexagon will be called a *regular* or *ideal embedding* in $\mathrm{PG}(d, q)$ if the following conditions are satisfied

(i) $\mathcal{H}$ is fully embedded in $\mathrm{PG}(d, q)$,

(ii) the points collinear (in $\mathcal{H}$) with any given point in $\mathcal{H}$ are coplanar in $\mathrm{PG}(d, q)$,

(iii) the points not opposite any given point in $\mathcal{H}$ are contained in a hyperplane of $\mathrm{PG}(d, q)$.

**Theorem 2 (Thas and Van Maldeghem[19**b]).** *A finite thick generalized hexagon $\mathcal{H}$ is regularly embedded in some projective space if and only if it is a natural embedding of a classical generalized hexagon.*

It should be mentioned that in the proof of Theorem 2 we rely on the following result on embeddings by Cameron and Kantor[1979].

**Theorem 3.** *Suppose that the finite thick generalized hexagon $\mathcal{H}$ is fully embedded in* $\mathrm{PG}(d, q)$ *and satisfies*

*(ii)′ the set of all points collinear (in $\mathcal{H}$) with any given point in $\mathcal{H}$ is a plane of* $\mathrm{PG}(d, q)$,

*(iii)′ the points not opposite any given point $x$ in $\mathcal{H}$ are contained in a hyperplane $\pi$ of* $\mathrm{PG}(d, q)$ *and $\pi$ does not contain points opposite $x$.*

*Then $\mathcal{H}$ is a natural embedding in* $\mathrm{PG}(5, q)$ *or* $\mathrm{PG}(6, q)$ *of a classical generalized hexagon.*

**Remark.** In Theorem 3 the order of $\mathcal{H}$ is necessarily $(q, q)$.

We motivate the term *regular* or *ideal embedding* by the following observations. The (finite) generalized quadrangles admitting a regular embedding are precisely those classical generalized quadrangles (that is, the generalized quadrangles in the Theorem of Buekenhout and Lefèvre) with all points regular (in the sense of Payne and Thas[1984]). Also, every generalized hexagon admitting a regular embedding must have ideal lines (in the sense of Ronan[1980]). The notions of "regular points" for generalized quadrangles and "ideal lines" for generalized hexagons were unified for generalized polygons by Van Maldeghem[1995] to "distance-2-regular" points in generalized polygons. He shows that for a thick generalized octagon not all points can be distance-2-regular. As Thas and Van Maldeghem[19**b] show that a finite thick generalized octagon admitting a regular embedding must have all its points distance-2-regular, we have the following result.

**Theorem 4.** *If a finite thick generalized $n$-gon $\mathcal{G}, n \geq 4$, is regularly embedded, then $n = 4$ or $6$ and we have one of the following cases*

(a) *$\mathcal{G}$ is the symplectic quadrangle $W(q)$ naturally embedded in* $\mathrm{PG}(3, q)$,

(b) *$\mathcal{G}$ is the hermitian quadrangle $H(3, q^2)$ naturally embedded in* $\mathrm{PG}(3, q^2)$,

(c) *$\mathcal{G}$ is the classical hexagon $H(q), q$ even, naturally embedded in* $\mathrm{PG}(5, q)$,

(d) *$\mathcal{G}$ is the classical hexagon $H(q)$ naturally embedded in* $\mathrm{PG}(6, q)$,

(e) *$\mathcal{G}$ is the classical hexagon $H(q, \sqrt[3]{q})$ naturally embedded in* $\mathrm{PG}(7, q)$.

Any representation $\mathcal{H}$ of any thick finite generalized hexagon will be called a *flat embedding* in $\mathrm{PG}(d, q)$ if conditions (i) and (ii) are satisfied.
Recently Thas and Van Maldeghem[19**d] proved the following theorem.

**Theorem 5.** *If a finite thick generalized hexagon $\mathcal{H}$ is flatly embedded in $\mathrm{PG}(d, q)$, then $4 \leq d \leq 7$. Also, if $d = 7$, or if $d = 6, s \neq t^3$, and $\mathcal{H}$ is isomorphic to a classical or dual classical generalized hexagon, or if $d = 5$ and $s = t$, then the hexagon is regularly embedded in $\mathrm{PG}(d, q)$, hence it is a natural embedding of a classical generalized hexagon.*

## 3.2 Polar spaces

The next theorem is due to Buekenhout and Lefèvre [1976] and Lefèvre-Percsy [1977].

**Theorem 6.** *Let $\mathcal{S}$ be a finite polar space of rank at least three (that is, which is not a generalized quadrangle) fully embedded in $\mathrm{PG}(d, q)$. Then one of the following holds:*

*(a)* $\mathcal{S}$ *is formed by the points and lines of a quadric of rank at least three in* $\mathrm{PG}(d,q)$;

*(b)* $q$ *is a square and* $\mathcal{S}$ *is formed by the points and lines of a hermitian variety of rank at least three in* $\mathrm{PG}(d,q)$;

*(c)* $\mathcal{S}$ *is formed by the points and lines of some symplectic polar space of rank at least three in* $\mathrm{PG}(d,q)$.

## 3.3 Partial geometries

Let $P$ be the set of all points of $\mathrm{PG}(d,q)$ which are not contained in a fixed subspace $\mathrm{PG}(d-2,q)$, with $d \geq 2$. Let $B$ be the set of all lines of $\mathrm{PG}(d,q)$ having no point in common with $\mathrm{PG}(d-2,q)$. Finally, let I be the natural incidence. Then $\mathcal{S} = (P,B,\mathrm{I})$ is a dual net with parameters $s = q, t = q^{d-1} - 1, \alpha = q$. This dual net will be denoted by $H_q^d$.

In $\mathrm{PG}(2,q)$ any nonempty set of $k$ points may be described as a $\{k;m\}$-arc, where $m$ $(m \neq 0)$ is the greatest number of collinear points in the set. For given $q$ and $m$ $(m \neq 0), k$ can never exceed $mq-q+m$, and a $\{mq-q+m;m\}$-arc will be called a *maximal arc*. Equivalently, a maximal arc may be defined as a nonempty set of points of $\mathrm{PG}(2,q)$ meeting every line in just $m$ points or in none at all. Trivial maximal arcs are the plane $\mathrm{PG}(2,q)$ $(m = q+1)$, the affine plane $\mathrm{AG}(2,q)$ obtained by deleting a line from $\mathrm{PG}(2,q)$ $(m = q)$, and a single point $(m = 1)$. A necessary condition for the existence of a maximal arc, with $m \leq q$, is that $m$ should be a factor of $q$. Denniston[1969] proved that for $q$ even the condition $m|q$ is also sufficient for the existence of a $\{mq-q+m;m\}$-arc. Recently Ball, Blokhuis and Mazzocca[19**] proved the longstanding conjecture that for $q$ odd and $1 < m < q$ no $\{mq-q+m;m\}$-arc exists in $\mathrm{PG}(2,q)$.

The following theorem is due to De Clerck and Thas[1978].

**Theorem 7.** *If* $\mathcal{S} = (P,B,\mathrm{I})$ *is a partial geometry with parameters* $s,t,\alpha$ *which is fully embedded in* $\mathrm{PG}(d,s)$, *then one of the following holds*

*(a)* $\alpha = s+1$ *and* $\mathcal{S}$ *is the design formed by all points and all lines of* $\mathrm{PG}(d,s)$,

*(b)* $\alpha = 1$ *and* $\mathcal{S}$ *is a nondegenerate generalized quadrangle fully embedded in* $\mathrm{PG}(d,s)$,

*(c)* $\alpha = t+1, d = 2$, $\mathrm{PG}(2,s) - P$ *is a maximal* $\{sm-s+m;m\}$*-arc* $K$ *of* $\mathrm{PG}(2,s)$, *with* $m = s/\alpha$ *and* $2 \leq m < s$, *and* $B$ *consists of all lines of* $\mathrm{PG}(2,s)$ *having an empty intersection with* $K$,

*(d)* $\alpha = s, d \geq 2$ *and* $\mathcal{S} = H_s^d$.

### 3.4 Semipartial geometries and $(0, \alpha)$-geometries

Let $M$ be a $(d+1) \times (d+1)$ skewsymmetric matrix over $\mathrm{GF}(q), d \geq 2$, with rank $M = 2k, k \neq 0$. Then $M$ defines a symplectic polarity $\theta$ of $\mathrm{PG}(d, q)$. The points of $\mathrm{PG}(d, q)$ together with the lines of $\mathrm{PG}(d, q)$ which are not totally isotropic for $\theta$ form a $(0, \alpha)$-geometry, which is denoted by $\overline{W(d, 2k, q)}$; here $s = \alpha = q$ and $t = q^{d-1} - 1$. If $k = 1$, then $\overline{W(d, 2, q)}$ is the dual net $H_q^d$. If $2k = d + 1$, then $\theta$ is nonsingular; in this case $\overline{W(d, d+1, q)}$ is a semipartial geometry, also denoted by $\overline{W(d, q)}$. In all other cases $\overline{W(d, 2k, q)}$ is not a semipartial geometry, that is, it is a *proper* $(0, \alpha)$-geometry.

Let $Q^+(3, 2^h)$ be a nonsingular hyperbolic quadric of $\mathrm{PG}(3, 2^h), h \geq 2$. Let $B$ be the set of lines skew to $Q^+(3, 2^h)$, let $P$ be the set of points of $\mathrm{PG}(3, 2^h) - Q^+(3, 2^h)$ and let I be the incidence of $\mathrm{PG}(3, 2^h)$. Then $\mathcal{S} = (P, B, \mathrm{I})$ is a proper $(0, 2^{h-1})$-geometry, denoted by $NQ^+(3, 2^h)$, with $s = 2^h, t = 2^{2h-1} - 2^{h-1} - 1$.

**Theorem 8 (Debroey and Thas[1978]).** *If $\mathcal{S}$ is a proper semipartial geometry with parameters $s, t, \alpha, \mu$ which is fully embedded in* $\mathrm{PG}(3, s)$, *then one of the following holds*

(a) $\alpha = s$ *and* $\mathcal{S} = \overline{W(3, s)}$,

(b) $\alpha = s = t = 2$ *and $\mathcal{S}$ consists of all points of* $\mathrm{PG}(3, 2)$ *not on a given elliptic quadric $O$ of* $\mathrm{PG}(3, 2)$ *together with all lines skew to $O$.*

The following theorem is due to De Clerck and Thas[1983], combined with a result by Thas (see Theorem 26.8.6 of Hirschfeld and Thas [1991]) and the allready mentioned theorem on maximal arcs by Ball, Blokhuis and Mazzocca[19**].

**Theorem 9.** *For $s$ odd there is no proper $(0, \alpha)$-geometry, $\alpha > 1$, fully embedded in* $\mathrm{PG}(3, s)$.

**Remark.** For $s$ even, several results on fully embedded proper $(0, \alpha)$-geometries, $\alpha > 1$, have been proved; see De Clerck and Thas[1983] and Thas (see Theorems 26.8.6 and 26.8.7 by Thas in Hirschfeld and Thas[1991]).

**Conjecture.** If $\mathcal{S}$ is a proper $(0, \alpha)$-geometry, $\alpha > 1$, fully embedded in $\mathrm{PG}(3, s)$, then $\mathcal{S} = NQ^+(3, s)$ with $s = 2^h, h \geq 2$.

**Theorem 10 (Thas, Debroey and De Clerck[1984]).** *If $\mathcal{S}$ is a $(0, \alpha)$-geometry fully embedded in* $\mathrm{PG}(d, s)$, *with $\alpha > 1, s > 2$ and $d \geq 4$, then $\mathcal{S}$ is either the design formed by all points and all lines of* $\mathrm{PG}(d, s)$, *or $\mathcal{S} = \overline{W(d, 2k, s)}$ with* $2k \in \{2, 4, \ldots, d+1\}$.

**Theorem 11 (Thas, Debroey and De Clerck[1984]).** *If $\mathcal{S}$ is a proper semipartial geometry fully embedded in* $\mathrm{PG}(d, s), d \geq 4, \alpha > 1$ *and $s > 2$, then $\alpha = s, d$ is odd, and $\mathcal{S} = \overline{W(d, s)}$.*

**Theorem 12 (De Clerck and Thas[1983]).** *If $\mathcal{S}$ is a dual semipartial geometry fully embedded in* $\mathrm{PG}(d,s), d \geq 3$ *and* $\alpha > 1$, *then it is one of the following*

*(a)* $\alpha = s+1$ *and $\mathcal{S}$ is the design formed by all points and all lines of* $\mathrm{PG}(d,s)$,

*(b)* $\alpha = s$ *and* $\mathcal{S} = H_s^d$,

*(a)* $\alpha = s = t = 2$ *and $\mathcal{S}$ consists of all points of* $\mathrm{PG}(3,2)$ *not on a given elliptic quadric $O$ of* $\mathrm{PG}(3,2)$ *together with all lines skew to $O$.*

**Open problems**

(a) Classify all proper $(0,\alpha)$-geometries, $\alpha > 1$, fully embedded in $\mathrm{PG}(3,s)$, $s$ even.

(b) Determine all partial quadrangles (here $\alpha = 1$) fully embedded in $\mathrm{PG}(d,s)$, $d \geq 4$.

(c) Determine all dual partial quadrangles fully embedded in $\mathrm{PG}(d,s), d \geq 3$ (for $d = 3$ an infinite class of fully embedded dual partial quadrangles, which are not generalized quadrangles, is known).

(d) Determine all $(0,2)$-geometries fully embedded in $\mathrm{PG}(d,2)$, with $d \geq 4$. This amounts to the classification of all semipartial geometries with $\alpha = 2$ in $\mathrm{PG}(d,2)$. In this connection several theorems have been proved; see Shult[1975], Hall[1983], Thas, Debroey and De Clerck[1984].

# 4 Weak embeddings

## 4.1 Generalized quadrangles

Let $x_1, x_2, x_3, x_4, x_5$ be the consecutive vertices of a proper pentagon in $W(2)$, with $W(2)$ the generalized quadrangle arising from a nonsingular symplectic polarity in $\mathrm{PG}(3,2)$. Let $K$ be any field and identify $x_i$, $i \in \{1,2,3,4,5\}$, with the point $(0,\ldots,0,1,0,\ldots,0)$ of $\mathrm{PG}(4,K)$, where the 1 is in the $i$th position. Identify the unique point $y_{i+3}$ of $W(2)$ on the line $x_i x_{i+1}$ and different from both $x_i$ and $x_{i+1}$, with the point $(0,\ldots,0,1,1,0,\ldots,0)$ of $\mathrm{PG}(4,K)$, where the 1's are in the $i$th and the $(i+1)$th position (subscripts are taken modulo 5). Finally, identify the unique point $z_i$ of the line $x_i y_i$ ($x_i y_i$ is indeed a line of $W(2)$) different from both $x_i$ and $y_i$, with the point whose coordinates are all 0 except in the $i$th position, where the coordinate is $-1$, and in the positions $i-2$ and $i+2$, where it takes the value 1 (again subscripts are taken modulo 5). It is an elementary excercise to check that this defines a weak embedding of $W(2)$ in $\mathrm{PG}(4,K)$. We call this the *universal weak embedding* of $W(2)$ in $\mathrm{PG}(4,K)$.

The following theorem is due to Thas and Van Maldeghem[19**c].

**Theorem 13.** *Let $\mathcal{S}$ be a finite thick generalized quadrangle of order $(s,t)$ weakly embedded in* $\mathrm{PG}(d,q)$. *Then either $s$ is a prime power,* $\mathrm{GF}(s)$ *is a subfield of* $\mathrm{GF}(q)$ *and $\mathcal{S}$ is fully embedded in some subspace* $\mathrm{PG}(d,s)$ *of* $\mathrm{PG}(d,q)$, *or $\mathcal{S}$ is isomorphic to $W(2)$ and the weak embedding is the universal one in a projective 4-space over an odd characteristic finite field.*

## 4.2 Polar spaces

**Theorem 14 (Thas and Van Maldeghem[1996]).** *Let $\mathcal{S}$ be a nonsingular finite polar space of rank at least three weakly embedded in* $\mathrm{PG}(d,q)$. *Then $\mathcal{S}$ is fully embedded in some subspace* $\mathrm{PG}(d,q')$ *of* $\mathrm{PG}(d,q)$, *for some subfield* $\mathrm{GF}(q')$ *of* $\mathrm{GF}(q)$.

If $\mathcal{S}$ is a polar space then *Rad* $\mathcal{S}$ is the set of all points of $\mathcal{S}$ which are collinear with every point of $\mathcal{S}$.

**Theorem 15 (Thas and Van Maldeghem[19**a]).** *If $\mathcal{S}$ is a polar space with rank $\mathcal{S} - dim(Rad\,\mathcal{S}) - 1 \geq 3$ which is weakly embedded in* $\mathrm{PG}(d,q)$, *then there is a projective space* $\mathrm{PG}(\overline{d},q), \overline{d} \geq d$, *containing* $\mathrm{PG}(d,q)$ *such that $\mathcal{S}$ is the projection from a* $\mathrm{PG}(\overline{d}-d-1,q) \subseteq \mathrm{PG}(\overline{d},q)$ *into* $\mathrm{PG}(d,q)$ *of a polar space $\widetilde{\mathcal{S}}$ which is fully embedded in some subspace* $\mathrm{PG}(\overline{d},q')$ *of* $\mathrm{PG}(\overline{d},q)$, *for some subfield* $\mathrm{GF}(q')$ *of* $\mathrm{GF}(q)$.

# 5 Lax embeddings

Surprisingly, also for the weakest form of embeddings, the lax embeddings, strong results were recently found by Thas and Van Maldeghem[19**e],[19**f].

## 5.1 Generalized quadrangles

The elements of PSL will be called special linear transformations, those of PGL linear transformations and those of **PΓL** semi-linear transformations.

**Theorem 16.** *If the thick generalized quadrangle $\mathcal{S}$ of order $(s,t)$ is laxly embedded in* $\mathrm{PG}(d,q)$, *then $d \leq 5$.*

(*i*) *If $d = 5$, then $\mathcal{S} \cong Q(5,s)$ and it is either weakly embedded (and hence fully and naturally embedded in some subspace* $\mathrm{PG}(5,s)$ *of* $\mathrm{PG}(5,q)$), *or $s = 2$, $q$ is odd and there exists, up to a special linear transformation of* $\mathrm{PG}(5,q)$, *a unique (non-weak) lax embedding, which is a full affine embedding if $q = 3$. If $s = 2$ and $q$ is a power of 3, then the (non-weak) lax embedding is a full embedding in some affine subspace* $\mathrm{AG}(5,3)$ *over the subfield* $\mathrm{GF}(3)$ *of* $\mathrm{GF}(q)$. *In all cases, the full automorphism group of $\mathcal{S}$ is induced by* $\mathrm{PGL}_6(q)$.

(*ii*) *If* $d = 4$, *then* $s \leq t$.

(*a*) *If* $s = t$, *then* $\mathcal{S} \cong Q(4, s)$ *and one of the following occurs:*

* $s \neq 2$ *for* $q$ *odd, and* $\mathcal{S}$ *is weakly embedded (and hence fully and naturally embedded in some subspace* $\mathrm{PG}(4, s)$ *in* $\mathrm{PG}(4, q)$*);*
* $s = 2$ *and* $q$ *is odd, and* $\mathcal{S}$ *is weakly embedded in* $\mathrm{PG}(4, q)$ *(the so-called universal weak embedding of* $W(2)$ *in* $\mathrm{PG}(4, q)$, $q$ *odd, see Thas and Van Maldeghem[19**a]);*
* $s = 3$, $q \equiv 1 \bmod 3$, *and there exists, up to a special linear transformation, a unique (non-weak) lax embedding (the case* $q = 4$ *corresponds to a full affine embedding; the case* $q$ *even corresponds to a full affine embedding in an affine subspace over the subfield* $\mathrm{GF}(4)$ *of* $\mathrm{GF}(q)$*). The group* $\mathrm{PSp}_4(3)$ *acting naturally as an automorphism group on* $W(3)$ *(which is dual to* $Q(4, 3)$*) is induced on* $\mathcal{S}$ *by* $\mathrm{PSL}_5(q)$. *If* $q$ *is a perfect square and if* $\sqrt{q} \equiv -1 \bmod 3$, *then the full automorphism group* $\mathrm{PGSp}_4(3)$ *of* $\mathcal{S}$ *is the group induced by* $\mathbf{P\Gamma L}_5(q)$*; otherwise,* $\mathbf{P\Gamma L}_5(q)$ *just induces* $\mathrm{PSp}_4(3)$.

(*b*) *If* $s > 2$, *then* $t \neq s + 2$.

(*c*) *If* $t^2 = s^3$, *then* $\mathcal{S} \cong H(4, s)$ *and it is weakly embedded (and hence fully and naturally embedded in some subspace* $\mathrm{PG}(4, s)$ *of* $\mathrm{PG}(4, q)$*).*

(*d*) *If* $\mathcal{S} \cong Q(5, s)$, *then there exists a* $\mathrm{PG}(5, q)$ *containing* $\mathrm{PG}(4, q)$ *and a point* $x \in \mathrm{PG}(5, q) \setminus \mathrm{PG}(4, q)$ *such that* $\mathcal{S}$ *is the projection from* $x$ *onto* $\mathrm{PG}(4, q)$ *of a generalized quadrangle* $\widetilde{\mathcal{S}} \cong Q(5, s)$ *which is laxly embedded in* $\mathrm{PG}(5, q)$ *(and hence determined by (i) above).*

(*iii*) $d = 3$.

- *If* $s = t^2$, *then* $\mathcal{S} \cong H(3, s)$ *and* $\mathcal{S}$ *is weakly embedded in* $\mathrm{PG}(3, q)$ *(and hence fully and naturally embedded in some subspace* $\mathrm{PG}(3, s)$ *of* $\mathrm{PG}(3, q)$*);*
- *if* $\mathcal{S}$ *is classical, but not isomorphic to* $W(s)$ *with* $s$ *odd, then we have*
  (*a*) $\mathcal{S}$ *is not dual to* $H(4, s^{2/3})$;
  (*b*) *if* $\mathcal{S} \cong H(4, s)$, *then there exists a* $\mathrm{PG}(4, q)$ *containing* $\mathrm{PG}(3, q)$ *and a point* $x \in \mathrm{PG}(4, q) \setminus \mathrm{PG}(3, q)$ *such that* $\mathcal{S}$ *is the projection from* $x$ *onto* $\mathrm{PG}(3, q)$ *of a generalized quadrangle* $\widetilde{\mathcal{S}} \cong H(4, s)$ *which is fully embedded in a subspace* $\mathrm{PG}(4, s)$ *of* $\mathrm{PG}(4, q)$, *for some subfield* $\mathrm{GF}(s)$ *of* $\mathrm{GF}(q)$, *with* $s$ *a perfect square;*
  (*c*) *if* $\mathcal{S} \cong Q(4, s)$, *then there exists a* $\mathrm{PG}(4, q)$ *containing* $\mathrm{PG}(3, q)$ *and a point* $x \in \mathrm{PG}(4, q) \setminus \mathrm{PG}(3, q)$ *such that* $\mathcal{S}$ *is the projection from* $x$ *onto* $\mathrm{PG}(3, q)$ *of a generalized quadrangle* $\widetilde{\mathcal{S}} \cong Q(4, s)$ *which is laxly embedded in* $\mathrm{PG}(4, q)$ *(and hence determined by (ii) above);*

(d) *if* $\mathcal{S} \cong Q(5,s)$, *then there exists a* $\mathrm{PG}(5,q)$ *containing* $\mathrm{PG}(3,q)$ *and a line* $L \in \mathrm{PG}(5,q)$ *skew to* $\mathrm{PG}(3,q)$ *such that* $\mathcal{S}$ *is the projection from* $L$ *onto* $\mathrm{PG}(3,q)$ *of a generalized quadrangle* $\widetilde{\mathcal{S}} \cong Q(5,s)$ *which is laxly embedded in* $\mathrm{PG}(5,q)$ *(and hence determined by (i) above).*

## 5.2 Polar spaces

**Theorem 17**. *Let* $\mathcal{S}$ *be a nonsingular finite polar space of rank at least three laxly embedded in* $\mathrm{PG}(d,q)$.

(a) *If* $\mathcal{S}$ *is not the polar space* $W(2m+1,s)$ *arising from a nonsingular symplectic polarity of* $\mathrm{PG}(2m+1,s)$, $s$ *odd, and if* $d \geq 3$, *then there exists a* $\mathrm{PG}(n,q)$ *containing* $\mathrm{PG}(d,q)$, *a* $\mathrm{PG}(n-d-1,q)$ *in* $\mathrm{PG}(n,q)$ *skew to* $\mathrm{PG}(d,q)$ *and a polar space* $\widetilde{\mathcal{S}} \cong \mathcal{S}$ *fully embedded in a subspace* $\mathrm{PG}(n,s)$ *of* $\mathrm{PG}(n,q)$, *such that* $\mathcal{S}$ *is the projection of* $\widetilde{\mathcal{S}}$ *from* $\mathrm{PG}(n-d-1,q)$ *onto* $\mathrm{PG}(d,q)$.

(b) *If* $\mathcal{S} \cong W(2m+1,s), m \geq 2, s$ *odd and* $d \geq 4$, *then there exists a* $\mathrm{PG}(2m+1,q)$ *containing* $\mathrm{PG}(d,q)$, *a* $\mathrm{PG}(2m-d,q)$ *in* $\mathrm{PG}(2m+1,q)$ *skew to* $\mathrm{PG}(d,q)$ *and a polar space* $\widetilde{\mathcal{S}} \cong \mathcal{S}$ *fully embedded in a subspace* $\mathrm{PG}(2m+1,s)$ *of* $\mathrm{PG}(2m+1,q)$, *such that* $\mathcal{S}$ *is the projection of* $\widetilde{\mathcal{S}}$ *from* $\mathrm{PG}(2m-d,q)$ *onto* $\mathrm{PG}(d,q)$.

**Remark**. At the moment we are working on the case $\mathcal{S} \cong W(2m+1,s)$, $m \geq 2$, $s$ odd and $d = 3$, and also on the case $d = 2$.

## 5.3 Generalized hexagons

In [19**d] Thas and Van Maldeghem also obtain several strong results on regularly lax and flatly lax embedded generalized hexagons.

# References

[19**] Ball, S., A. Blokhuis and F. Mazzocca, Maximal Arcs in Desarguesian planes of odd order don't exist, preprint.

[1974] Buekenhout, F. and C. Lefèvre, Generalized quadrangles in projective spaces, *Arch. Math.* (Basel), **25**, 540–552.

[1976] Buekenhout, F. and C. Lefèvre, Semi-quadratic sets in projective spaces, *J. Geom.*, **7**, 17–42.

[1979] Cameron, P. J. and W. M. Kantor, 2-transitive and antiflag transitive collineation groups of finite projective spaces, *J. Algebra*, **60**, 384–422.

[1978] Debroey, I. and J. A. Thas, Semipartial geometries in $\mathrm{PG}(2,q)$ and $\mathrm{PG}(3,q)$, *Atti Accad. Naz. Lincei Rend.*, **64**, 147-151.

[1978] De Clerck, F. and J. A. Thas, Partial geometries in finite projective spaces, *Arch. Math.* (Basel), **30**, 537–540.

[1983] De Clerck, F. and J. A. Thas, The embedding of $(0,\alpha)$-geometries in $\mathrm{PG}(n,q)$. Part I, *Ann. Discrete Math.*, **18**, 229–240.

[1969] Denniston, R. H. F., Some maximal arcs in finite projective planes, *J. Combin. Theory*, **6**, 317–319.

[1964] Feit, W. and G. Higman, The non-existence of certain generalized polygons, *J. Algebra*, **1**, 114–131.

[1983] Hall, J. I., Linear representations of cotriangular spaces, *Linear Algebra Appl.*, **49**, 257–273.

[1991] Hirschfeld, J. W. P. and J. A. Thas, *General Galois Geometries.* Oxford : Oxford University Press.

[1977] Lefèvre-Percsy, C., Sur les semi-quadriques en tant qu'espaces de Shult projectifs, *Acad. Roy. Belg. Bull. Cl. Sci.*, **63**, 160–164.

[1984] Payne, S. E. and J. A. Thas, *Finite Generalized Quadrangles.* Boston: Pitman.

[1980] Ronan, M. A., A geometric characterization of Moufang hexagons, *Invent. Math.*, **57**, 227–262.

[1975] Shult, E. E., Groups, polar spaces and related structures, *Proceedings of the Advanced Study Institute on Combinatorics, Breukelen*, **55**, 130–161, Dordrecht, Reidel.

[1984] Thas, J. A., I. Debroey and F. De Clerck, The embedding of $(0,\alpha)$-geometries in $\mathrm{PG}(n,q)$. Part II. *Discrete Math.*, **51**, 123–137.

[1996] Thas, J. A. and H. Van Maldeghem, Orthogonal, symplectic and unitary polar spaces sub-weakly embedded in projective space, *Compos. Math.*, **103**, 1–19.

[19**a] Thas, J. A. and H. Van Maldeghem, Sub-weakly embedded singular and degenerate polar spaces, *Geom. Dedicata* (to appear).

[19**b] Thas, J. A. and H. Van Maldeghem, Embedded thick finite generalized hexagons in projective space, *J. London Math. Soc.* (to appear).

[19**c] Thas, J. A. and H. Van Maldeghem, Generalized quadrangles weakly embedded in finite projective space, *J. Stat. Plan. Inf.* (to appear).

[19**d] Thas, J. A. and H. Van Maldeghem, Flat lax and weak lax embeddings of finite generalized hexagons, preprint.

[19**e] Thas, J. A. and H. Van Maldeghem, Lax embeddings of polar spaces in finite projective spaces, preprint.

[19**f] Thas, J. A. and H. Van Maldeghem, Lax embeddings of generalized quadrangles in finite projective spaces, preprint.

[1959] Tits, J., Sur la trialité et certains groupes qui s'en déduisent, *Publ. Math. IHES*, **2**, 14–60.

[1995] H. Van Maldeghem, The non-existence of certain regular generalized polygons, *Arch. Math.* **64**, 86–96, Basel : Birkhäuser Verlag.

Josef A. Thas,
University of Gent,
Department of Pure Mathematics and Computer Algebra,
Galglaan 2, B–9000 Gent, Belgium
e-mail: `jat@cage.rug.ac.be`

Trends in Mathematics, © 1998 Birkhäuser Verlag Basel/Switzerland

# Affine Extensions of Near Hexagons Related to the Spin Module of Type $B_3$

John van Bon and Hans Cuypers

**Abstract**

We give a geometric characterization of two classes of geometries related to the spin representation of the groups of type $B_3$. These geometries appear as quotient geometries of point-line spaces obtained from an $F_4$-building by removing a geometric hyperplane.

## 1 Introduction

In [9] the second author of this paper started the investigation of circle extensions of (near) hexagons in connection with lines systems in $\mathbb{R}^n$. The extensions considered in [9] are all extensions of generalized hexagons and near hexagons of order 2, i.e. all lines contain 3 points. Such extensions can also be considered to be extensions by affine planes, in which case we call them *affine extensions of (near) hexagons.* While in [2] affine extensions of generalized hexagons are studied, this paper is devoted to the study of affine extensions of *near hexagons with quads*, i.e. near hexagons in which any two points at distance two are contained in a subgeometry, called *quad*, isomorphic to a nondegenerate generalized quadrangle.

Such affine extensions may be constructed as follows. Suppose $\Delta$ is a metasymplectic space of type $F_4$ in the sense of [13]. If $H$ is a geometric hyperplane of $\Delta$, then, after removing all the subspaces of $\Delta$ whose point shadows are completely contained in $H$, we are left with an affine extension of a dual polar space belonging to the diagram:

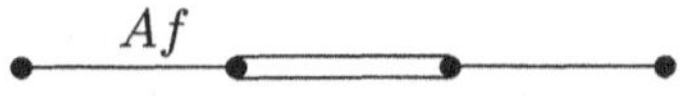


Since the point-line spaces $\Delta$ admit several different types of geometric hyperplanes (see [6]), we obtain several non isomorphic examples of affine extensions of near hexagons. Moreover, some of the constructed affine extensions of near hexagons admit quotients which lead us to even more examples. Other examples can possibly be obtained by taking covers of the geometries constructed above, however, we do not no whether this really leads to new geometries.

The affine extensions of near hexagons studied in [9] include examples related to the groups $PSU_4(2)$, $Co_2$, $2^{11}{:}M_{24}$ and $2^8{:}Sp_6(2)$. The first three examples seem to be sporadic examples, while the last one appears in a series of examples of affine extensions of dual polar spaces obtained as a quotient of the affine extensions of dual polar spaces described in the previous paragraph.

Indeed, consider the case where $\Delta$ is a metasymplectic space of type $F_{4,4}(k)$ over some field $k$. The collinearity graph of $\Delta$ has diameter 3. If one fixes a point $\infty$, then all the points at distance at most 2 from $\infty$ form a geometric hyperplane, see Section 3. Let $\widehat{\mathcal{G}}$ be the affine extension of a near hexagon obtained by removing this hyperplane from $\Delta$.

The stabilizer in the group $G = F_4(k)$ of $\infty$ is the maximal parabolic $P$ of type $B_3$. This group admits a quotient group $P/K$ isomorphic to $k^8{:}B_3(k)$, the split extension of the 8-dimensional spin module for the group $B_3(k)$ by that group itself. The orbits of the kernel $K$ of the quotient map induce an equivalence relation on $\widehat{\mathcal{G}}$. Let $\mathcal{G}$ be the quotient geometry of $\widehat{\mathcal{G}}$ obtained by factoring out this equivalence relation. The example related to the group $2^8{:}Sp_6(2)$ appears in this way if we start with the $F_4(2)$ metasymplectic space. The following result gives a characterization of the geometries constructed in this way in terms of their points, lines and planes:

**Theorem 1.1** *Let $\mathcal{G} = (\mathcal{P}, \mathcal{L}, \mathcal{A})$ be a connected geometry satisfying:*

*i. $(\mathcal{P}, \mathcal{L})$ is a partial linear gamma space with at least 3 points per line;*

*ii. $\mathcal{A}$ is a set of subspaces of $(\mathcal{P}, \mathcal{L})$ isomorphic to affine planes;*

*iii. For any point $p \in \mathcal{P}$ the residue at $p$, $Res(p) = (\mathcal{L}_p, \mathcal{A}_p)$, is a near hexagon with quads and $m, l \in \mathcal{L}_p$ are not opposite in $Res(p)$ if and only if $m \subseteq l^\perp$.*

*Then $\mathcal{G}$ is isomorphic to the affine extension of a dual polar space associated to the group $k^8 : B_3(k)$, with $k$ a field.*

In Section 4 we will give a more detailed description of the examples appearing in the above theorem. In the proof of Theorem 1.1 we encounter partial linear spaces which carry the structure of an affine extension of a generalized quadrangle. In the finite case such geometries are studied in [8] and [11]. In Section 5 we will indicate how the proof of [8] extends to the infinite case as well. The proof of Theorem 1.1 will be completed in Section 6.

Combining the above theorem with the results of [9] we obtain the following:

**Theorem 1.2** *Let $\mathcal{G} = (\mathcal{P}, \mathcal{L}, \mathcal{A})$ be a connected geometry satisfying:*

*i. $(\mathcal{P}, \mathcal{L})$ is a partial linear gamma space;*

*ii. $\mathcal{A}$ is a set of subspaces of $(\mathcal{P}, \mathcal{L})$ isomorphic to affine planes;*

iii. *For any point* $p \in \mathcal{P}$ *the residue at* $p$, $Res(p) = (\mathcal{L}_p, \mathcal{A}_p)$, *is a finite regular near hexagon with quads and* $m$, $l \in \mathcal{L}_p$ *are not opposite in* $Res(p)$ *if and only if* $m \subseteq l^{\perp}$.

*Then* $\mathcal{G}$ *is isomorphic to the affine extension of a dual polar space associated to the group* $k^8 : B_3(k)$, $k$ *a finite field, or lines in* $\mathcal{L}$ *contain only two points and* $\mathcal{G}$ *is isomorphic to the affine extension of a near hexagon associated with* $PSU_4(2)$, $Co_2$ *or* $2^{11}{:}M_{24}$.

## 2 Metasymplectic spaces, polar and dual spaces

In this section we give some definitions and properties of metasymplectic spaces that are useful in the construction of some affine extension of dual polar spaces as will be given in the next section.

Consider a point-line space $(\mathcal{P}, \mathcal{L})$, where lines in $\mathcal{L}$ consist of subsets of the point set $\mathcal{P}$. If $x$ and $y$ are points from $\mathcal{P}$, then by $x \perp y$ we denote that $x$ and $y$ are collinear, i.e. there is a line $l \in \mathcal{L}$ containing both $x$ and $y$. The set of all points collinear to $x$ is denoted by $x^{\perp}$. If $X$ is a subset of $\mathcal{P}$, then $X^{\perp}$ is the intersection of all the sets $x^{\perp}$, where $x$ runs through $X$.

A subset $X$ of $\mathcal{P}$ is called a *subspace* of $(\mathcal{P}, \mathcal{L})$ if it meets every line in 0, 1 or all points. We usually identify a subspace $X$ of $(\mathcal{P}, \mathcal{L})$ with the point-line space with as points the points of $X$ and as lines those lines of $\mathcal{L}$ that are contained in $X$.

Now we consider metasymplectic spaces. As a starting point we take the axiom system for metasymplectic spaces as given by Tits in [13]:

A *metasymplectic space* is a connected point-line space $(\mathcal{P}, \mathcal{L})$ containing distinguished sets of subspaces called *planes* and *symplecta* satisfying the following axiom:

(M1) The intersection of any two distinct symplecta is empty, a point or a plane.

(M2) A symplecton together with its points, lines and planes is a (nondegenerate) rank 3 polar space.

(M3) The symplecta, planes and lines on a fixed but arbitrary point $p$ of $\mathcal{P}$ form a rank 3 polar space. (Here incidence is symmetrized inclusion.)

In [5] Cohen has given a characterization of metasymplectic spaces in terms of points and lines only. We will make use of some of the properties of metasymplectic spaces that are listed in [5] (here $d$ is the distance function in the collinearity graph of $(\mathcal{P}, \mathcal{L})$):

(P1) Each noncollinear pair $x, y \in \mathcal{P}$ for which $x^{\perp} \cap y^{\perp}$ contains at least two points is contained in a symplecton which also contains $x^{\perp} \cap y^{\perp}$. In this case the pair $x, y$ is called a *polar pair.* If $x$ and $y$ are noncollinear points and $x^{\perp} \cap y^{\perp}$ contains just one point, then the pair $x, y$ is called *special.*

(P2) The *pentagon property*: any proper 5-gon $x_1, \ldots, x_5$ from $\mathcal{P}$ (i.e. with $x_i$ only collinear with $x_{i-1}$, itself and $x_{i+1}$) is contained in a (unique) symplecton.

(P3) Let $(z, x_1)$ and $(z, x_2)$ be special pairs and let $y_i \in x_i^\perp \cap z^\perp$, $i = 1, 2$. If $x_1$ and $x_2$ are collinear, then so are $y_1$ and $y_2$.

(P4) If $x, y$ is a polar pair and a point $z \in y^\perp$, then $z^\perp \cap x^\perp$ is nonempty. In particular, if $d(x, y) = 2$ and there is a point $z$ in $y^\perp$ and at distance 3 from $x$, then the pair $x, y$ is special.

(P5) $(\mathcal{P}, \mathcal{L})$ is a partial linear space of diameter 3. Moreover, for $x \in \mathcal{P}$ and $l \in \mathcal{L}$, either one or all points $y \in l$ satisfy $d(x, y) = d(x, l)$.

(P6) Given a symplecton $Sym$ and a point $z \notin Sym$ with $z^\perp \cap Sym \neq \emptyset$, we have $z^\perp \cap Sym$ is a line. Moreover any point in $Sym \setminus (z^\perp \cap Sym)^\perp$ makes a special pair with $z$.

We notice that (P1) is a direct consequence of Proposition 3.8 of [5]. Property (P2) is Lemma 4.1 of of [5]. Also (P3) follows from (P4). Indeed, if $z, x_1, x_2, y_1$ and $y_2$ are as in (P3), they cannot form a proper pentagon as such a pentagon is contained in a symplecton so that all pairs of noncollinear points would have to be polar pairs. Hence $y_1$ and $y_2$ are collinear. Property (P6) is Lemma 4.2 of [5] and (P4) follows from it. Finally (P5) is (F6) of [5].

# 3 Geometric hyperplanes

In this short section we present some useful result on geometric hyperplanes of polar spaces, dual polar spaces and metasymplectic spaces.

Let $\Delta = (\mathcal{P}, \mathcal{L})$ be a point-line space. Then a subset $X$ of the point set $\mathcal{P}$ is called a *geometric hyperplane* if it is a proper subspace of $\mathcal{P}$ that meets each line in $\mathcal{L}$ nontrivially.

If $\Delta$ is a nondegenerate polar space of rank at least 2 or a dual polar space or near polygon, then it is straightforward to check that the set of points that are not at maximal distance of a fixed point $p \in \mathcal{P}$ is a geometric hyperplane of $\Delta$. A similar result also holds for metasymplectic spaces:

**Proposition 3.1** *Let $x$ be a point of the metasymplectic space $\Delta = (\mathcal{P}, \mathcal{L})$. Then the set $\mathcal{H}_x$ of points at distance at most two from $x$ forms a proper geometric hyperplane of $(\mathcal{P}, \mathcal{L})$.*

*Proof.* If $l$ is a line, then by (P5) either 0, 1 or all points of the line are in $\mathcal{H}_x$. We have to show that 0 cannot occur.

Suppose $x, y, z, u$ is a path of length 3 in $\Delta$, and let $l$ be a line through $u$ coplanar with the line $\langle z, u \rangle$. Then inside the residue of $z$, which is a classical near hexagon, we see that there is a quadrangle containing $\langle y, z \rangle$ and some line $m$

on $z$ inside the plane through $z, u$ and $l$. Let $Sym$ be the symplecton containing the quadrangle, then $Sym$ contains $y$ and $m$. Since planes in polar spaces are projective, this implies that there is a point $v = m \cap l$, which is inside $Sym$. Thus, $(y, v)$ is a polar pair, and by (P4) the point $v$ is contained in $\mathcal{H}_x$. Now by connectedness of the residue of $u$, we see that each line on $u$ contains some point in $\mathcal{H}_x$. Hence we see that the diameter of $\Delta$ is at most 3, and $\mathcal{H}_x$ meets every line nontrivially. This proves the proposition. □

In general the geometric hyperplanes we consider are maximal subspaces of $\Delta$. However, there are exceptions as has been proven by Blok and Brouwer [1]. Here we give that part of their results that we need in the sequel:

**Proposition 3.2** (Blok & Brouwer) *Let $p$ be a point of $\Delta$ and suppose $\mathcal{H}_p$ is the set of points not at maximal distance from $p$.*

*i. If $\Delta$ is a nondegenerate polar space then $\mathcal{H}_p$ is maximal subspace of $\Delta$.*

*ii. If $\Delta$ is a nondegenerate dual polar space of rank 3, then $\mathcal{H}_p$ is a maximal subspace of $\Delta$, except when $\Delta$ is a $C_3(2)$ dual polar space.*

# 4 Some examples of affine extensions of dual polar spaces

In this section we construct a class of affine extensions of dual polar spaces appearing in the conclusion of Theorem 1.1. We will describe them as quotients of a metasymplectic space from which a geometric hyperplane is removed. Due to some technical difficulties caused by the exceptional case in 3.2 (ii), we will not cover the case of a metasymplectic space of type $F_4(2)$, whose point residue is a $C_3(2)$ dual polar space. However, a description of the quotient geometry obtained in this particular case can be found in [9], where also a description of the sporadic examples occurring in Theorem 1.2 is given.

Let $\Delta$ be a metasymplectic space, not of type $F_4(2)$, and let $x$ be a point in $\Delta$. Then we denote by $\mathcal{H}_x$ the geometric hyperplane of $\Delta$ consisting of all points at distance at most two from $x$. Fix a point $\infty$ of $\Delta$. By $\widehat{\mathcal{G}}$ we denote the affine extension of the dual polar space consisting of all the points, lines, planes and symplecta of $\Delta$ not contained in $\mathcal{H}_\infty$.

**Lemma 4.1** *Suppose $Sym$ is a symplecton of $\Delta$. Then $Sym$ is either contained in $\mathcal{H}_\infty$ or contains a point $p$ such that $Sym \cap \mathcal{H}_\infty = Sym \cap p^\perp$.*

*Proof.* Let $Sym$ be a symplecton not contained in $\mathcal{H}_\infty$. Fix a point $q \in Sym$ at distance 3 from $\infty$ and let $\pi$ be a singular plane of $Sym$ containing $q$. Since $\mathcal{H}_\infty$ is a geometric hyperplane, it intersects $\pi$ in a line, $l$ say. Fix two points $x_1$ and $x_2$ on $l$. Since these two points are at distance 2 from $\infty$ we can find points $y_i \in x_i^\perp \cap \infty^\perp$. By (P4) the pairs $(\infty, x_i)$ and $(q, y_i)$ are special. Thus $y_i^\perp \cap l$ contains only the point $x_i$. This also implies that $y_1 \neq y_2$. By (P3) applied to $\infty, y_1, y_2, x_1, x_2$ we find that $y_1 \perp y_2$ and $y_1, y_2, x_1, x_2$ form a quadrangle in $\mathcal{H}_\infty$.

The unique symplecton $Sym'$ on this quadrangle intersects $Sym$ in a at least a line and thus a plane, see (M1). Let $p$ be the point in this plane collinear with both $y_1$ and $y_2$. Then $p$ and $\infty$ form a polar pair, and, by (P4), all lines on $p$ are contained inside $\mathcal{H}_\infty$. Thus $p^\perp \cap Sym \subseteq Sym \cap \mathcal{H}_\infty$. However, since $p^\perp \cap Sym$ is a maximal subspace of $Sym$, see Proposition 3.2, they are the same. This proves the lemma. □

If $Sym$ is a symplecton not contained in $\mathcal{H}_\infty$, then the point $p$ appearing in the above lemma will be called the *deep point* of $Sym$.

On the point set of $\widehat{\mathcal{G}}$ we define a relation $\sim$ as follows: $x \sim y$ if and only if $\mathcal{H}_x \cap \infty^\perp = \mathcal{H}_y \cap \infty^\perp$. The relation $\sim$ is an equivalence relation. The equivalence class of the point $x$ will be denoted by $[x]$. For a point $x$ and line $l$ not in $\mathcal{H}_\infty$, let $\pi(x) = \mathcal{H}_x \cap \infty^\perp$, $\pi(l) = \cap_{x \in l \setminus (l \cap \mathcal{H}_\infty)} \pi(x)$ and $\rho(l)$ be the unique point in $\infty^\perp$ collinear with a point on $l$. We notice that the induced geometry on $\pi(x)$ is isomorphic to the residue of $\Delta$ at the point $\infty$. In particular, by (M3) we can identify $\pi(x)$ with a rank 3 dual polar space. Inside this space we see that two points are at distance 2 if and only if they form a polar pair, and at distance 3 if they form a special pair.

**Lemma 4.2** *Suppose $x$ and $y$ are two collinear points of $\widehat{\mathcal{G}}$. Let $l$ be the line of $\Delta$ through $x$ and $y$. Then $\pi(x)$ meets $\pi(y)$ in all the points of $\pi(x)$ at distance at most two of $\rho(l)$.*

*Proof.* The line $l$ through $x$ and $y$ contains a point $z$ at distance 2 from $\infty$. Let $m$ be a line through $\infty$ containing a point $u$ at distance 3 from $\rho(l)$ inside $\pi(x)$. This point $u$ is at distance 3 from $z$. For, if $v$ is a common neighbor of $u$ and $z$, then the five points $u, v, z, \rho(l)$ and $\infty$ form a pentagon with special pairs $(z, \infty)$ and $(u, \rho(l))$, which is impossible by (P2) and (P3).

As $z$ is at distance 3 from $u$, (P5) implies that also $y$ is at distance 3 from $u$. Thus any point in $\pi(x)$ which is at distance 3 (inside $\pi(x)$) from $\rho(l)$ is not in $\pi(y)$. Since $\pi(x) \cap \pi(y)$ is a geometric hyperplane of $\pi(x)$, all points of $\pi(x)$ which are at distance 2 (inside $\pi(x)$) from $\rho(l)$ are contained in $\pi(y)$. But then it easily follows that all points of $\pi(x)$ at distance at most 2 from $\rho(l)$ are in $\pi(y)$. This proves the lemma. □

**Lemma 4.3** *Suppose $Sym$ is a symplecton of $\Delta$ not contained in $\mathcal{H}_\infty$. Then for any two points $x$ and $y$ of $Sym \cap \widehat{\mathcal{G}}$ we have $x \sim y$ if and only if $x$ and $y$ are on the same hyperbolic line of $Sym$ through the deep point of $Sym$.*

*Proof.* By Lemma 4.2 two collinear points are never $\sim$- equivalent.

Let $x$ and $y$ be two noncollinear points of $Sym$, both at distance 3 from $\infty$. Let $w$ be a common neighbor of $x$ and $y$. Consider a plane $\pi$ on $x$ and $w$ but not in $Sym$. Then, as $\pi$ meets $Sym$ in at most a line on $x$, the line $\pi \cap \mathcal{H}_\infty$ is not contained in *Sym*. Now choose a point $z$ on $\pi \cap \mathcal{H}_\infty$ but outside *Sym*. Clearly $z$ is collinear with both $x$ and $w$, and, by (P6), makes a special pair with $y$. Moreover, $z^\perp \cap y^\perp = \{w\}$.

Let $u$ be the common neighbor of $z$ and $\infty$. The point $u$ is at distance 2 from $x$. If $x \sim y$, then $u$ is also at distance 2 from $y$. Let $v$ be the common neighbor of $y$ and $u$. Consider the 5 points $u, v, w, y$ and $z$. Since $(y, u)$ and $(y, z)$ are special, (P3) implies that $w$ and $v$ have to be collinear. But that implies that $(w, u)$ is a polar pair, and the distance of $w$ to $\infty$ equals 2.

So $x \sim y$ implies that $x^\perp \cap y^\perp \subseteq \mathcal{H}_\infty$, from which we deduce that $x$ and $y$ are on a hyperbolic line of *Sym* through the deep point of *Sym*.

Now suppose that $x$ and $y$ are on a hyperbolic line of *Sym* through the deep point of *Sym*. Then $x^\perp \cap y^\perp \subseteq \mathcal{H}_\infty$. Let $u$ be a point in $\pi(x)$ and $z$ a common neighbor of $x$ and $u$. If $z \in y^\perp$, then $u \in \pi(y)$. If $z \notin y^\perp$, then by (P6) there exists a point $w \in x^\perp \cap y^\perp \cap z^\perp$. But then $w$ is at distance 2 from $\infty$. Let $v$ be the common neighbor of $\infty$ and $w$. Applying (P3) to $\infty, u, z, w$ and $v$ yields that $u$ and $v$ are collinear and that $(u, w)$ is a polar pair. But then (P4) implies that the distance between $y$ and $u$ equals 2 and $u \in \pi(y)$. Hence $\pi(x) \subseteq \pi(y)$. Similarly we find that $\pi(y) \subseteq \pi(x)$. This proves the lemma. □

Now we can construct a geometry $\mathcal{G} = \widehat{\mathcal{G}}/\sim$ as follows. As points we take the equivalence classes of points of $\widehat{\mathcal{G}}$. If $l$ is a line, respectively $\pi$ is a plane, of $\widehat{\mathcal{G}}$, then $[l]$, respectively $[\pi]$, denotes the set of equivalence classes $[x]$ where $x$ runs through the point set of $l$, respectively $\pi$. The lines (resp. planes) of $\mathcal{G}$ are the sets $[l]$ (resp. $[\pi]$) where $l$ (resp. $\pi$) runs through the set of lines (resp. planes) of $\widehat{\mathcal{G}}$. In the following lemmas we will show that in this quotient space the residue of a point $[x]$ is isomorphic to the residue of $x$ in $\Delta$.

**Lemma 4.4** *Let $x$ be a point and $l$ a line. If $l$ meets $[x]$ nontrivially, then there is a line $m$ on $x$ with $[m] = [l]$. In particular, $m$ is the unique line on $x$ with $\rho(m) = \rho(l)$.*

*Proof.* If $m$ is a line on $x$ with $[m] = [l]$, then by 4.2 we have $\rho(l) = \rho(m)$.

So let $m$ be the unique line through $x$ with $\rho(m) = \rho(l)$. Then by Lemma 4.2 we have $\pi(l) = \pi(m)$. Let $y$ be a point on $l$, at distance 3 from $\infty$ and not in $[x]$. Fix a point $z \in \pi(y) \setminus \pi(x)$. The point $z$ is at distance 3 from $x$ and therefore makes a special pair with some point $y' \in m$. Inside $\pi(y)$ it is at distance 3 from $\rho(l)$, and thus makes a special pair with $\rho(l)$. Let $u$ be the common neighbor of $z$ and $y'$. If $y'$ is in $\mathcal{H}_\infty$, then $y'$ and $\rho(l)$ are collinear. But then (P3) implies that $\infty$ and $y'$ form a polar pair, contradicting (P4). Hence $y'$ is at distance 3 from $\infty$. Moreover, $\pi(y')$ contains both $\pi(m)$ and $z$. But, by Proposition 3.2 (ii), $\pi(m)$ and $z$ generate the whole of $\pi(y)$. So $\pi(y')$ contains $\pi(y)$, from which it easily follows that $\pi(y') = \pi(y)$ and $y \sim y'$. Hence $[m] \subseteq [l]$ and by symmetry of the argument $[m] = [l]$. □

**Lemma 4.5** *Let $x$ be a point and $\pi$ a plane. If $\pi$ meets $[x]$ nontrivially, then there is a unique plane $\tau$ on $x$ with $[\pi] = [\tau]$.*

*Proof.* Suppose $y \in \pi \cap [x]$. Let $l$ be a line on $y$ inside $\pi$. By Lemma 4.4 there is a unique line $l'$ on $x$ with $[l'] = [l]$. It is the unique line on $x$ containing a point collinear with $\rho(l)$.

Let $m \neq l$ be a line of $\pi$ through $y$. Using (P3) we find that $\rho(m)$ is collinear with $\rho(l)$. Moreover, as $\pi(l)$ does not contain planes, all the $\rho(m)$ where $m$ runs through the lines of $\pi$ on $y$ are collinear. They are all on a line called $k$. But then again by (P3) also all the lines $m'$ on $x$ with $\rho(m')$ on $k$ are inside a plane, $\tau$ say. Lemma 4.4 now easily implies that $[\pi] = [\tau]$. □

**Lemma 4.6** *If $x \sim y$ then $x$ and $y$ are not at distance one or two in $\widehat{\mathcal{G}}$.*

*Proof.* Notice that if $x$ and $y$ are collinear then the result follows from 4.2.

Suppose that $x \sim y$ are at distance 2 and let $z$ be a common neighbor of $x$ and $y$ inside $\widehat{\mathcal{G}}$. Consider the line $l$ of $\Delta$ through $x$ and $z$. Let $x_0$ be $l \cap \mathcal{H}_\infty$ and $u = \rho(l)$ be the unique point in $\infty^\perp$ collinear with $x_0$. Then $u$ is also at distance 2 from $y$. Let $w$ denote the common neighbor of $y$ and $u$. Applying (P3) shows that $x_0$, $z$ and $y$ must be in some symplecton, whence $x$ and $y$ are in some symplecton. But inside a symplecton we see that they are not on a hyperbolic line with the deep point, a contradiction to Lemma 4.3. □

Let $Sym$ be a symplecton of $\Delta$ containing a point at distance 3 from $\infty$, then the geometry obtained by removing $\mathcal{H}_\infty \cap Sym$ from $Sym$ and taking the quotient with respect to $\sim$ is a tangent geometry as described in [8, 10]. We will call these subspace of $\widehat{\mathcal{G}}/\sim$ *affine symplecta*. The above results imply:

**Proposition 4.7** *The geometry of points, lines, planes and affine symplecta of $\widehat{\mathcal{G}}/\sim$ is an affine extension of a dual polar space with diagram*

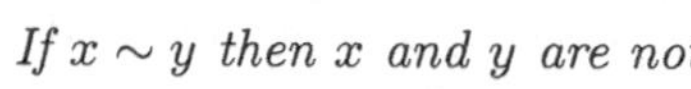

We end this section with the observation that the geometry $\widehat{\mathcal{G}}/\sim$ satisfies the conditions of Theorem 1.1 if and only if the affine symplecta are linear spaces. This is only the case when we start with a metasymplectic space in which all symplecta are symplectic rank 3 polar spaces, see [8, 10], and $\Delta$ is a metasymplectic space of type $F_{4,4}(k)$ for some field $k$.

The stabilizer in the group $G = F_4(k)$, which is contained in $Aut(\Delta)$, of $\infty$ is the maximal parabolic $P$ of type $B_3$. This group admits a normal subgroup $K$ which is the kernel of the action of $P$ on the $\sim$-equivalence classes. The quotient group $P/K$ is isomorphic to $k^8{:}B_3(k)$, the split extension of the 8-dimensional spin module for the group $B_3(k)$ by that group itself. For this reason we call the geometry $\widehat{\mathcal{G}}/\sim$ in this particular case the *affine extension of a dual polar space associated to the group $k^8{:}B_3(k)$.*

# 5 The affine extension of a symplectic quadrangle

Suppose $f$ is a nondegenerate symplectic form on a 4-dimensional vector space $V$ over a field $k$. Let $\mathcal{P}$ denote the set of vectors of $V$. By $\mathcal{L}$ we denote the set affine lines in $V$, and by $\mathcal{A}$ the set of affine planes of $V$ which are parallel to a linear plane from $V$ on which $f$ vanishes. Then $(\mathcal{P}, \mathcal{L}, \mathcal{A})$ is called an *affine extension of the symplectic quadrangle* and denoted by $ASp(4, k)$. It is characterized by the following:

**Theorem 5.1** *Let $\mathcal{G} = (\mathcal{P}, \mathcal{L}, \mathcal{A})$ be a connected geometry satisfying:*

*i. $(\mathcal{P}, \mathcal{L})$ is a linear space with at least 3 points per line;*

*ii. $\mathcal{A}$ is a set of subspaces of $(\mathcal{P}, \mathcal{L})$ isomorphic to affine planes;*

*iii. For any point $P \in \mathcal{P}$ the residue at p, $Res(p) = (\mathcal{L}_p, \mathcal{A}_p)$, is a nondegenerate generalized quadrangle.*

*Then $\mathcal{G}$ is isomorphic to $ASp(4, k)$, $k$ a field.*

For finite geometries $\mathcal{G}$ the above result follows from the results of [8, 10] and [11].

In this section we prove Theorem 5.1. We will indicate how the proof of [8], which only works in the finite case, can be extended to obtain a proof of 5.1.

Let us assume that $\mathcal{G} = (\mathcal{P}, \mathcal{L}, \mathcal{A})$ is a geometry satisfying the hypothesis of 5.1. As in [8], we use the convention that points, lines and other subspaces of the residue of some point $p$ of $\mathcal{P}$ are referred to as 'points', 'lines' or 'subspaces'. This is done in order to distinguish them from points, lines and subspaces of $\mathcal{G}$. (This convention will also be used in the next section.) Let $p$ be a point and $l$ a line of $\mathcal{G}$ such that $p$ and $l$ are not coplanar. Let $H$ denote the set of lines on $p$ meeting $l$ nontrivially. The set $H$ is a set of all or all but one of the 'points' of a 'hyperbolic line' of $Res(p)$ as follows by Proposition 4.5 of [8].

Let $\pi$ denote the set of all the points that are on a line of $\mathcal{L}$ which is in $H$ or meets at least two of the lines of $H$ in some point different from $p$. The results of Section 4 of [8] show that if $H$ is a full 'hyperbolic line' of $Res(p)$, then $\pi$ is a subspace of $(\mathcal{P}, \mathcal{L})$ in which any two lines intersect. In particular, any two lines in $H$ generate this subspace which is isomorphic to a projective plane. Whereas in case $H$ is the set of all but one 'points' of a hyperbolic line, $\pi$ is a subspace isomorphic to an affine plane.

In any case we find that two intersecting lines of $\mathcal{G}$ generate a subspace of $(\mathcal{P}, \mathcal{L})$ isomorphic to an affine or projective plane.

As $(\mathcal{P}, \mathcal{L})$ contains some affine plane, Teirlinck's results, see [12], imply that we only encounter affine planes and no projective ones. The results of [8, 10] also imply that being parallel in an affine plane determines an equivalence relation on $\mathcal{L}$, from which we derive that $(\mathcal{P}, \mathcal{L})$ is an affine space. Of course, we can also appeal to Buekenhout's well known result, see [3], that any linear space in which any two intersecting lines generate an affine plane is a an affine space, provided all lines contain at least 4 points. (The case where lines contain 3 points being handled by the above remark, see [8].)

The theorem follows now easily. Indeed, for each point $p$ of $\mathcal{G}$ the residue $Res(p)$ at $p$ is a generalized quadrangle in the projective space on all the lines through $p$. All the points of this projective space are in $Res(p)$. But then $Res(p)$ is a symplectic quadrangle, which clearly implies Theorem 5.1.

# 6 Affine extensions of near hexagons

This section is devoted to the proof of Theorem 1.1. Throughout this section let $\mathcal{G} = (\mathcal{P}, \mathcal{L}, \mathcal{A})$ be a connected geometry satisfying:

i. $(\mathcal{P}, \mathcal{L})$ is a partial linear gamma space with at least 3 points per line;

ii. $\mathcal{A}$ is a set of subspaces of $(\mathcal{P}, \mathcal{L})$ isomorphic to affine planes;

iii. For any point $p \in \mathcal{P}$ the residue at $p$, $Res(p) = (\mathcal{L}_p, \mathcal{A}_p)$, is a near hexagon with quads and $m, l \in \mathcal{L}_p$ are not opposite in $Res(p)$ if and only if $m \subseteq l^\perp$.

**Lemma 6.1** *i. Let $\pi$ be a plane and $p$ be a point, then either $\pi \cap p^\perp = \emptyset$, $\pi \cap p^\perp \in \mathcal{L}$ or $\pi \subseteq p^\perp$.*

*ii. Any two different planes that meet non trivially, meet either in a point or a line.*

*Proof.* See Lemma 3.2 and Lemma 3.3 of [2]. □

**Lemma 6.2** *Suppose $l$ and $m$ are two lines on the point $p$ that are not coplanar. If $n$ is a line meeting both $l$ and $m$ at points different from $p$, then all the lines through $p$ meeting $n$ are on the hyperbolic line of $Res(p)$ through $l$ and $m$.*

*Proof.* Suppose $l$, $m$ and $n$ are lines as in the hypothesis of the lemma. Suppose $r$ is also a line through $p$ meeting $n$ nontrivially. If $k$ is a line on $p$, which is, as a 'point' of $Res(p)$, at distance at most 2 from both $l$ and $m$, then $l$, $m$ and by the gamma space property also $n$ are contained in $k^\perp$. But then at least two points of $r$ and hence all points of $r$ are in $k^\perp$. So, in $Res(p)$ the 'point' $r$ is also at distance at most 2 from $k$. Hence $r$ is contained in the hyperbolic 'line' of $Res(p)$ defined by $l$ and $m$. □

Let $\mathcal{Q}$ be a quad in the residue of a point $p$. Then the union of all the points on a line on $p$ belonging to $\mathcal{Q}$ will be denote by $\mathcal{AQ} = \mathcal{AQ}(p, \mathcal{Q})$ and is called an *affine quad* of $\mathcal{G}$ defined by $p$ and $\mathcal{Q}$.

**Proposition 6.3** *$\mathcal{AQ}$ is a subgeometry of $\mathcal{G}$ isomorphic to an affine symplectic quadrangle. Moreover, if $q$ is a point in $\mathcal{AQ}$, then there is a quad $\mathcal{Q}'$ in $Res(q)$ with $\mathcal{AQ} = \mathcal{AQ}(q, \mathcal{Q}')$.*

*Proof.* We prove that $\mathcal{AQ}$ is a subgeometry of $\mathcal{G}$. First we notice that any two points in $\mathcal{AQ}$ are collinear in $\mathcal{G}$ by the second part of axiom (iii). Now fix two such points, $a$ and $b$ say, and let $l$ be the line through them. If $a$, $b$ and $p$ are collinear or coplanar, then clearly $l$ is contained in $\mathcal{AQ}$. Thus suppose that they are not collinear nor coplanar. Let $m_a$, respectively, $m_b$ be the line through $p$ and $a$, respectively, $b$. Then $m_a$ and $m_b$ are two 'points' of $\mathcal{Q}$, and all the lines through $p$ meeting $l$ form a subset of a 'hyperbolic line' of $Res(p)$. Any 'point' $r$ of $\mathcal{Q}$ is at distance at most 2 from both $m_a$ and $m_b$. Thus the 'hyperbolic line' through $m_a$ and $m_b$ is contained in the set of 'points' which are at distance at most 2 from every point of $\mathcal{Q}$. But this set equals just $\mathcal{Q}$. So all the lines through $p$ meeting $l$ are contained in $\mathcal{AQ}$, and so is $l$.

This shows that $\mathcal{AQ}$ is a subspace of $(\mathcal{P}, \mathcal{L})$. As any plane in $\mathcal{A}$ is generated by any three noncollinear points in it, we easily obtain, with the help of the above, that $\mathcal{AQ}$ contains all planes of which it contains at least 3 noncollinear points. Thus $\mathcal{AQ}$ is a subgeometry of $\mathcal{G}$.

Now we show that inside $\mathcal{G}$ the residue of a point is a generalized quadrangle. For the point $p$ this is by definition the case. Thus consider a point $q \neq p$ of $\mathcal{AQ}$. Let $m \in \mathcal{L}$ be a line on $q$ and $\pi \in \mathcal{A}$ be a plane on $q$ both contained in $\mathcal{AQ}$, the line $m$ not in $\pi$.

Since any two points in $\mathcal{AQ}$ are collinear, we find that $\pi \subseteq m^\perp$. Thus there is a unique plane on $m$ meeting $\pi$ in a line. Since $\mathcal{AQ}$ is a subspace this plane is in $\mathcal{AQ}$. This implies that $Res(q)$ in $\mathcal{AQ}$ is a possibly degenerate generalized quadrangle. Let $k$ be the line through $p$ and $q$. The line $k$ is contained in at least two planes from $\mathcal{A}$. So, if $Res(q)$ in $\mathcal{AQ}$ is degenerate, its radical equals $k$. On $p$ there is a line $k'$ which is not coplanar with $k$. But then any line through $q$ different from $k$ which meets $k'$ is also not coplanar with $k$. This shows that the residue of $q$ restricted to $\mathcal{AQ}$ is indeed a nondegenerate generalized quadrangle.

We can apply Theorem 5.1 to $\mathcal{AQ}$ to find that $\mathcal{AQ}$ is indeed isomorphic to an affine symplectic quadrangle. This proves the first part of the proposition.

Now consider a point $q$ in $\mathcal{AQ}$ different from $p$. The residue of $q$ inside $\mathcal{AQ}$ is contained in a unique 'quad' $\mathcal{Q}'$ of $Res(q)$. Let $r$ be a line through $q$, which is a 'point' of $\mathcal{Q}'$. Then every line on $p$ which is inside $\mathcal{AQ}$ is contained in $r^\perp$. But that implies that the lines through $p$ meeting $r$ are 'points' of $Res(p)$ which are at distance at most 2 from every point of $\mathcal{Q}$. But then, as $Res(p)$ is a near hexagon, they are 'points' of $\mathcal{Q}$. In particular, $r$ is contained in $\mathcal{AQ}$, and hence $\mathcal{AQ} = \mathcal{AQ}(q, \mathcal{Q}')$. □

**Corollary 6.4** *For each point $p$ of $\mathcal{G}$, the residue $Res(p)$ is isomorphic to the dual polar space of type $B_3$.*

*Proof.* To prove this corollary it suffices to show that for any point $p$ the local near hexagon $Res(p)$ is a classical near hexagon. For this purpose we have to show that each 'point-quad'-pair of $Res(p)$ is classical, i.e., given a 'quad' $\mathcal{Q}$ and a 'point' $l$, then there is a unique 'point' $m$ in $\mathcal{Q}$ closest to $l$. See [4].

Suppose $p$, $l$, and $\mathcal{Q}$ are as above. Then let $\mathcal{AQ}$ be the affine quad defined by $p$ and $\mathcal{Q}$. We may assume that $l$ is not contained in $\mathcal{AQ}$. But then, by the gamma space property, all 'points' in $Res(p)$ that are inside $l^{\perp} \cap \mathcal{AQ}$ form a 'geometric hyperplane' of $\mathcal{Q}$ closed under 'hyperbolic lines'. Such a 'geometric hyperplane' in $\mathcal{Q}$ is a 'point' $m$ with all the 'lines' through that 'point'. But then $m$ must be collinear with $l$ inside $Res(p)$, and it is the unique 'point' of $\mathcal{Q}$ closest to $l$. □

The above also implies that the geometry of points, lines, affine planes and affine quads of $\mathcal{G}$ is a diagram geometry with diagram:

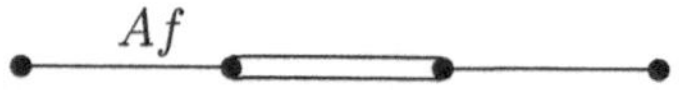


Consider the space $(\mathcal{P}, \mathcal{L})$. As follows from 6.3, any two intersecting lines $l$ and $m$ with $l \subseteq m^{\perp}$ generate a subspace of $(\mathcal{P}, \mathcal{L})$ isomorphic to an affine plane. Let $\mathcal{B}$ be the set of all these affine planes not already contained in $\mathcal{A}$.

**Proposition 6.5** *The geometry $\mathcal{AS} = (\mathcal{P}, \mathcal{L}, \mathcal{A} \cup \mathcal{B})$ is an affine polar space of type $Af.D_4$.*

*Proof.* It is well known that after adding all the hyperbolic lines of the quads in a dual polar space of type $B_3$ one obtains a polar space of type $D_4$. Thus inside $\mathcal{AS}$ we find for any point $p$, that $Res(p)$ is isomorphic to the polar space of type $D_4$. Since $\mathcal{AS}$ is a gamma space we have proved the proposition. □

From [7] we know that $\mathcal{AS} = (\mathcal{P}, \mathcal{L}, \mathcal{A} \cup \mathcal{B})$ can be obtained from a polar space $\mathcal{S}$ of type $D_5$, by removing a hyperplane $\mathcal{H}$. We identify $\mathcal{AS}$ with this complement of the geometric hyperplane $\mathcal{H}$ of a fixed polar space $\mathcal{S}$ of type $D_5$. The geometric hyperplanes of the $D_5$ polar space $\mathcal{S}$ are classified in [7]. There are two types of them: nondegenerate subspaces, which are polar spaces of type $B_4$, and degenerate ones consisting of a point and all the lines through a point. Below (6.7) we shall see that $\mathcal{H}$ is a degenerate hyperplane.

Let $l$ be a line and $\pi$ a plane of $\mathcal{AS}$, then $l$ and $\pi$ can be obtained from a line $\hat{l}$, respectively a plane $\hat{\pi}$, of $\mathcal{S}$ by removing the intersection point $[l]$ of $\hat{l}$, respectively the intersection line $[\pi]$ of $\hat{\pi}$ with $\mathcal{H}$. The point $[l]$ is called the *point at infinity* of $l$ or the *projection* of $l$ in $\mathcal{H}$. For a plane $\pi$ we call $[\pi]$ the *line at infinity* of $\pi$ or the *projection* of $\pi$ in $\mathcal{H}$.

In this way the hyperplane $\mathcal{H}$ provides us with a parallelism on the lines and planes of $\mathcal{G}$. Two lines $l$ and $m$ (resp. planes $\pi$ and $\rho$) are called *parallel* if and only if $[l] = [m]$ (respectively $[\pi] = [\rho]$). We identify the point $[l]$ of $\mathcal{H}$ with the corresponding parallel class of $\mathcal{L}$. In the notation of [7] we recognize this parallelism as follows: a line $l$ is called parallel to $m$ if and only if $\{p \in \mathcal{P} \mid p^{\perp} \cap l = \emptyset \text{ or } l\} = \{p \in \mathcal{P} \mid p^{\perp} \cap m = \emptyset \text{ or } m\}$.

**Lemma 6.6** *If $\pi$ is a plane in $\mathcal{G}$ then the whole parallel class of $\pi$ in $\mathcal{AS}$ is in $\mathcal{G}$.*

*Proof.* This follows from the proof of Lemma 5.2 of [2], since near hexagons do not contain triangles. □

Let $\mathcal{P}_\infty$ be the set $\{[l] \mid l \in \mathcal{L}\}$, and $\mathcal{L}_\infty$ be the set $\{[\pi] \mid \pi \in \mathcal{A}\}$. By $\mathcal{G}_\infty = (\mathcal{P}_\infty, \mathcal{L}_\infty)$ we denote the *space at infinity* of $\mathcal{G}$. It follows from the above lemma that $\mathcal{G}$ is isomorphic to the *tangent space* of its projection $\mathcal{G}_\infty$ on $\mathcal{H}$. This means, $\mathcal{G}$ consists of all the lines and planes of $\mathcal{AS}$ whose point respectively line at infinity is in $\mathcal{G}_\infty$. Thus to prove our main result Theorem 1.1, we still have to determine $\mathcal{G}_\infty$ and its embedding in $\mathcal{H}$.

For a point $p$ the geometry at infinity of $Res(p)$ is a dual polar space $\mathcal{DS}$ of type $B_3$. Indeed, the projection of $Res(p)$ into $\mathcal{H}$ is injective and incidence preserving. Similarly, the projection on $\mathcal{H}$ of the residue of $p$ in $\mathcal{AS}$ is a subgeometry $\mathcal{D}$ of $\mathcal{H}$ isomorphic to a polar space of type $D_4$. Clearly $\mathcal{DS}$ is contained in $\mathcal{D}$ and $\mathcal{D}$ is a geometric hyperplane of $\mathcal{H}$, which is simply the intersection of $\mathcal{H}$ with the geometric hyperplane $p^\perp$ of the polar space $\mathcal{S}$.

A 'quad' of $Res(p)$ is mapped onto maximal singular subspaces of this $D_4$ polar space $\mathcal{D}$. Elements of the other type of maximal singular subspaces of $\mathcal{D}$ can be obtained as the projection of a 'point' of $Res(p)$ and all the 'lines' through it. So, if $l$ is a line of $\mathcal{G}$, then the projection of the set of all planes of $\mathcal{A}$ on $l$ on $\mathcal{H}$ is contained in a 3-space of $\mathcal{H}$, which will be called the *special* 3-space of $l$ and be denoted by $\Pi_l$. Moreover, all lines in $\Pi_l$ through the point $[l]$ are projections of planes on $l$.

**Lemma 6.7** *If $l$ and $m$ are parallel lines, then the special 3-spaces $\Pi_l$ and $\Pi_m$ are contained in some singular subspace of $\mathcal{H}$. In particular, $\mathcal{H}$ contains singular 4-spaces and is degenerate.*

*Proof.* Let $\pi_l, \pi_m \in \mathcal{A}$ be affine planes on $l$, respectively $m$, contained in the singular plane $\hat{\pi}_l$, respectively $\hat{\pi}_m$, of $\mathcal{S}$. Notice that $[\pi_l]$ and $[\pi_m]$ intersect in $[l] = [m]$. Suppose that $[\pi_l]$ and $[\pi_m]$ are not together in a singular subspace of $\mathcal{H}$. In the polar space $\mathcal{S}$ we see that $[\pi_m]^\perp \cap \pi_l$ contains a line $k$ of $\pi_l$. Let $\pi$ be the affine plane of $\mathcal{AS}$ defined by the $\mathcal{S}$ plane through $k$ and $[\pi_m]$. By 6.6 the plane $\pi$ is in $\mathcal{A}$; furthermore both $[\pi_l]$ and $[\pi_m]$ are contained in the special 3-space $\Pi_k$; a contradiction with our assumption. We have now proved he first part of the statement of the lemma.

Now take two distinct but parallel lines $l$ and $m$ which are not coplanar. Then the two special 3-spaces $\Pi_l$ and $\Pi_m$ are distinct but contained in a singular subspace of $\mathcal{H}$, which therefore contains singular 4-spaces. As a $B_4$ polar space contains no singular 4-spaces, we can conclude that $\mathcal{H}$ is degenerate. □

**Theorem 6.8** *The geometry $\mathcal{G}$ is isomorphic to the affine extension of a dual polar space associated to $k^8{:}B_3(k)$, where $k$ is some field.*

*Proof.* The hyperplane $\mathcal{H}$ is degenerate and of the form $\infty^\perp$, where $\infty$ is a point of $\mathcal{S}$, see [7]. For each point $p$ in $\mathcal{P}$, which is simply $\mathcal{S} \setminus \mathcal{H}$, the parallel classes of lines and planes of $\mathcal{G}$ on $p$ form the points and lines of a dual polar space $D_p$ isomorphic to $\mathcal{DS}$ embedded in $p^{\perp}\cap\infty^{\perp}$. The lines and singular planes of $\mathcal{S}$ through the point $\infty$ that meet $D_p$ in a point, respectively, line, also form a dual polar space which will be denoted with $\overline{D}_p$.

We claim that for any two points $p$ and $q$ in $\mathcal{P}$ we have $\overline{D}_p = \overline{D}_q$. Let $\pi$ be a plane on $q$. Since $\overline{D}_p$ is a hyperplane of $\mathcal{H}$, $[\pi]$ must intersect it in a point $r$. Let $m$, respectively $n$, be the line on $p$, respectively $q$, whose point at infinity equals $r$.

The two special 3-spaces $\Pi_m$ and $\Pi_n$ are contained in a singular subspace of $\mathcal{H}$, but do not contain the point $\infty$. Since $\mathcal{H}$ is degenerate and modulo its radical is isomorphic to a polar space of type $D_4$ we must have that $\infty$ is also contained in this singular subspace. Moreover, both special 3-spaces are mapped onto each other by projection from $\infty$. Thus the singular plane on $\infty$ and $[\pi]$ is not only a line of $\overline{D}_q$ but also of $\overline{D}_p$ and we can conclude that $\overline{D}_q \subseteq \overline{D}_p$. By symmetry of the argument we have also $\overline{D}_p \subseteq \overline{D}_q$, and we can identify $\mathcal{G}$ with the affine extension of the dual polar space associated to $\overline{D}_q = \overline{D}_p$. Since there is, up to isomorphism, only one embedding of the dual polar space of type $B_3$ into the polar space $D_4$, we see that there is, up to isomorphism, also just one embedding of $\mathcal{G}_\infty$ into $\mathcal{H}$. As already noticed before, this proves our proposition and therefore also Theorem 1.1. □

**Acknowledgment.** The authors want to thank Sergey Sphectorov for very useful remarks on a first version of this paper.

# References

[1] R. Blok & A.E. Brouwer, The geometry far away from a residue, these Proceedings.

[2] J. van Bon & H. Cuypers, Affine extensions of generalized polygons, preprint 1996.

[3] F. Buekenhout, Une caractérization des espaces affines basée sur la notion de droite, *Math. Z.* **111** (1969), 367–371.

[4] P. Cameron, Dual polar spaces, *Geom. Dedicata* **12** (1982), 75–85.

[5] A.M. Cohen, An axiom system for metasymplectic spaces, *Geom. Dedicata* **12** (1982), 417–433.

[6] A.M. Cohen & B. Cooperstein, The 2-spaces of the standard $E_6$-module, *Geom. Dedicata* **25** (1988), 467–480.

[7] A.M. Cohen & E. Shult, Affine polar spaces, *Geom. Dedicata* **35** (1990), 43–76.

[8] H. Cuypers, Finite locally generalized quadrangles with affine planes, *Europ. J. of Comb.* **13** (1992), 439–453.

[9] H. Cuypers, Extended near hexagons and line systems, preprint 1993.

[10] H. Cuypers & A. Pasini, Locally polar geometries with affine planes, *Europ. J. of Comb.* **13** (1992), 39–57.

[11] A. Pasini, On locally polar geometries whose planes are affine, *Geom. Dedicata* **34** (1990), 35–56.

[12] L. Teirlinck, On linear spaces in which every plane is either projective or affine, *Geom. Dedicata* **4** (1975), 39–44.

[13] J. Tits, *Buildings of spherical type and finite BN-pairs*, Lecture Notes in Mathematics **386**, Springer, 1974.

John van Bon,
Dipartimento di Matematica,
Università della Calabria,
Arcavacata di Rende, 87036 Rende (CS), Italy.
e-mail: `vanbon@ccuws4.unical.it`

Hans Cuypers,
Department of Mathematics,
Eindhoven University of Technology,
P.O. BOX 513, 5600 MB Eindhoven, The Netherlands.
e-mail: `hansc@win.tue.nl`

Trends in Mathematics, © 1998 Birkhäuser Verlag Basel/Switzerland

# Ovoids and Spreads Arising from Involutions

Hendrik Van Maldeghem*

**Abstract**

In this paper, we give a new construction of the hermitian spreads in $H(q)$ without using the standard embedding in $\mathbf{PG}(6,q)$, without using the group $U_3(q)$, but using some geometric properties of the hexagon and an involution. Remarking that a similar construction holds in certain quadrangles of order $s$, with $s$ a power of 2, we obtain ovoids in quadrangles of type $T_2(O)$.

## 1 Introduction and definitions

A *generalized polygon* or *generalized n-gon*, $n \in \mathbf{N}$, $n \geq 2$, is a point-line incidence geometry with an incidence graph of diameter $n$ and girth $2n$ (or gonality $n$). For finite generalized quadrangles, we refer to PAYNE & THAS [4]. The only known examples of finite generalized hexagons (6-gons) are defined in TITS [8] and they satisfy the so-called Moufang condition, see TITS [9]. They arise from the Chevalley groups $G_2(q)$ and ${}^3D_4(q)$. We will be concerned with the class arising from $G_2$ and sometimes called the *split Cayley hexagons*, because they can be constructed using a split Cayley algebra. We will use another construction below: the one due to TITS [8].

It is common to call a generalized polygon *thick* if every element is incident with at least three other elements. It is well-known that for thick generalized polygons the number $s+1$ of points on a line is a constant, and, dually, the number $t+1$ of lines incident with a point is a constant. In this case, the pair $(s,t)$ is called the *order* of the polygon.

An *ovoid* of a generalized quadrangle $\Gamma$ is a set $\mathcal{O}$ of points such that every line is incident with a unique element of $\mathcal{O}$. It follows readily that all points of $\mathcal{O}$ are mutually at distance 4 (distances measured in the incidence graph) and also, $|\mathcal{O}| = 1 + q^2$ if the quadrangle has order $(q,q)$. An *ovoid* in a generalized hexagon is a set of points such that every point is at distance $\leq 2$ from a unique element of the ovoid (distances again measured in the incidence graph). It follows readily that all points of an ovoid are at distance 6 from each other, and that there are $1 + q^3$ elements in an ovoid if the hexagon has order $(q,q)$. A *spread* is the dual notion of an ovoid.

*Senior Research Associate of the Fund for Scientific Research – Flanders (Belgium) (F.W.O.).

Let $\Gamma = H(q)$ be the generalized hexagon of order $(q,q)$ arising from $G_2(q)$. For an element $u$ of $\Gamma$, we denote by $\Gamma_i(u)$ the set of points and lines of $\Gamma$ at distance $i$ from $u$. We fix the duality class of $H(q)$ by requiring that all points of $H(q)$ are *regular*, i.e., for every three points $x, y, z$ such that $y, z \in \Gamma_6(x)$, the inequality $|\Gamma_i(x) \cap \Gamma_{6-i}(y) \cap \Gamma_{6-i}(z)| \geq 2$ implies $|\Gamma_i(x) \cap \Gamma_{6-i}(y) \cap \Gamma_{6-i}(z)| = q+1$, for $i = 2, 3$ (see RONAN [5]). We will use that property along with a certain involution to construct a spread $\mathcal{S}$ in a subhexagon $H(\sqrt{q})$ of $H(q)$, and we show that $\mathcal{S}$ is isomorphic to the so-called hermitian spread, as contructed by THAS [6]. Then we apply this method to quadrangles and give some non-classical examples.

# 2 Hermitian spreads of $H(q)$

The generalized hexagon $H(q)$ can be constructed as follows (see TITS [8]). Consider in $\mathbf{PG}(6,q)$ the quadric $Q(6,q)$ with equation $X_0X_4 + X_1X_5 + X_2X_6 = X_3^2$. The points of $H(q)$ are all the points of $Q(6,q)$ and the lines of $H(q)$ are the lines of $Q(6,q)$ the Grassmann coordinates of which satisfy the following six linear equations:

$$p_{12} = p_{34}, \qquad p_{54} = p_{32}, \qquad p_{20} = p_{35},$$
$$p_{65} = p_{30}, \qquad p_{01} = p_{36}, \qquad p_{46} = p_{31}.$$

One can deduce all above equations from the first one by consecutively applying the following rule: if $p_{ij} = p_{3k}$ is in the list, then so are $p_{(i\pm 4)k} = p_{3j}$ and $p_{k(j\pm 4)} = p_{3i}$, where in $\pm 4$, one should choose the appropriate sign in order to obtain a number between 0 and 7. Incidence is inherited from $\mathbf{PG}(6,q)$. Now consider a hyperplane $H$ of $\mathbf{PG}(6,q)$ that intersects $Q(6,q)$ in an elliptic quadric. Then the lines of $H$ which also belong to $H(q)$ form a spread $\mathcal{S}$ in $H(q)$, called the *hermitian spread*, see THAS [6]. The spread $\mathcal{S}$ has the following property. Let $L, M$ be 2 lines of $\mathcal{S}$, then

$\mathrm{H}_{L,M}$ every line of $H(q)$ at distance 3 from every point of $H(q)$ which is itself at distance 3 from both $L$ and $M$, is contained in $\mathcal{S}$.

By the regularity mentioned above, the number of lines in $H(q)$ at distance 3 from all points at distance 3 from both $L$ and $M$ is equal to $q+1$. Note that BLOEMEN, THAS & VAN MALDEGHEM [1] show that, whenever a spread of $H(q)$ has the property $\mathrm{H}_{L,M}$, for all lines $L$ and at least 2 lines $M$, then the spread is a hermitian spread.

Now consider $H(q)$ and embed $H(q)$ in $H(q^2)$. Let $\theta$ be an involution in $H(q^2)$ fixing $H(q)$ pointwise. Such an involution always exists (apply the field automorphism $x \mapsto x^q$ on the above representation of $H(q^2)$ in $\mathbf{PG}(6,q^2)$). Let $L$ and $M$ be two opposite lines of $H(q)$. Let $p$ be a point of $H(q^2)$ incident with $L$, but not fixed by $\theta$. Let $p'$ be the projection of $p^\theta$ onto $M$ (the point of $M$ nearest to $p^\theta$). By RONAN [5](6.12), there exists a unique subhexagon $\Gamma$ of order $(1,q^2)$ through $p$ and $p'$, and $\Gamma$ is isomorphic to the incidence graph of the projective plane $\mathbf{PG}(2,q^2)$.

Let $\mathcal{S}$ be the set of lines of $\Gamma$ fixed by $\theta$, or in other words, $\mathcal{S}$ is the intersection of the set of lines of $\Gamma$ with the set of lines of $H(q)$. Then we claim:

**Theorem 1** *With the above notation, the set $\mathcal{S}$ of lines is a spread of $H(q)$.*

*Proof.* Since $\theta$ fixes $L$ and $M$, it maps $p'$ to the projection of $p$ onto $M$. Hence both $p^\theta$ and $p'^\theta$ belong to $\Gamma$ and hence $\theta$ preserves $\Gamma$. Note that no point of $\Gamma$ is a point of $H(q)$. Indeed, every point of $\Gamma$ is either at distance 4 from $p$ or at distance 4 from $p^\theta$. Hence if a point $w$ of $H(q)$ would belong to $\Gamma$, then, since $L$ belongs to $H(q)$, also the point $p$ or $p^\theta$ would belong to $H(q)$, a contradiction. Since $\Gamma$ is the incidence graph of $\mathbf{PG}(2,q^2)$, the involution $\theta$ induces in $\mathbf{PG}(2,q^2)$ a polarity (which we also denote by $\theta$). Let $x$ be the unique point of $\Gamma$ collinear with both $p$ and $p'^\theta$, and let $y$ be the unique point collinear with both $p^\theta$ and $p'$. Note that $x^\theta = y$ and that $\{L,M\} \subseteq \Gamma_3(x) \cap \Gamma_3(y)$. Let $u_i$, $i = 1,2$, $u_1 \neq u_2$, be any two points of $H(q)$, both at distance 3 from both $L$ and $M$. By the regularity in $H(q^2)$, there are $q^2+1$ lines of $H(q^2)$ at distance 3 from both $x$ and $y$, and hence also at distance 3 from both $u_1$ and $u_2$. But in $H(q)$, there are $q+1$ lines at distance 3 from both $u_1$ and $u_2$. Hence there are $q+1$ lines of $H(q)$ at distance 3 from both $x$ and $y$. Without loss of generality, we may assume that $x$ represents a point of $\mathbf{PG}(2,q^2)$, and $y$ represents a line of $\mathbf{PG}(2,q^2)$ not incident with $x$. Then we have shown that the polarity $\theta$ in $\mathbf{PG}(2,q^2)$ contains exactly $1+q$ absolute points incident with $y$ (namely, the points on the $q+1$ lines of $H(q)$ at distance 3 from both $x$ and $y$, collinear in $H(q^2)$ with $y$), and equivalently, $1+q$ absolute lines incident with $x$. Hence $\theta$ is a unitary polarity in $\mathbf{PG}(2,q^2)$ and hence it contains $1+q^3$ absolute points. If $z$ is such a point, then $\{z, z^\theta\}$ represents a collinear pair of points in $H(q^2)$ and the line $zz^\theta$ is fixed by $\theta$, hence it belongs to $H(q)$. So we have found $1+q^3$ lines in the intersection of $\Gamma$ and $H(q)$. Clearly, no two of these lines are at distance $\leq 4$ from each other, because this would imply that the shortest path connecting these lines lies in both $\Gamma$ and $H(q)$, and hence $\Gamma$ and $H(q)$ would share at least one point, a contradiction. So $\mathcal{S}$ is a set of $1+q^3$ lines mutually at distance 6 from each other. By CAMERON, THAS & PAYNE [2], $\mathcal{S}$ is a spread of $H(q)$. □

It is clear that $\mathcal{S}$ is a hermitian spread. Indeed, if two lines belong to this spread (and we may take $L$ and $M$), then all lines at distance 3 from two points at distance 3 from $L$ and $M$ belong to $\mathcal{S}$, as follows directly from the above proof. At the same time, $\mathcal{S}$ can be viewed as a hermitian curve in $\mathbf{PG}(2,q^2)$, motivating the name for this spread.

## 3 Some ovoids of non-classical quadrangles

We now apply the method of the previous section to generalized quadrangles. Dualizing the situation, there is the following result.

**Theorem 2** *Let $\Gamma$ be a generalized quadrangle having a subquadrangle $\Gamma'$ with the following properties:*

(*i*) *every point of $\Gamma'$ is incident with exactly two lines of $\Gamma'$;*

(*ii*) *every point of $\Gamma$ incident with a line of $\Gamma'$ belongs to $\Gamma'$;*

(*iii*) *every line of $\Gamma$ is incident with at least one point of $\Gamma'$.*

*Suppose moreover that there is an involution $\theta$ of $\Gamma$ which preserves $\Gamma'$ and which has the following properties:*

(*a*) *there exist two points $x_1, x_2$ of $\Gamma'$ such that $\theta$ interchanges the two lines through $x_i$, for each $i = 1, 2$;*

(*b*) *$\theta$ fixes a thick subquadrangle $\Gamma''$.*

*Then the set of points of $\Gamma'$ fixed under $\theta$ forms an ovoid of $\Gamma''$, or in other words, the intersection of the point sets of $\Gamma'$ and $\Gamma''$ is an ovoid in $\Gamma''$.*

*Proof.* By (*iii*), every line $L$ of $\Gamma''$ is incident with a unique point $x$ of $\Gamma'$ (unique indeed because otherwise $L$ lies in $\Gamma'$, contradicting (*a*), which asserts that $L$ is not fixed in this case). Since $\theta$ fixes $L$ and $\Gamma'$, it fixes $x$, hence $x$ belongs to $\Gamma''$. The result follows. □

In the finite case, conditions (*i*), (*ii*) and (*iii*) are equivalent with saying that the order of $\Gamma$ is $(s, s)$ and that the order of $\Gamma'$ is $(s, 1)$, for some integer $s \geq 2$ (see Payne & Thas [4](2.2.1)). Putting $\Gamma \cong Q(4, q^2)$, the generalized quadrangle arising from a non-degenerate quadric in $\mathbf{PG}(4, q^2)$, and $\Gamma'' \cong Q(4, q)$, we obtain an ovoid isomorphic to $Q^-(3, q)$ in $Q(4, q)$. So for $q$ even, the two known ovoids in $Q(4, q)$, $q = 2^{2h+1}$, arise either from a polarity (Suzuki-Tits ovoid), or from an involution. So one could say that they are both phenomena related to order 2 elements of the correlation group of $Q(4, q)$ (a similar remark holds for the Ree-Tits spreads and hermitian spreads in $H(3^{h+1})$ above).

Now we apply the above theorem to non-classical quadrangles of type $T_2(O)$. We describe a certain class of them algebraically. Let $\Gamma$ be a geometry whose points are $(\infty)$, $(a)$, $(k, b)$ and $(a, l, a')$, for $a, a', k, l \in \mathbf{GF}(2^{2e})$, whose lines are $[\infty]$, $[k]$, $[a, l]$ and $[k, b, k']$, for $k, k', a, b \in \mathbf{GF}(2^{2e})$, and incidence is given by

$$[k, b, k'] \; \mathbf{I} \, (k, b) \; \mathbf{I} \, [k] \; \mathbf{I} \, (\infty) \; \mathbf{I} \, [\infty] \; \mathbf{I} \, (a) \; \mathbf{I} \, [a, l] \; \mathbf{I} \, (a, l, a'),$$

for all $a, a', b, k, k', l \in \mathbf{GF}(2^{2e})$, and by

$$(a, l, a') \; \mathbf{I} \, [k, b, k']$$

$$\Updownarrow$$

$$\begin{cases} a' &= k^{2^h} a + b, \\ k' &= ka + l. \end{cases}$$

It is an elementary calculation to verify that this defines a generalized quadrangle, using the results of HANSSENS & VAN MALDEGHEM [3], if and only if $(h, 2e) = 1$. Since in this case $(h, e) = 1$, we see that restricting coordinates to $\mathbf{GF}(q)$, we obtain a subquadrangle $\Gamma''$ which can be seen as the fix point structure of the involution $\theta$ obtained by applying the field automorphism $x \mapsto x^{2^e}$ on each coordinate of each element (and fixing $(\infty)$ and $[\infty]$). It is also an elementary calculation, using the description above of $\Gamma$ to verify that there is a subquadrangle $\Gamma'$ (necessarily unique) of order $(2^{2e}, 1)$ through any pair of lines $\{[k], [k]^\theta\}$, for which $k \in \mathbf{GF}(2^{2e}) \setminus \mathbf{GF}(2^e)$. Hence we can apply the previous theorem and obtain an ovoid $\mathcal{O}$ of $\Gamma''$. The explicit form of the ovoid is, after calculation,

$$\mathcal{O} = \{(\infty)\} \cup \{(a, l, l(k + k^{2^e})^{2^h-1} + a\frac{k^{2^e+2^h} + k^{1+2^{e+h}}}{k + k^{2^e}})|a, l \in \mathbf{GF}(2^e)\}.$$

**Remark 3** The construction of ovoids via involutions is in fact inspired by the situation in the classical case: the intersection of a standard embedded quadrangle with a non-tangent hyperplane yields either a subquadrangle or an ovoid. But in a quadratic extension, we always get a subquadrangle. This is the quadrangle $\Gamma'$ of the last theorem. The idea is to reverse the procedure, and start with $\Gamma'$, then restrict coordinates in $\Gamma$ with the aid of an involution and obtain an ovoid in the subquadrangle $\Gamma''$ over the subfield. A similar argument holds in case of hexagons.

# References

[1] **I. Bloemen, J. A. Thas & H. Van Maldeghem**, Translation ovoids of Generalized Quadrangles and Hexagons, *preprint*, 1996.

[2] **P. J. Cameron, J. A. Thas & S. E. Payne**, Polarities of generalized hexagons and perfect codes, *Geom Dedicata* **5** (1976), 525–528.

[3] **G. Hanssens & H. Van Maldeghem**, Algebraic properties of quadratic quaternary rings, *Geom. Dedicata* **30** (1989), 43–67.

[4] **S. E. Payne & J. A. Thas**, *Finite Generalized Quadrangles*, Pitman, Boston London Melbourne, 1984.

[5] **M. A. Ronan**, A geometric characterization of Moufang hexagons, *Invent. Math.* **57** (1980), 227–262.

[6] **J. A. Thas**, Polar spaces, generalized hexagons and perfect codes, *J. Combin. Theory Ser* A **29** (1980), 87–93.

[7] **J. A. Thas & S. E. Payne**, Spreads and ovoids in finite generalized quadrangles, *Geom. Dedicata* **52** (1994), 227–253.

[8] **J. Tits**, Sur la trialité et certains groupes qui s'en déduisent, *Inst. Hautes Études Sci. Publ. Math.* **2** (1959), 14–60.

Trends in Mathematics, © 1998 Birkhäuser Verlag Basel/Switzerland

# The Borel-Tits Property for Finite Groups

Satoshi Yoshiara

**Abstract**

A recursive method to determine $p$–radical subgroups is illustrated with the simple groups of characteristic–$p$ type, especially the Mathieu group $M_{24}$ for $p = 2$.

## 1 Introduction

An exposition will be given on a method for determining a certain simplicial complex $\mathcal{B}_p(G)$, associated with a finite group $G$ and each prime divisor $p$ of $|G|$, which can be used to obtain the additive structure of modulo $p$ cohomology ring of $G$. For a finite group $G$ of Lie type in characteristic $p$, a theorem by Borel and Tits implies that $\mathcal{B}_p(G)$ coincides with the complex of chains of $p$-unipotent radicals, i.e., the building associated with $G$. Suggested by this fact, an inductive method to compute $\mathcal{B}_p(G)$ for any finite group $G$ is proposed (see 3.6): It contains as its main part the verification of a property $(BT/\Delta)$ (see 3.4), which is an interpretation of a part of the conclusion of the Borel-Tits theorem in terms of complexes.

The aim of the paper is to explain how naturally we are led to the method and to illustrate how to apply it effectively, rather than to give a complete description of $\mathcal{B}_p(G)$ for each group $G$. For the latter purpose, I choose the simple groups of characteristic-$p$ type, especially the Mathieu group $M_{24}$ for $p = 2$, because the method works well when all $p$-local groups are $p$-costrained.

The main idea 3.5,3.6 was already contained in my recent joint paper with S. D. Smith [SY, 1.8], in which we verified the property $(BT/\Delta)$ for some sporadic groups and their geometries $\Delta$, but did not attempt to determine $\mathcal{B}_p(G)$ completely. In [SY], we rely on the classification of the maximal subgroups of each sporadic group to verify $(BT/\Delta)$, although more elementary direct proof is desirable for a unified treatment. A proof of such nature is given in this paper for $M_{24}$ using its "natural" module, the Golay code.

## 2 Alternating-sum decomposition of cohomology

The central object considered in this exposition is the simplicial complex associated with a well-known partially ordered set $\mathcal{B}_p(G)$. Formally it can be defined for any finite group $G$ and any prime divisor $p$ of $|G|$ as follows:

**Definition 2.1** $\mathcal{B}_p(G) := \{1 \neq U \leq G | U : p\text{-subgroup}, O_p(N_G(U)) = U\}$.

We regard $\mathcal{B}_p(G)$ as a poset with respect to inclusion in $G$. Abusing notation, we also use the same symbol $\mathcal{B}_p(G)$ to denote the associated simplicial complex (sometimes called the *order complex*) with this poset. That is, a simplex of $\mathcal{B}_p(G)$ is defined to be a subset $\{U_1, \ldots U_n\}$ of $p$-subgroups in the poset $\mathcal{B}_p(G)$ with $U_1 < \ldots < U_n$, and the face relation is defined by inclusion.

The simplicial complex $\mathcal{B}_p(G)$ is called the *p-radical* or *p-stubborn* complex of $G$. The second name is usually used by topologists, who have recognized this complex as an important object concerning its close relation with recent developements in group cohomology. As we can find, for example, in the book of Adem and Milgram [AM], many works have been recently done on determination of ring structure of modulo $p$ cohomology of sporadic finite simple groups, because they show exotic aspects of classifying spaces of the groups.

Usually the calculation of group cohomology is very complicated. However, on its additive structure, sometimes one can reduce the problem to those for smaller groups: For a finite group $G$ and a prime divisor $p$ of $|G|$, if we can find some *nice* simplicial complex $\Delta$ on which $G$ acts simplicially (that is, if $g \in G$ stabilizes a simplex $\sigma$ then $g$ fixes each vertex in $\sigma$), then the reduced cohomology $\tilde{H}(G, \mathbf{Z}_p)$ of the whole group $G$ with coefficients in the ring $\mathbf{Z}_p$ of $p$-adic integers can be decomposed as the alternating sum of cohomologies of the stabilizers of simplices (in the Grothendieck group of $p$-torsion abelian groups):

**2.2** *([We] Alternating-Sum Decomposition)*

$$\tilde{H}^n(G, \mathbf{Z}_p) = \sum_{\sigma \in \Delta/G} (-1)^{\dim(\sigma)} \tilde{H}^n(G_\sigma, \mathbf{Z}_p),$$

*where $n$ is any non-negative integer, and $G_\sigma$ is the stabilizer of a simplex $\sigma$, which ranges over the set $\Delta/G$ of representatives of $G$-orbits on the simplexes of $\Delta$.*

Once such a decomposition, and hence enough information on the additive structure of cohomology of the whole group $G$, is obtained from those for its subgroups $G_\sigma$, it seems not so much difficult for experts to guess the ring structure of cohomology by observing behavior of the squaring operator, etc. See [We, 7] for some simple examples of computations of cohomology based on the decomposition. Adem and Milgram applied the above decomposition to the 2-local geometries $\Delta$ of the sporadic groups $M_{22}$ and $McL$ in order to determine their cohomology rings over $\mathbf{Z}_2$ [AM2] [AM3]. For practical purposes, smaller complexes $\Delta$ are evidently easier to handle. Hence the following problems may be interesting to consider:

**Problem 2.3** *(a) Given an arbitrary finite group $G$ and a prime divisor $p$ of $|G|$, find a standard (hopefully small) simplicial complex $\Delta$ for which the decomposition of cohomology 2.2 holds, or give a sufficient condition for $(G, \Delta)$ to allow the decomposition.*

(b) *Find a smallest possible simplicial complex $\Delta$ for an explicit finite group $G$ and a prime divisor $p$ of its order so that the alternating-sum decomposition 2.2 holds.*

As for Problem (a) above, Webb [We] gave a nice sufficient condition by extracting the essence of arguments in the pioneering work by Brown and Quillen [Qu].

**Theorem 2.4** *[Th, 2.8], [We, Theorem A] Under the above notation, assume that*

> $(*)$ *for each element $g \in G$ of order $p$, the subcomplex $\Delta^g$ of $\Delta$ fixed by $g$ is acyclic (i.e. the rational reduced cohomology $H^n(\Delta^g, \mathbf{Z}) = 0$ for any integer $n \geq 0$).*

*Then we have the alternating-sum decomposition 2.2 of the cohomology $\tilde{H}(G, \mathbf{Z}_p)$.*

There are several known standard complexes for which the condition $(*)$ and hence the decomposition 2.2 holds: Consider the following posets and their ordered complexes (denoted by the same symbols) for a finite group $G$ and a prime divisor $p$ of $|G|$:

**Definition 2.5** $\mathcal{S}_p(G) := \{U | 1 \neq U : p\text{-subgroup of } G\}$,
$\mathcal{A}_p(G) := \{E \in \mathcal{S}_p(G) | E : \text{elementary abelian } p\text{-group}\}$,
$\mathcal{Z}_p(G) := \{E \in \mathcal{A}_p(G) | E = \langle \text{all elements of order } p \text{ of } Z(C_G(E)) \rangle\}$.

It is proved by Quillen, Bouc, etc. [Qu],[Be, 6.6] that their geometric realizations together with that of the $p$-stubborn complex $\mathcal{B}_p(G)$ are $G$-homotopy equivalent, and that the complex $\mathcal{S}_p(G)$ satisfies the condition $(*)$ above (see [Be, 6.7.3] and the remark after the statement of Theorem A in [We]). Hence all these complexes satisfy the condition $(*)$ and we can use any of them to obtain the alternating-sum decomposition of mod $p$ cohomology of $G$.

Apparently, it is quite complicated to determine $\mathcal{S}_p(G)$ or $\mathcal{A}_p(G)$. For example, if $G = SL_n(p^e)$, the dimension of $\mathcal{S}_p(G)$ is $(en(n+1)/2) - 1$, the number of composition factors of its Sylow group, and that of $\mathcal{A}_p(G)$ is $em(n-m)$ for $n = 2m$ or $n = 2m + 1$. Smaller complexes $\mathcal{B}_p(G)$ and $\mathcal{Z}_p(G)$ seem more convenient to handle. So we are naturally led to the following question:

**Problem 2.6** *Determine $\mathcal{B}_p(G)$ (or $\mathcal{Z}_p(G)$) for any finite (especially simple) group $G$ and a prime divisor $p$ of $|G|$ (particularly $p = 2$) up to conjugacy.*

This might not be a tractable problem for general finite groups. For example, we can verify by straightforward but tedious calculations that $\mathcal{Z}_2(G)$ for $G = M_{24}$ has exactly 19 classes, but the complications of the result (see [SY, 2.8]) seem to suggest that the size of $M_{24}$ may represent the practical limit for such simple methods. In contrast, it is well known that $\mathcal{B}_p(G)$ for a finite group of Lie type in characteristic $p$ can be neatly described by virtue of a theorem of Borel-Tits.

The main aim of this expository is to provide a reccursive method to solve the question about $\mathcal{B}_p(G)$ above for some groups $G$ in nature of characteristic $p$.

Let me conclude this section with mentioning recent contributions to the problems in 2.3: During the last few years, there has been much progress towards Problem(a) 2.3 by topologists using techniques in $p$-adic homotopy theory (see, e.g. [SY, Section 4]). Towards Problem (b), there is a joint work by S.D. Smith and the author [SY], which verifies for many sporadic simple groups $G$ that some ($p$-local) geometries (which are of small dimension, less than 4) are homotopy equivalent to $\mathcal{S}_p(G)$, and hence their cohomologies allow the alternating-sum decomposition 2.2.

# 3 The Borel-Tits property

Recall that the *Borel-Tits property* is a property for finite groups of Lie type, which can be stated as follows ([BW], [BT] for the original result, also see [Be, 6.8.4]).

**Theorem 3.1** *Let $\Gamma$ be a simple linear algebraic group defined over an algebraically closed field of characteristic $p \neq 0$, and $\rho$ be an endomorphism of $\Gamma$ onto itself for which the fixed subgroup $G := \Gamma^\rho$ is finite. Then for each non-trivial p-subgroup $U$ of $G$, there exists a parabolic subgroup $P$ of $G$ (that is, the fixed subgroup $\Pi^\rho$ under $\rho$ of a parabolic subgroup $\Pi$ of $\Gamma$ stabilized by $\rho$) such that $N_G(U) \leq P$ and $U \leq O_p(P)$.*

As a corollary of this theorem, we can easily obtain the $p$-stubborn complex $\mathcal{B}_p(G)$ for $G$ a finite group of Lie type. Observe that if $U \neq O_p(P)$ in the conclusion above, then $N_{O_p(P)}(U)$ is a normal $p$-subgroup of $N_G(U)$ properly containing $U$. This implies:

**Corollary 3.2** *(See also [Be, 6.8.4]) For a group $G$ above, the poset $\mathcal{B}_p(G)$ consists of the unipotent radicals $O_p(P)$ for all parabolics $P$.*

The above property 3.1 can be thought of as a property of a finite group $G$ of Lie type, concerning its action on the associated building $\Delta := \cup(G/P)$, where $P$ ranges over the parabolic subgroups containing a fixed Borel subgroup $B$. Namely, as a parabolic subgroup is just the stabilizer of certain simplex of the building $\Delta$, we can rephrase 3.1 as follows:

**3.3** *For each non-trivial p-subgroup $U$ of a finite group $G$ of Lie type in characteristic p, there exists a simplex $\sigma$ of the building $\Delta$ of $G$ such that $N_G(U) \leq G_\sigma$ and $U \leq O_p(G_\sigma)$.*

As we will observe later (see 3.5 below): in order to check the two inclusions for $U$ above, it suffices to consider $U$ in $\mathcal{B}_p(G)$; and furthermore, the former inclusion for $U \in \mathcal{B}_p(G)$ is equivalent to $O_p(G_\sigma) \leq U$, the opposite to the latter inclusion. Thus we will consider the following property, corresponding to the former inclusion.

**Definition 3.4** $(BT/\Delta)$: *For each non-trivial p-subgroup U of G, there exists a simplex σ of Δ such that* $N_G(U) \leq G_\sigma$.

The statement makes sense for any finite group $G$ acting on a simplicial complex (or geometry) $\Delta$. We will refer to this property as the *Borel-Tits property* for a finite group $G$. Note that we can always find such a complex $\Delta$ for any $G$, though they are huge in general: the ordered complex $\Delta$ of the poset of all maximal (or maximal $p$-local) subgroups clearly satisfies $(BT/\Delta)$. So the problem is whether we can find such a complex of reasonably small size. For a sporadic simple group $G$, we can in fact verify in many cases that a suitable ($p$-local) geometry $\Delta$ of small dimension satisfies the Borel-Tits property $(BT/\Delta)$ and allows the alternating sum decomposition of cohomology 2.2 [SY].

In general, there are several complexes $\Delta$ satisfying $(BT/\Delta)$ for a given $G$. For example, for $G = S_5 \cong P\Gamma L_2(4)$, besides the usual building for $L_2(4)$, which is just a 0-dimensional disconnected complex with 5 points, we have the following geometry $\bar{C}(5)$ of rank 2 for which $(BT/\Delta)$ is satisfied: The vertices of $\bar{C}(5)$ has two types: 5 letters and 10 triples of letters, with incidence given by inclusion. A simplex of $\bar{C}(5)$ is a pair of incident letter and triple. It is straightforward to check $(BT/\Delta)$ for this geometry $\Delta = \bar{C}(5)$. Note that $\bar{C}(5)$ is connected, so it is not homotopy equivalent to the building for $L_2(4)$.

Now we give some elementary but important remarks on the property $(BT/\Delta)$:

**Lemma 3.5** *Let G be a finite group acting simplicially on a complex Δ, and let p a prime divisor of* $|G|$.

(1) *For each non-trivial p-subgroup V of G, there is a p-subgroup* $U \in \mathcal{B}_p(G)$ *such that* $V \leq U$ *and* $N_G(V) \leq N_G(U)$. *In particular, it suffices to consider p-groups in* $\mathcal{B}_p(G)$ *for verifying the Borel-Tits property* $(BT/\Delta)$.

(2) *If* $(BT/\Delta)$ *is satisfied, then* $O_p(G_\sigma) \leq U$ *for each* $U \in \mathcal{B}_p(G)$, *where σ is a simplex of Δ with* $N_G(U) \leq G_\sigma$. *Furthermore, if* $U \neq O_p(G_\sigma)$, $U/O_p(G_\sigma)$ *is a member of the p-stubborn poset* $\mathcal{B}_p(G_\sigma/O_p(G_\sigma))$ *for the section* $G_\sigma/O_p(G_\sigma)$ *of G.*

*Proof.* (1) Let consider the following chain of subgroups: $U_0 := V$, $N_0 := N_G(V)$, and define $U_i$ and $N_i$ inductively by $U_i := O_p(N_{i-1})$ and $N_i := N_G(U_i)$ for $i = 1, 2, \ldots$. Evidently $U_{i-1} \leq U_i$ as $U_{i-1} \trianglelefteq N_{i-1}$, and $N_{i-1} \leq N_i$. Since $G$ is finite, the increasing chain of subgroups $V = U_0 \leq U_1 \leq \cdots$ terminates at some $U_n =: U$. This implies that $U = O_p(N_G(U))$, or equivalently $U \in \mathcal{B}_p(G)$. As $N_G(V) = N_0 \leq N_n = N_G(U)$, the claim follows.

(2) Set $P := O_p(G_\sigma)$. As $N_G(U) \leq G_\sigma$, $UP$ is normalized by $N_G(U)$. Then $U(N_G(U) \cap P)$ is a normal $p$-subgroup of $N_G(U)$, and so $U(N_G(U) \cap P) = U$, since $U = O_p(N_G(U))$. This implies $N_{UP}(U) = N_G(U) \cap UP = U(N_G(U) \cap P) = U$, and so $UP = U$, or equivalently $P \leq U$. This proves the former part of the claim. Now the latter is easy to see, because $N_{G_\sigma/P}(U/P) = N_{G_\sigma}(U)/P = N_G(U)/P$ and so $O_p(N_{G_\sigma/P}(U/P)) = O_p(N_G(U))/P = U/P$. □

Remark (2) above shows that we can determine the $p$-stubborn complex $\mathcal{B}_p(G)$ inductively, if we find a suitable simplicial complex $\Delta$ for which the Borel-Tits property $(BT/\Delta)$ holds. Note that even when we only find such a complex $\Delta$ with $O_p(G_\sigma) = 1$ for some simplex $\sigma$, it suffices to work with a non $p$-local but smaller group $G_\sigma$. Hence we have the following reccursive method to determine $\mathcal{B}_p(G)$.

**3.6** *(Method) To determine the p-stubborn complex $\mathcal{B}_p(G)$ for a finite group $G$,*
*(i) Find a (small) simplicial complex $\Delta$ for which the Borel–Tits property $(BT/\Delta)$ holds (and $O_p(G_\sigma) \neq 1$ for each simplex $\sigma$ of $\Delta$, if possible).*
*(ii) For each simplex $\sigma$ of $\Delta$, determine the p-stubborn complex $\mathcal{B}_p(G_\sigma/O_p(G_\sigma))$ of a section $G_\sigma/O_p(G_\sigma)$ of $G$.*
*(iii) Check whether each candidate in fact lies in $\mathcal{B}_p(G)$.*

## 4 Examples and some technical details

In this section, I explain how to apply the method 3.6 to obtain $\mathcal{B}_p(G)$ and how to verify that a given complex $\Delta$ satisfies the Borel-Tits property $(BT/\Delta)$, taking simple groups $G$ of characteristic-$p$ type (that is, the generalized Fitting subgroup $F^*(H)$ is a $p$-group for every $p$-local subgroup $H$ of $G$) as examples. For such a group, we can in fact always find a complex $\Delta$ which satisfies $(BT/\Delta)$ and homotopy equivalent to $\mathcal{B}_p(G)$ (see [SY, Section 2,3]). The most typical example of groups of characteristic-$p$ type is a finite simple group of Lie type in characteristic $p$ for which the $p$-stubborn complex is completely determined by 3.2. Thus we may assume that $G$ is not of that type.

In each exposition of [SY, Section 2,3], we (sometimes very briefy) verified the Borel-Tits property $(BT/\Delta)$ for suitable complex $\Delta$, but did not try to determine $\mathcal{B}_p(G)$ completely. Here I summarize the distributed information on $(BT/\Delta)$ in the form of the table below, and give a brief description of $\mathcal{B}_p(G)$ to each simple group $G$:

In the first and second columns of Table, we give a prime $p$ and the name of a simple group $G$ of characteristic-$p$ type. In the last two columns, a description of $\mathcal{B}_p(G)$ is given: the number of $G$-conjugacy classes on the poset $\mathcal{B}_p(G)$ (the vertices of $\mathcal{B}_p(G)$ as a simplicial complex) in the fourth column, and the dimension (the maximal length of chains) of the complex $\mathcal{B}_p(G)$ in the fifth column. The third column gives a description of a geometry $\Delta$ (usually with $p$-local simplex stabilizers) for which the Borel-Tits property $(BT/\Delta)$ is satisfied. Many of them are described in [RS], but for more references consult the relevant subsections in [SY, Section 2,3]. Here we give a structure of the stabilizer $G_v$ for each typical vertex $v$ of $\Delta$ in the form $A\backslash B$, which means $A$ is isomorphic to the kernel of the action of $G_v$ on the residue $Res(v)$ at $v$ and $B \cong G_v/K_v$.

We also sometimes describe the structure of the residue $Res(v)$ at a vertex $v$, under the follwoing convention: $GQ(s,t)$, $GH(s,t)$ and $PG(n,s)$ are a generalized quadrangle, hexagon and $d$-dimensional desarguesian projective space of order

$(s,t)$, $(s,t)$ and $s$, respectively; $PG(3,2)$/plane is a projective space with planes truncated; $\tilde{Q}(2,2)$ is the "2-split cover" of $Q(2,2)$; $C(k)$ means the circle geometry, that is, it consists of vertices and edges of the complete graph on $k$ vertices with incidence given by inclusion; $\bar{C}(5)$ is the geometry we met in §3; $X * Y$ means the join of two geometries $X$ and $Y$; and the 0-dimensional complex of just $n$ points is denoted by $n$.

Note that for a general simplex $F$ we can easily recognize the kernel $K_F$ on the residue $Res(F)$, because for a vertex $v$ in $F$, we have $K_v \trianglelefteq K_F$ and the factor group $K_F/K_v$ coincides with the kernel on the residue at $F$ in $Res(v)$, under the action of $G_F/K_v$.

Observe for each $\Delta$ in Table below and each vertex $v$ of $\Delta$, we have $O_p(G_v) = O_p(K_v)$. Except for a few cases, $O_p(K_v)$ is non-trivial.

As is proved in [SY], each geometry of $G$ appearing in this list is homotopy equivalent to $\mathcal{B}_p(G)$, and it is the smallest possible complex with such property and the Borel-Tits property. In that sense, these geometries can be thought of as "buildings" for these sporadic groups with respect to their nature in $p$, though some have trivial kernel or do not contain a Sylow $p$-subgroup.

## 4.1 Table of geometries $\Delta$ with $(BT/\Delta)$, and $\mathcal{B}_p(G)$

| $p$ | $G$ | $\Delta$ | | | | $\mathcal{B}_p(G)$ | |
|---|---|---|---|---|---|---|---|
| 2 | $L_3(3)$ | $2\backslash S_4$ | $2^2\backslash S_3$ | | | 3 | 1 |
| 2 | $U_4(3)$ | $2^4\backslash A_6$<br>$GQ(2,2)$ | $2_+^{1+4}\backslash S_3 \times S_3$<br>$3 * 3$ | $2^4\backslash A_6$<br>$GQ(2,2)$ | | 7 | 2 |
| 2 | $G_2(3)$ | $1\backslash G_2(2)$<br>$GH(2,2)$ | $2_+^{1+4}\backslash 3^2.2$<br>$3 * 3$ | $2^3\backslash L_3(2)$<br>$PG(2,2)$ | | 7 | 2 |
| 2 | $M_{11}$ | $2\backslash S_4$ | $2^2\backslash S_3$ | | | 3 | 1 |
| 2 | $M_{22}$ | $2^3\backslash L_3(2)$<br>$PG(2,2)$ | $2^4\backslash A_6$<br>$\tilde{Q}(2,2)$ | $2^4\backslash S_5$<br>$\bar{C}(5)$ | | 7 | 2 |
| 2 | $M_{23}$ | $2^3\backslash L_3(2)$ | $2^4\backslash A_7$ | $2^4.3\backslash S_5$ | | 7 | 2 |
| 2 | $M_{24}$ | $2^4\backslash L_4(2)$<br>$P(3,2)$/plane | $2^6\backslash S_3 \times L_3(2)$<br>$3 * PG(2,2)$ | $2^6 3\backslash S_6$<br>$GQ(2,2)$ | | 11 | 3 |
| 2 | $J_3$ | $2^4\backslash 3 \times L_2(4)$ | $2^{2+4}\backslash 3 \times S_3$ | $2_+^{1+4}\backslash A_5$ | | 4 | 2 |
| 2 | $J_4$ | $2^{10}\backslash L_5(2)$ | $2^{3+12}\backslash S_5 \times L_3(2)$ | $2_+^{1+12}.3\backslash M_{22}.2$ | $2^{11}\backslash M_{24}$<br>$\Delta$ for $M_{24}$ | 23 | 4 |
| 2 | $Co_2$ | $2_+^{1+6} \times 2^4\backslash L_4(2)$ | $2_+^{1+8}\backslash Sp_6(2)$ | $2^{4+10}\backslash(S_5 \times S_3)$ | $2^{10}\backslash M_{22}.2$<br>$\Delta$ for $M_{22}$ | 15 | 3 |
| 2 | $Th$ | $2_+^{1+8}\backslash A_9$ | $2^5\backslash L_5(2)$ | | | 16 | 4 |
| 3 | $U_5(2)$ | $3\backslash PSp_4(3)$<br>$GQ(3,3)$ | $(3 \times 3^{1+2}).2\backslash 2 \times A_4$<br>$2 * 4$ | $3^4\backslash S_5$<br>$C(5)$ | | 4 | 2 |
| 3 | $Ru$ | $3\backslash A_6.2^2$ | $3^2 2\backslash PGL_2(3)$ | | | 3 | 2 |
| 3 | $J_4$ | $3 \times 2\backslash M_{22}.2$ | $3^2.2 \times 2^2\backslash PGL_2(3)$ | | | 3 | 2 |
| 3 | $O'N$ | $3^2 4\backslash S_6$ | $3^4 2\backslash 2^4 D_{10}$ | | | 2 | 1 |
| 3 | $McL$ | $1\backslash \Omega_6^-(3)$<br>$GQ(3,9)$ | $3^4\backslash M_{10}$<br>$2 * 10$ | $3^{1+4}2\backslash S_5$<br>$C(5)$ | | 3 | 1 |
| 3 | $Ly$ | $3\backslash McL.2$ | $3^{2+4}.2\backslash A_5.D_8$ | $3^5.2\backslash M_{11}$<br>$C(11)$ | | 4 | 2 |
| 5 | $Ly$ | $1\backslash G_2(5)$<br>$GH(5,5)$ | $5^{1+4}.4\backslash S_6$<br>$6 * 6$ | $5^3\backslash L_3(5)$<br>$PG(2,5)$ | | 4 | 2 |
| 5 | $Th$ | $5^{1+2}.4\backslash S_4$ | $5^2.4\backslash L_2(5)$ | | | 2 | 1 |

## 4.2 How to obtain $\mathcal{B}_p(G)$ from column for $\Delta$ in Table

From the column for $\Delta$ in Table 4.1, we can find the $p$-stubborn complex $\mathcal{B}_p(G)$ for each $G$. I will explain this for $G = M_{24}$. Table 4.1 shows that the 2-local geometry $\Delta$ of $M_{24}$ satisfies the Borel-Tits property $(BT/\Delta)$. (We will give a proof of this fact later in 4.4.) Recall that $\Delta$ has three types of vertices, *octads*, *trios* and *sextets* in the Steiner syetem $S(5,8,24)$, with incidence among them defined by natural subdivision, so that a simplex of $\Delta$ consists of mutually incident vertices. (I assume some familiarity with these neat combinatorial objects: e.g. refer to [As, Chap.6].)

Take any $P \in \mathcal{B}_2(M_{24})$. As $(BT/\Delta)$ is satisfied and $O_2(G_F) = O_2(K_F)$ for every flag $F$ of $\Delta$, it follows from 3.5(2) that $P$ contains $O_2(K_F)$ for some simplex $F$, and so $O_2(K_v)$ for some $v$ which is an octad, trio, or sextet. For brevity, we use the letters $O$, $T$ and $\Sigma$ to denote typical octad, trio, and sextet, respectively, which form a maximal simplex.

If $O_2(K_O) = K_O \trianglelefteq P$, $P/K_O \in \mathcal{B}_2(L_4(2))$ or $K_O = P$ by 3.5(2). Now we know $\mathcal{B}_2(L_4(2))$ by 3.2, as $L_4(2) \cong G_O/K_O$ is a group of Lie type. Thus $P/K_O$ is conjugate to one of the $2^3 - 1$ unipotent radicals of standard parabolics of $L_4(2)$. They can be described as $O_2(K_X)/K_O$ for a simplex $X$ with $O \in X \subseteq \{O, T, \Sigma, \square\}$, where $\square$ denotes a typical vertex of the truncated type in $Res(O)$, the truncated building for $L_4(2)$. (It corresponds to the "square" node in the usual diagram of $\Delta$.)

If $P$ does not contain $K_O$, either $O_2(K_T) = K_T \trianglelefteq P$ or $O_2(K_\Sigma) \trianglelefteq P$. Since $G_\Sigma/K_\Sigma \cong Sp_4(2)$ is a group of Lie type acting on the generalized quadrangle $Res(\Sigma)$ of order $(2,2)$, in the latter case $P = O_2(K_\Sigma)$ or $P$ is conjugate to $K_{\{T,\Sigma\}}$. (Note the other possible unipotent radical contains $K_O$.) Now in the remaining case, we have $P = K_T$, as each non-trivial 2-subgroup of $G_T/K_T = G_{T,O}G_{T,\Sigma}/K_T \cong L_2(2) \times L_3(2)$ is conjugate to the quotient groups of the 2-groups we already obtained, and so they contain $K_O$ or $K_\Sigma$.

We can verify that every 2-subgroup we obtained in fact lies in $\mathcal{B}_2(G)$. Thus,

> $\mathcal{B}_2(M_{24})$ consists of the $8 + 3 = 11$ conjugacy classes with representatives $U_F := O_2(K_F)$ for a non-empty subset $F$ of $\{O, T, \Sigma, \square\}$ with $F \cap \{O, \square\} \neq \{\square\}$.

Similary we can easily obtain the candidates from Table 4.1, but the last step of 3.6 usually requires a bit messy calculation of their normalizers (and their inclusions).

## 4.3 How to check the result

The complex $\Delta = \mathcal{B}_p(G)$ satisfies the condition $(*)$ in 2.4, which implies the projectivity of the *reduced Lefschetz module* $\tilde{L}(\Delta)$ of $\Delta$, given by the alternating sum of chain spaces of $\Delta$ (e.g.[Th]). Then the dimension $\tilde{\chi}(\Delta)$ of the vertical $G$-module $\tilde{L}(\Delta)$ (called the *reduced Euler characteristic* of $\Delta$) can be divisible by the

$p$-part $|G|_p$ of $|G|$. On the other hand, it is by definition just the alternating sum of the numbers of $i$-simplices of $\Delta$:

For $\Delta$ with condition $(*)$ in 2.4, $\tilde{\chi}(\Delta)$ is divisible by $|G|_p$, where

$$\tilde{\chi}(\Delta) = -1 + \sum_{i=0}^{\dim(\Delta)} (-1)^i |\{ i\text{-simplices of } \Delta\}|$$

Independently calculating $\tilde{\chi}(\Delta)$ can help to detect possible errors in our determination of $\mathcal{B}_p(G)$, though it does not guarantee that the answer is correct. Let check our earlier list $\mathcal{B}_2(M_{24})$. As a poset, it consists of 11 classes, with certain representatives $U_F$. We can see the following inclusion relations hold among them (see Figure 1 below). Note $U_{F_1} \subset \cdots \subset U_{F_n}$ is a chain iff $F_1 \subset \cdots \subset F_n$. In Figure 1, for each $U_F$, $N_G(U_F)$ and the index $[G : N_G(U_F)]$ are given, where $G = M_{24}$, $a := 3^2 \cdot 5 \cdot 11 \cdot 23$, and the symbol $\frac{B}{A}$ means that $U_F$ is of order indicated by $A$ and $B \cong N_G(U_F)/U_F$. Note that $G_F = N_G(U_F)$ in this case, and so the index $[G : N_G(U_F)]$ is easily obtained as the number of flags of type $F$. For example, if $F = \{O, \Sigma, \square\}$, the index is given by $|(\text{octads})| \cdot |(\text{lines in } Res(O) \cong PG(3,2))| \cdot |(\text{planes in } Res(O) \text{ containing a line})| = 759 \cdot 35 \cdot 3 = 3a$.
Other calculation can be similarly performed.

Figure 1: Inclusions of $U_F$, representatives of $\mathcal{B}_2(M_{24})$.

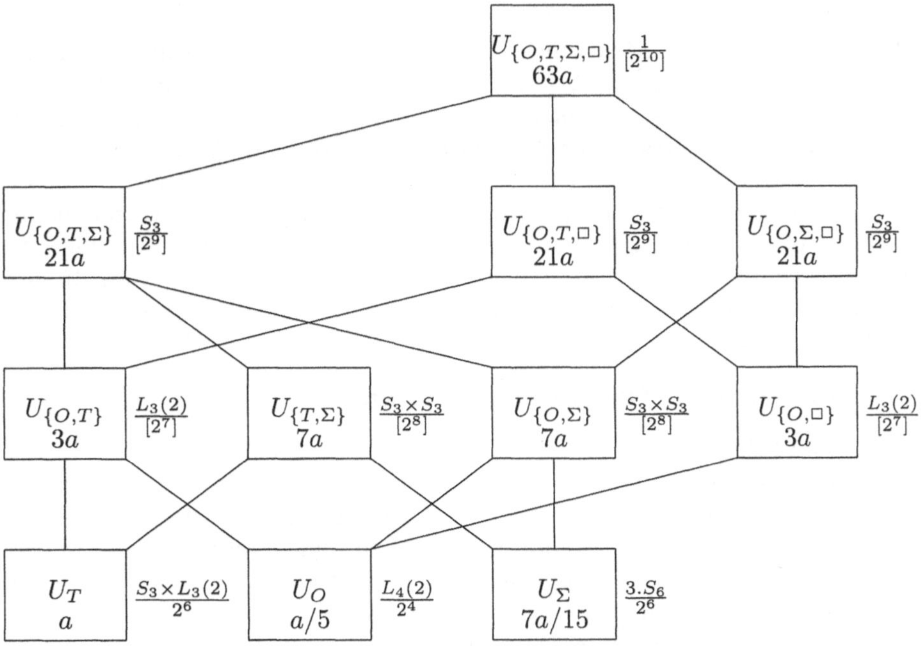

The simplicial complex $\mathcal{B}_2(M_{24})$ consists of the conjugate of possible chains of the lattice above. Note the stabilizer of a chain with top term $U_{F_n}$ is given by the index $[G : N_G(U_{F_n})]$, as we can verify that $U_{F_1} \subset \cdots \subset U_{F_n}$ implies $G_{F_1} = N_G(U_{F_1}) \supset \cdots G_{F_n}$ in this case. From Figure 1 above we can see the longest chain is of length 3, and so $\mathcal{B}_2(M_{24})$ is of dimension 3. Moreover, for example, there are 6 possible types $F$ included in $\{O, T, \Sigma\}$ and each type of size 2 contains 2 types of size 1, and so there are 6 types of 1-chains and $2+2+2$ types of 2-chains with top term conjugate to $U_{\{O,T,\Sigma\}}$.

Similarly, we can verify the following: There are 2 (resp. 1) 1-chains of types with top $\{O, T\}$, $\{T, \Sigma\}$, $\{O, \Sigma\}$ (resp. $\{O, \square\}$), 6 (resp. 4) 1-chains of types with top $\{O, T, \Sigma\}$ (resp. $\{O, T, \square\}$ and $\{O, \Sigma, \square\}$); there are $2+2+2$ (resp. $2+1$ and $2+1$) 2-chains of types with top $\{O, T, \Sigma\}$ (resp. $\{O, T, \square\}$ and $\{O, \Sigma, \square\}$). Therefore there are 10 1-chains, $2+2+2+1+6+4+4 = 21$ 2-chains, and $6+3+3$ 3-chains of types with top $\{O, T, \Sigma, \square\}$.

Thus there are $2 \cdot 3a + 2 \cdot 7a + 2 \cdot 7a + 1 \cdot 3a + 6 \cdot 21a + 4 \cdot 21a + 4 \cdot 21a + 10 \cdot 63a = 331a + 10 \cdot 63a$ 1-simplices, $(6+3+3) \cdot 21a + 21 \cdot 63a =$ 2-simplices, and $12 \times 63a$ 3-simplices of $\mathcal{B}_2(M_{24})$. Also the number of 0-chains (vertices) is $63a + 21a + 21a + 21a + 3a + 7a + 7a + 3a + a + a/5 + (7/15)a = 84a + (2/3)a + 63a$.

Hence the reduced Euler characteristic $\tilde{\chi}(\mathcal{B}_2(M_{24}))$ is given by $-1 + 331a - 252a + 84a + (2/3)a + (1 - 10 + 21 - 12) \cdot 63a = 21504 = 2^{10} \cdot 21$, which is certainly divisible by $2^{10} = |M_{24}|_2$.

As we saw above, the complex $\mathcal{B}_2(M_{24})$ has 11 classes of vertices and is of dimension 3. So, though this complex is one of the standard complex and can be completely determined, it is still large, especially when it is compared with the 2-local geometry $\Delta$ for $M_{24}$, which has just 3 classes of vertices and of dimension 2. However, they turn out to be quite similar from topological point of view.

Remark that the above Euler characteristic of $\mathcal{B}_2(M_{24})$ coincides with that for the 2-local geometry $\Delta$ for $M_{24}$, because the simplex $\Delta$ is a subcomplex of $\mathcal{B}_2(M_{24})$ with vertices of type $F$ without including $\square$, and then we can easily check in the computation of $\tilde{\chi}(\mathcal{B}_2(M_{24}))$ above that the terms correponding to simpleces of type including $\square$ are cancelled out to 0. This strongly suggests the homotopy equivalence of $\mathcal{B}_2(M_{24})$ with the 2-local geometry $\Delta$ of $M_{24}$. In fact, it is true as we verified in [SY, 2.8]. The equivalence was first observed by Ronan.

## 4.4 How to verify the Borel-Tits property

In the rest of this exposition, we give some technical details for the verification of the Borel-Tits property $(BT/\Delta)$. One simple way is to quote the classification of maximal subgroups of each simple group $G$, and examine each possible inclusion of a $p$-local. This method was adopted in [SY] to some larger sporadics, though it is not so neat. For smaller groups, a method based on [SY, Lemma 1.7] is used.

In fact, as an elementary proof of the theorem of Borel-Tits given in [Be, 6.8.4] suggests, we could use a "natural" module for some sporadics. Here I give an elementary verification of the Borel-Tits property $(BT/\Delta)$ for the Mathieu group $M_{24}$ and its 2-local geometry $\Delta$.

**Lemma 4.5** *The 2-local geometry $\Delta$ of $M_{24}$ satisfies the Borel–Tits property $(BT/\Delta)$.*

*Proof.* Let $\Omega$ be a set of 24 letters with a set $\mathcal{O}$ of 8-subsets, called *octads*, so that $(\Omega, \mathcal{O})$ forms a Steiner system $S(5, 8, 24)$, and set $G := Aut(\Omega, \mathcal{O}) \cong M_{24}$. The *Golay code* is the subspace of the power set $\mathcal{P}(\Omega)$ (recognized as a vector space with respect to the symmetric difference) generated by $\Omega$. We denote by $V$ the quotient module of the Golay code by the trivial $G$-subspace generated by $\Omega$. The group $G$ has two orbits on the nonzero vectors of $V$, one corresponds to the octads and the other to the pairs of complementary dodecads. We sometimes identify an octad (in fact the pair of an octad and the complementary 16-ad) with the corresponding vector of $V$.

Let $P$ be a non-trivial 2-subgroup of $G$. Since $N_G(P)$ normalizes the subgroup generated by the involutions of $Z(P)$, we may assume that $P$ is an elementary abelian group in order to show the Borel-Tits property. Note that $G$ has two classes of involutions of permutation types $1^8 2^8$ and $2^{12}$ on $\Omega$.

Assume that $P$ contains an involution $\sigma$ of the latter type. It is a standard fact that there is a unique sextet $\Sigma = \{T_i | i = 1, \ldots, 6\}$ for which $\sigma$ fixes all the teterads $T_i$ of $\Sigma$, and that the 15 octads $T_i \cup T_j$ exhaust the octads stabilized by $\sigma$. As $P$ centralizes $\sigma$, it permutes these 15 octads, so it fixes at least one such octad, and hence the subspace $V(P)$ of $V$ generated by the octads fixed by $P$ is a non-trivial subspace of the 4-subspace $V(\Sigma) := \langle T_i \cup T_j | 1 \leq i < j \leq 6 \rangle$ corresponding to the sextet $\Sigma$. We may introduce a non-degenerate symplectic bilinear form $f$ on $V(\Sigma)$ by $f(X, Y) := |X \cap Y|/4 \bmod 2$, and isotropic subspaces with respect to $f$ are conjugate under the action of $G_\Sigma / K_\Sigma \cong Sp_4(2)$ to a 1(resp. 2)-space corresponding to an octad (resp. trio). Moreover, $V(\Sigma)$ is the unique sextet space containing any of its non-degenerate 2-subspace, because its generating octads intersects in a tetrad which uniquely determines such a sextet. Since $N_G(P)$ acts on the subspace $V(P)$ and so on its radical, it is conjugate to a subgroup of $G_\Sigma$, if $V(P)$ is a non-degenerate 2 or 4-subspace of $V(\Sigma)$, and to a subgroup of $G_O$ or $G_T$ otherwise. The property $(BT/\Delta)$ is verified in this case.

Thus we may assume that all involutions of $P$ are of type $1^8 2^8$. Let $\Phi(\sigma)$ be the octad of letters fixed by an involution $\sigma$ of type $1^8 2^8$, and $V(P)$ be the subspace of $V$ generated by $\Phi(\sigma)$ for all involutions $\sigma$ in $P$. For any distinct involutions $\sigma$, $\tau$ of $P$, the commutativity implies $\tau$ acts on $\Phi(\sigma)$, and so $\tau$ induces on $\Phi(\sigma)$ a permutation of type $1^8$, $1^4 2^2$ or $2^4$, as $G_O/K_O$ induces $A_8$ on an octad $O$. Then $|\Phi(\sigma) \cap \Phi(\tau)| = 0$, 4 or 8. As this holds for any pair of octads generating $V(P)$, all nonzero vectors of $V(P)$ are octads, and $|X \cap Y| = 0$ or 4 for any distinct octads $X, Y$ of $V(P)$. That is, each pair of octads in $V(P)$ is contained in a trio, or a sextet. If $V(P)$ is contained in a sextet, as we saw in the above paragraph, $V(P)$ is an octad, a trio, or it is contained in a unique sextet, and hence $N_G(P)$ lies in the stabilizer of an octad, a trio, or a sextet, as $N_G(P)$ acts on $V(P)$.

Thus we may assume that there is no sextet space containing $V(P)$. In particular, $V(P)$ contains octads $x, y$ with $|x \cap y| = 4$. Let $\Sigma = \{T_i | i = 1, \dots, 6\}$ be the sextet determined by $x \cap y$. We may assume $x = T_1 \cup T_2$ and $y = T_1 \cup T_3$. As $V(P)$ is not contained in the sextet space $V(\Sigma)$, there is an octad $z \in V(P) - V(\Sigma)$. Since $|z \cap x| = 0$ or 4, and $z \notin V(\Sigma)$, it intersects the tetrads of $\Sigma$ in $\{2^4, 0^2\}$ and $|x \cap z| = 4$. Similarly $|y \cap z| = 4$. Thus $|z \cap T_i| = 2$ for all $i = 1, 2, 3$, and there is a unique index $j(z) \in \{4, 5, 6\}$ such that $|z \cap T_{j(z)}| = 2$. Note that for any $z \in V(P) - V(\Sigma)$, $j(z)$ is constant, for if $j(w) \neq j(z)$ for some octad $w$ of $V(P)$ outside $V(\Sigma)$ we conclude that $z + w$ (or its complement) would intersect the tetrads of $\Sigma$ in $\{42^2 0^3\}$, applying the conclusion on $z$ above to $w$ and noticing $|z \cap w| = 0$ or 4. Thus we may take $j(z) = 4$ for all $z \in V(P) - V(\Sigma)$. Then $O := T_5 \cup T_6$ is the unique octad (not necesarily in $V(P)$) intersecting $x$, $y$ and $z$ trivially for every octad $z \in V(P) - V(\Sigma)$. Note that there is no octad $w \in V(P) \cap V(\Sigma)$ with $|\Phi(w) \cap O| = 4$, for otherwise $|z \cap w| = 2$ for each $z \in V(P) - V(\Sigma)$. Hence $O$ is the octad uniquely determined by $V(P)$ with the property that $O \cap \Phi(z) = \emptyset$ or $O = \Phi(z)$ for every nonzero $z \in V(P)$. Since $N_G(P)$ acts on $V(P)$, it stabilizes the octad $O$. Thus $(BT/\Delta)$ is verified in this case too. □

As for geometries of larger groups like $J_4$ and $Co2$, no illuminating explanation is available so far about the reason why they satisfy the Borel-Tits poperty. Furthermore, for several larger sporadics their $p$-stubborn complexes are not completely determined, though we have some conjectures [SY, Section 4].

# References

[AM] A. Adem and R. J. Milgram, *The Cohomology of Finite Groups*, *Grundlehren der mathematischen Wissenschaft* **309**, Springer-Verlag, Berlin–New York, 1994.

[AM2] A. Adem and R. J. Milgram, The cohomology of the Mathieu group $M_{22}$, *Topology* **34** (1995), 389–410.

[AM3] A. Adem and R. J. Milgram, The cohomology of the McLaughlin group and some associated groups, to appear in *Math. Z.*.

[As] M. Aschbacher, *Sporadic Groups*, Cambridge tracts in Math. **104**, Cambridge U. Press, Cambridge, 1994.

[Be] D. Benson, *Representations and Cohomology: Cohomology of Groups and Modules*, *Cambridge studies in advanced mathematics* **31**, Cambridge U. Press, Cambridge, 1991.

[BT] A. Borel and J. Tits, Eléments unipotents et sousgroupes paraboliques des groups réductives, *Inv. Math.*, **12** (1971), 97–104.

[BW] N. Burgoyne and C. Williamson, On theorem of Borel and Tits for finite Chevalley groups, *Arch.Math.* **27** (1976), 489–491.

[Qu] D. G. Quillen, Homotopy properties of the poset of non-trivial $p$-subgroups of a group, *Adv. in Math.*, **28** (1978), 101–128.

[RS] M. A. Ronan and S. D. Smith, 2-local geometries for some sporadic groups, pp.283–289, *The Santa Cruz Conference on Finite Groups*, eds. B. Cooperstein and G. Mason, *Proc. Symp. Pure Math.*, **37** (1980), Providence RI, Amer. Math. Soc.

[RSY] A. Ryba, S. Smith, and S. Yoshiara, Some projective modules determined by sporadic geometries, *J. Algebra*, **129** (1990), 279–311.

[SY] S.D. Smith and S. Yoshiara, Some homotopy equivalences for sporadic geometries, to appear in *J. Algebra*.

[Th] J. Thévenaz, Permutation representations arising from simplicial complexes, *J. Comb. Th. A*, **46** (1987), 121–155.

[We] P. Webb, A local method in group cohomology, *Comment. Math. Helv.*, **62** (1987), 135–167.

Satoshi Yoshiara,
Division of Mathematical Sciences,
Osaka Kyoiku University,
Kashiwara, Osaka 582, Japan.
e-mail: `yoshiara@cc.osaka-kyoiku.ac.jp`

Trends in Mathematics, © 1998 Birkhäuser Verlag Basel/Switzerland

# Generalized Reflection Groups I. Constructing the Algebra

François Zara


**Abstract**

Let $G$ be a group generated by a set $X$ of involutions such that the product of two elements of $X$ is finite. We define a graph structure on $X$ and call the corresponding graph $\Gamma$ by: $\{x, y\}$ is an edge of $\Gamma$ if the order of $xy$ is $\geq 3$. We suppose that $\Gamma$ is connected. Let $K$ be a ring and let $G$ act on the $K$–module $M$. Suppose that the following condition is satisfied:

($MG$) let $x, y$ in $X$, $r := xy$ of order $n$ and let $V := [M, x] + [M, y]$; if $n \geq 3$, then $P_n(r)_{|V} = 0$; if $n = 2$, then $(Id_M + r)_{|V} = 0$, where $P_n(Y) := \frac{Y^n-1}{(Y-1)(Y+1)^f}$ and $(Y+1)^f$ is the greatest power of $Y+1$ which divides $(Y^n - 1)/(Y - 1)$.

We show that if $x, y$ are in $X$, then $[M, x]$ and $[M, y]$ are isomorphic $K$–modules and there is a canonical $K$–algebra $L$ such that for each $x$ in $X$, $[M, x]$ is an $L$–module in a natural way.


## 1 Introduction

Let $K$ be a field and $E$ be a $K$–vector space. An element $x$ of $GL(E)$ is a "reflection" of $E$ if $x^2 = Id_E$ and if $\ker(x - Id_E)$ is a hyperplane of $E$ (equivalently if $\dim(\mathrm{Im}(x - Id_E)) = 1$). A subgroup $G$ of $GL(E)$ is a "reflection group" on $E$ if $G$ is generated by reflections.

If E is finite dimensional and if $Q : E \to K$ is a non degenerate quadratic form, then the orthogonal group $O(Q)$ (group of isometries of $Q$) is a reflection group (with a unique exception: $K = \mathbb{F}_2$, $\dim(E) = 4$ and the Witt index of $Q$ is 2, theorem of Cartan-Dieudonné).

If $K$ is one of the fields $\mathbb{R}$, $\mathbb{C}$ or $\mathbb{H}$ (quaternions field), the finite reflection groups have been classified by :

- if $K = \mathbb{R}$, Coxeter ([2], 1934)
- if $K = \mathbb{C}$, Shephard and Todd ([4], 1953)
- if $K = \mathbb{H}$, Cohen ([1], 1980)

Let G be a group generated by a set $X$ of involutions, such that if $x$ and $y$ are in $X$, then $xy$ *is* of finite order. Let $R$ be a ring and let $M$ be an $RG$–module. We impose an axiom pertaining to the action of $G$ on $M$ (axiom $(MG)$). The goal of this work is to show that, at least in particular cases, in $End([M,x]), (x \in X)$ we can find an $R$–algebra $L$ such that $G$ becomes a reflection group when $M$ is considered as an $L$–module : here a reflection is a linear map $x$ of the $L$–module $M$ into itself such that $x^2 = Id_M$ and $\mathrm{Im}(x - Id_M)$ is of rank one, that is, generated by *one element* as an $L$–module. This algebra is given by generators and relations. Of course, in most of the cases, the conditions imply that $M = \{0\}$. Now, starting from a representation of $L$ we can construct an $(R, L)$–bimodule $M$ which is a free $R$–module, by defining the way each $x \in X$ acts, so that $G$ becomes a reflection group when M is considered as an $L$–module. Applied when $K = \mathbb{C}$ or $\mathbb{H}$ we may reobtain some of the results of Shephard and Todd or Cohen (if we know a good presentation for the group).

If the order of $xy$, $x, y$ in $X$ is $1, 2$ or $3$, the axiom $(MG)$ has been applied with great success to the study of Fischer groups (see Zara [5] and [6] and below).

## 2 Some properties of reflections

**Definition 1** *Let $K$ be a field and let $E$ be a $K$–vector space.*

1. *An element $x$ of $GL(E)$ is a "reflection" of $E$ if $x^2 = Id_E$ and if $\ker(x - Id_E)$ is a hyperplane of $E$, equivalently if $\dim(\mathrm{Im}(x - Id_E)) = 1$.*
2. *A subgroup $G$ of $GL(E)$ is a "reflection group" on $E$ if $G$ is generated by reflections.*

**Notation.** (a) If a group $G$ acts on an abelian group $E$ and if $g$ is in $G$, we define $[E, g] := \{v - g(v) \mid v \in E\}$, and $[E, G] =< [E, g] \mid g \in G >= \sum_{g \in G}[E, g]$.
(b) If $G =< X >$, it is easy to see that $[E, G] = \sum_{x \in X}[E, x]$.
(c) Let $n$ be an integer and let $K$ be a field. Let $(X + 1)^f$ be the greatest power of $(X + 1)$ which divides $\frac{X^n - 1}{X - 1}$ in $K[X]$. We define:

$$P_n(X) := \frac{X^n - 1}{(X - 1)(X + 1)^f}$$

By construction $-1$ is not a root of $P_n(X)$.
(d) If $K$ is a field, we call $\chi$ $(= \chi(K))$ the characteristic of $K$.

**Proposition 2** *Let $n$ be an integer $\geq 3$ and let $K$ be a field. The following two conditions are equivalent:*

1. *$\chi(K) = 2$ and $n$ is a power of $2$;*
2. *$P_n(X) = 1$.*

**Proof.** An easy exercise. □

**Remark 1** • *If $n$ is odd, then*

$$P_n(X) = X^{n-1} + X^{n-2} + \cdots + X + 1.$$

- *If $\chi(K) = p$ with $p$ odd and if $n = 2p$, then*

$$P_n(X) = X^{p-1} + X^{p-2} + \cdots + X + 1.$$

We study now the product of two reflections.

**Proposition 3** *Let $K$ be a field and let $E$ be a $K$-vector space. Let $x$ and $y$ be two distinct reflections of $E$ and suppose that $r := xy$ is of finite order $n$. Let $V := [E, x] + [E, y]$ and let $r\prime := r_{|V}$. Then we have $P_n(r\prime) = 0$ except when $\chi(K) = p \neq 0$ and $n = 2p$.*

**Proof.** There exist $a$ and $b$ in $E$ and two linear forms $\varphi$ and $\psi$ such that $\varphi(a) = \psi(b) = 2$ and

$$\forall v \in E,\ x(v) = v - \varphi(v)a,\ y(v) = v - \psi(v)b$$

We put $\alpha := \psi(a)$, $\beta := \varphi(b)$ and $\gamma := \alpha\beta$. We have $V =< a, b >$ and $r(a) = (\gamma - 1)a - \alpha b$, $r(b) = \beta a - b$ and if $w$ is in $E$, $r(w) = w + (-\varphi(w) + \beta\psi(w))a - \psi(w)b$. Let us distinguish two cases.

- $V$ is one dimensional. We can choose $b = a$. We have $\alpha = \beta = 2$ and if $w$ is in $E$, $r(w) = w - (\varphi - \psi)(w)a$. As $x \neq y$ we have $\varphi \neq \psi$ and $(\varphi - \psi)(a) = 0$ so $r$ is a transvection: $\forall m \in \mathbb{Z}$,

  $$r^m(w) = w - m(\varphi - \psi)(w)a.$$

  If $\chi(K) = 0$, $r$ is of infinite order and this case is excluded by hypothesis; If $\chi(K) = p$, $r$ is of order $p$. If $p$ is odd, $\forall w \in E$,

  $$(Id_E + r + \cdots + r^{p-1})(w) = pw - \frac{p(p-1)}{2}(\varphi - \psi)(w)a = 0;$$

  if $p = 2$, $(Id_E + r)(a) = (\varphi - \psi)(a) = 0$.

- $V$ is two dimensional. The characteristic polynomial of $r'$ is $Q(Y) = Y^2 - (\gamma - 2)Y + 1$. Let us call $\theta$ a root of $Q(Y)$ in an extension of $K$. The other root of $Q(Y)$ is $\theta^{-1}$ and $\theta + \theta^{-1} = \gamma - 2$. $Q(Y)$ has a double root if and only if $\gamma \in \{0, 4\}$. If $\gamma = 0$, then $\theta = \theta^{-1} = -1$ and if $\gamma = 4$, then $\theta = \theta^{-1} = 1$. We have now three cases depending on the value of $\gamma$.

- $\gamma \notin \{0, 4\}$. An easy calculation shows that, $\forall m \in \mathbb{N}$,

  $$r'^m = \frac{\theta^m - \theta^{-m}}{\theta - \theta^{-1}} r' - \frac{\theta^{m-1} - \theta^{1-m}}{\theta - \theta^{-1}} Id_V,$$

and, $\forall m \in \mathbb{N}, \forall v \in E$,

$$\begin{aligned} r^m(v) &= v + \left[\frac{1}{4-\gamma}(\theta^m + \theta^{-m} - 2)\varphi(v)\right] a \\ &+ \left[\frac{\beta}{\gamma(4-\gamma)}(\theta^m(\theta+1) + \theta^{-m}(\theta^{-1}+1) - \gamma)\psi(v)\right] a \\ &+ \left[\frac{\alpha}{\gamma(4-\gamma)}(\theta^m(\theta^{-1}+1) + \theta^{-m}(\theta+1) - \gamma)\varphi(v)\right] b \\ &+ \left[\frac{1}{4-\gamma}(\theta^m + \theta^{-m} - 2)\psi(v)\right] b. \end{aligned}$$

We have, clearly, $r'^n = Id_V \iff \theta^n = 1 \iff r^n = Id_E$ as $\gamma = \theta + \theta^{-1} - 2$. If we suppose, as we may, that $\theta$ is in $K$, then there is a basis $(c, d)$ of $V$ such that $r'(c) = \theta c$ and $r'(d) = \theta^{-1} d$. We have $\theta^n - 1 = (\theta - 1)(\theta + 1)^f P_n(\theta) = 0 = (\theta^{-1} - 1)(\theta^{-1} + 1)^f P_n(\theta^{-1})$ so, as $\theta \neq 1$ and $\theta \neq -1$, we obtain $P_n(\theta) = P_n(\theta^{-1}) = 0$. We see, in this case, that $P_n(r) = 0$.

- $\gamma = 4 = \alpha\beta$. We have, $\forall m \in \mathbb{N}$,

$$r'^m = mr' - (m-1)Id_V$$

and $\forall m \in \mathbb{N}, \forall v \in E$,

$$\begin{aligned} r^m(v) &= v + \left[-m^2\varphi(v) + \beta\frac{m(m-1)}{2}\psi(v)\right] a \\ &+ \left[\alpha\frac{m(m-1)}{2}\varphi(v) - m^2\psi(v)\right] b \end{aligned}$$

so, if $\chi(K) = 0$, $r$ is of infinite order; if $\chi(K) = p$ $(p \neq 2)$, then $r'^p = Id_V$, $r^p = Id_E$, $r$ is of order $p$ and it is clear that $P_p(r) = 0$ in this case; if $\chi(K) = 2$, then $r'^2 = Id_V$ and $r^2(v) = v + \beta\psi(v)a + \alpha\varphi(v)b$, $r^4 = Id_E$. As $\gamma = 0 = \alpha\beta$, $\alpha$ or $\beta$ is 0. If $0 = \alpha = \beta$, $r^2 = Id_E$; if $(\alpha, \beta) \neq (0, 0)$, then $r^2$ is a reflection of $E$.

- $\gamma = 0 = \alpha\beta$ and we can suppose that $\chi(K) \neq 2$. We have, $\forall m \in \mathbb{N}$,

$$r'^m = (-1)^{m-1}mr' + (-1)^{m-1}(m-1)Id_V$$

and $\forall m \in \mathbb{N}, \forall v \in E$,

$$\begin{aligned} r^m(v) &= v + \left[-\frac{1}{2}(1 + (-1)^{m+1})\varphi(v) + \frac{\beta}{4}(1 + (-1)^{m+1}(2m+1))\psi(v)\right] a \\ &+ \left[\frac{\alpha}{4}(1 + (-1)^m(2m-1))\varphi(v) - \frac{1}{2}(1 + (-1)^{m+1})\psi(v)\right] b \end{aligned}$$

so, if $\chi(K) = 0$, $r$ is of infinite order; if $\chi(K) = p$ $(p \neq 2)$, $r'^p = -Id_V$, $r^{2p} = Id_E$. In this case, $r^p$ is not a reflection of $E$ and $P_{2p}(r') \neq 0$ as $r'^p + Id_V = 0 = (r' + Id_V)^p$. □

**Remark 2** *With the notation and hypothesis of the preceding proposition, we have* $Id_V + r' + \cdots + r'^{n-1} = 0$. *This is clear except when* $\chi(K) = p$ *and* $n = 2p$. *If* $p > 2$ *and if* 1 *is an eigenvalue of* $r'$, *there exists* $v \in V - \{0\}$ *such that* $r'(v) = v$, *so* $r'^p(v) = v$; *as* $r'^p = -Id_V$, *we obtain* $v = -v$ *so* $v = 0$; *a contradiction. As* $r' - Id_V$ *is invertible we obtain the result in this case. If* $p = 2$, *then* $r'^2 = Id_V$ *and* $r^4 = Id_E$ *so we have* $Id_V + r' + r'^2 + r'^3 = 0$.

**Proposition 4** *Let* $K$ *be a field and let* $E$ *be a* $K$*-vector space. Let* $x$ *and* $y$ *be two distinct involutions of* $GL(E)$ *whose product* $r$ *is of finite order* $n$. *Define* $V := [E,x] + [E,y]$. *Then the following two conditions are equivalent:*

1. $(Id_E + r + \cdots + r^{n-1})_{|V} = 0$;
2. $\forall e \in [E,x]$, $\forall f \in [E,y]$, $((e,x)(f,y))^n = 1$ *where* $(e,x)$ *and* $(f,y)$ *are two elements of the semidirect product* $E : GL(E)$.

**Proof.** Let $c$ be in $E$. We have the formula:

$$\forall m \in \mathbb{N},\ (c,r)^m = (c + r(c) + \cdots + r^{m-1}(c), r^m).$$

Suppose that the first condition is satisfied. We have $((e,x)(f,y)) = (e + x(f), r)$. But $x(f) = f + e'$ with $e'$ in $[E,x]$, so $((e,x)(f,y)) = (e + e' + f, r)$ and $c = e + e' + f$ is in $V$. We have $((e,x)(f,y)) = (c,r)$, so $((e,x)(f,y))^n = (c,r)^n = (c + r(= c) + \cdots + r^{n-1}(c), r^n) = (0,1)$.

Suppose that the second condition is satisfied. Let $c$ be in $V$, $c = e + f, e$ in $[E,x]$, $f$ in $[E,y]$. We have $r(f) = xy(f) = -x(f) = -f + e'$ with $e'$ in $[E,x]$, so $(c,r)(f,y) = (e - e', x)$ and $(c,r) = (e - e', x)(f,y)$. From the hypothesis we deduce that $(c,r)^n = (0,1)$, so $c + r(c) + \cdots + r^{n-1}(c) = 0$. As $c$ was arbitrary in $V$, we obtain $(Id_E + r + \cdots + r^{n-1})_{|V} = 0$. □

# 3 Generalized reflection groups

In all the rest of this paper, I make the following hypothesis:

(H) $G$ **is a group generated by a set** $X$ **of involutions with the property that for all** $x$ **and** $y$ **in** $X$ , $xy$ **is of finite order.**

On $X$ we put a graph structure $\Gamma$ (the Coxeter graph $\Gamma(G,X) = \Gamma$): $\{x,y\}$ is an edge of $\Gamma$ if $xy$ is of order $\geq 3$. Let $E(\Gamma)$ be the set of edges of $\Gamma$. We can also consider the associated labelled graph: on the edge $\{x,y\}$ we put the number $m_{xy}$ = *order of* $xy$. Without loss of generality we can suppose that $\Gamma$ *is connected* (if not, $G$ decomposes into a central product of groups having the property (H)).

**Definition 5** *Let us assume hypothesis (H). Let* $K$ *be a ring and let* $G$ *act on the left* $K$*-module* $M$. *We call* $(G,X)$ *a generalized reflection group if the following is satisfied:*

*(MG): Let* $x,y$ *be in* $X$, $r := xy$ *be of order* $n$. *Let* $V = [M,x] + [M,y]$. *If* $n \geq 3$, *then* $P_n(r)_{|V} = 0$; *if* $n = 2$ *then* $(Id_M + r)_{|V} = 0$.

# 4 Constructing the algebra

Let $K$ be a ring, $M$ a left $K$–module and $(G, X)$ a generalized reflection group acting on $M$. As we want to apply the Bezout identity to the polynomials $Y+1$ and $P_n(Y)$ we must add an hypothesis if some $n$ is even $\geq 4$: *If the product of two elements of $X$ is an even number $\geq 4$ then we suppose that the characteristic of $K$ is a prime number or 0. If the characteristic of $K$ is 0 then we suppose that $\mathbb{Q}$ is a subring of $K$. We suppose also that for each $x$ in $X$, $[M, x]$ is a free submodule of $M$.* As everything important is happening in $[M, G]$, *we suppose also in all the following that $M = [M, G]$.*

**Theorem 6** *(Constructing the algebra) With the preceding hypothesis and notation, we have:*

1. *If $x$ and $y$ are two distinct elements of $X$, $[M, x]$ and $[M, y]$ are isomorphic $K$–modules. Call $\rho$ the common rank of all the $[M, x]$ $(x \in X)$.*
2. *Let $T$ be a spanning tree of $\Gamma$. Let $J$ be the set of edges of $\Gamma$ which are not edges of $T$. To each $j$ in $J$ is associated an element $l_j$ of $GL_\rho(K)$. We consider the subalgebra $L$ of $M_\rho(K)$ generated by the $l_j$. For each $x$ in $X$, $[M, x]$ is an $L$–module in a natural way.*

**Proof.** Let $x$ and $y$ be two distinct elements of $X$. Put $r := xy$ and suppose that $r$ is of order $n$. Let $V := [M, x] + [M, y]$. Let $u \in [M, x]$. Then $u + r(u) + \cdots + r^{n-1}(u) = 0$. As $r^n = Id_M$, we have $r^{n-1}(u) = yx(u) = -y(u)$ so $r(u) + \cdots + r^{n-2}(u) = -u + y(u) \in [M, y]$ and we have constructed, if $n > 2$, a linear map $\sigma_{xy} : [M, x] \to [M, y]$ where $\sigma_{xy} = (r + r^2 + \cdots + r^{n-2})_{|[M,x]} = -(Id_M + r^{-1})_{|[M,x]}$. If $n = 2$ we have $-u + y(u) = 0$ and we obtain $[M, x] \subset C_M(y)$ and also $[M, y] \subset C_M(x)$, this is condition (MG) for $n = 2$. In the same way, $\sigma_{yx} = (yx + (yx)^2 + \cdots + (yx)^{n-2})_{|[M,y]} = -(Id_M + r)_{|[M,y]} : [M, y] \to [M, x]$. We also have $\sigma_{xy} \circ \sigma_{yx} = (Id_M + r)(Id_M + r^{-1})_{|[M,y]} : [M, y] \to [M, y]$ and $\sigma_{yx} \circ \sigma_{xy} = (Id_M + r)(Id_M + r^{-1})_{|[M,x]} : [M, x] \to [M, x]$. The way we construct $P_n(Y)$ from $Y^n - 1$ shows that $G.C.D.(Y+1, P_n(Y)) = 1$ and so there exist polynomials $A(Y)$ and $B(Y)$ such that $A(Y)(Y+1) + B(Y)P_n(Y) = 1$ (here, we are working in the prime subfield of $K$). We obtain $(A(r)(Id_V + r) + B(r)P_n(r))_{|[M,y]} = Id_{[M,y]}$. As $P_n(r)_{|[M,y]} = 0, (A(r)(Id_V + r))_{|[M,y]} = Id_{[M,y]}$ so the linear maps $\sigma_{xy}$ and $\sigma_{yx}$ are isomorphisms.

Now we make *three choices*. First we choose $x_0$ in $X$, then we choose a spanning tree $T$ of $\Gamma$. Let $E(T)$ be the set of edges of $T$ and let $J := E(\Gamma) - E(T)$, finally we choose a basis $a_0 = (a_{0,i} \mid i \in I)$ of $[M, x_0]$. Let $\{x_0, x\} \in E(\Gamma)$, then $\sigma_{x_0 x}$ induces an isomorphism between $[M, x_0]$ and $[M, x]$ and so $\sigma_{x_0 x}(a_0)$ is a basis of $[M, x]$. Next, let $x$ be in $X$ and let $(x_0, x_1, \ldots, x_s = x)$ be the unique path in $T$ between $x_0$ and $x$. Then $\theta_x = \sigma_{x_{s-1}x_s} \circ \sigma_{x_{s-2}x_{s-1}} \circ \ldots \circ \sigma_{x_0 x_1}$ induces an isomorphism between $[M, x_0]$ and $[M, x]$, and so $\theta_x(a_0) = a_x$ $(= (a_{x,i} \mid i \in I)$ is a basis of $[M, x]$ called *the* basis of $[M, x]$. In this way, for each $x$ in $X$, we have a canonical basis for

$[M,x]$. As $G =< X >$ and $M = [M,G]$, $M$ is the sum (in general non direct) of the $[M,x]$, $x$ in $X$, and $A = \cup_{x\in X} a_x$ is a generating set of $M$.

Let now $\{x,y\} \in J$. We have the basis $a_x$ of $[M,x]$ and the basis $a_y$ of $[M,y]$. Then $\sigma_{xy}(a_x)$ is a basis of $[M,y]$ and there exists an invertible linear transformation $l_{xy} \in GL([M,y])$ such that $\sigma_{xy}(a_x) = a_y l_{xy}$ (if $a_y = (a_i \mid i \in I)$, $a_y l_{xy} = (l_{xy}(a_i) \mid i \in I)$) (we denote by the same symbol the linear map $l_{xy}$ and its matrix in the basis $a_y$ of $[M,y]$ In $a_y l_{xy}$ we view $a_y$ as a matrix whose columns are the basis vectors).

(In the standard representation of a Coxeter group $l_{xy} = Id_{|[M,y]}$.)

We have $(\sigma_{xy} \circ \sigma_{yx})(a_y) = (Id_M + r)(Id_M + r^{-1})_{|[M,y]}(a_y) = \sigma_{xy}(a_x)\ l_{yx} = a_y(l_{xy}l_{yx})$. If $n = 3$, then $(Id_M + r)(Id_M + r^{-1}) = Id_M$, so in this case $l_{xy}l_{yx} = l_{yx}l_{xy} = Id : l_{yx} = l_{xy}^{-1}$.

Let $(x_0, x_1, \ldots, x_s = x)$ (resp. $(x_0, y_1, \ldots, y_t = y)$) be the unique path in $T$ between $x_0$ and $x$ (resp. $x_0$ and $y$). We have $\theta_x(a_0) = a_x$ and $\theta_y(a_0) = a_y$. We deduce that

$$(\theta_y^{-1}\sigma_{xy}\theta_x)(a_0) = a_0 l_{xy}. \tag{1}$$

To each $j$ in $J$ is associated an element $l_j$ of $GL_{|I|}(K)$. We can consider the subalgebra $L$ of $M_{|I|}(K)$ generated by the $l_j$, $j$ in $J$, taking into account the relations which exist between the elements of $X$ in order to obtain a presentation of $G$. □

**Proposition 7** *With the preceding notation and hypothesis, suppose that $X$ is finite of cardinality $v$. If $\Gamma$ has $w$ edges, then $|J| = w-v+1$ and the algebra $L$ is generated by $w - v + 1$ $l_j$, $j$ in $J$.* □

**Theorem 8** *With the hypothesis and notation of §4, we have the following properties:*

- *If we change the basis $a_0$ of $[M,x_0]$ into the basis $a_0'$, then we just conjugate the algebra $L$ in $M_{|I|}(K)$ by an element of $GL_{|I|}(K)$.*
- *The algebra $L$ is independent of the choice of $x_0$ in $X$.*
- *If $X$ is a finite set, the algebra $L$ is independent of the spanning tree $T$.*

**Proof.** The first point is clear. In the second point, it is equation (1) which shows that we don't change $L$.

For the third point, we have the following theorem which can be found in [3] (theorem 6.6.4): "When $T_1$ and $T_2$ are spanning trees in a finite connected graph, then one can be derived from the other by a finite number of singular interchanges." (We add an edge to $T_1$ and then delete another edge to obtain another spanning tree).

From this theorem, we deduce that it suffices to prove that the algebra $L$ is the same if we make a singular cyclic interchange. Let $T$ a spanning tree of $\Gamma$

and let $x_0$ be in $X$. Let $e = \{x, y\}$ be in $J$. Then if $\{x_0, x_1, \ldots, x_s = x\}$ (resp. $\{x_0, y_1, \ldots, y_t = y\}$) is the path joining $x_0$ to $x$ (resp. $x_0$ to $y$) then we have a circuit $\{x_0, x_1, \ldots, x_s, y_t, y_{t-1}, \ldots, y_1, x_0\}$ and we must remove one edge ( $\neq e$) of this circuit to obtain another spanning tree of $\Gamma$. Our problem now is to show that $l_e$ is independent of the edge we remove. We can suppose that $\Gamma = \{x_0, x_1, \ldots, x_n\}$ is a circuit: $E(\Gamma) = \{\{x_i, x_{i+1}\}, 0 \leq i \leq n \text{ (indices taken mod } n)\}$. Let $T$ be a spanning subgraph of $\Gamma : J = \{\{x_n, x_0\}\} = E(\Gamma) - E(T)$ and let $T'$ be another spanning subgraph of $\Gamma : J' = \{\{x_0, x_1\}\} = E(\Gamma) - E(T')$.

First we work with $T$. Let $a_0$ be a basis of $[M, x_0]$. Define $\sigma_i = \sigma_{x_i x_{i+1}}$ $(0 \leq i \leq n)$ an $a_i = \sigma_{i-1}(a_{i-1})$ $(1 \leq i \leq n)$. Then $L =< l >$, where $\sigma_n(a_n) = a_0 l$.

Now we work with $T'$. We start with $x_1$ with $a_1$ the basis of $[M, x_1]$, and we obtain $(\sigma_0 \sigma_n)(a_n) = \sigma_0(a_0 l) = a_1 l$, so we obtain the same algebra: $L$ is independent of the spanning tree $T$. □

I show now how we can consider the group $G$ as a reflection group. As I have not yet worked out the general case, I consider only the following situation: $X$ is a finite set, $|X| = n$, and for all $x$ and $y$ in $X$, $xy$ is of order $\leq 3$. We keep the hypothesis and notation of theorem 6.

Let $x_0$ be an element of $X$ and let $T$ be a spanning tree of $\Gamma$. Let $N$ be a free $K$–module of finite rank $p$ and let $\pi : L \to End(N)$ be a representation of $L$ such that $N$ becomes a cyclic right $KL$–module.

For each $x$ in $X$, let $N_x$ be a right $KL$–module isomorphic to $N$ and let $M$ be the direct sum of the $N_x$, $x \in X : M' = \oplus_{x \in X} N_x$. It is clear that $M'$ is a $KL$–module.

Let $a_x = (a_{xi} \mid 1 \leq i \leq p)$ be a basis of $N_x$ so that $A = \cup_{x \in X} a_x$ is a basis of $M'$. To each $x$ in $X$ we associate a linear application $\widetilde{x}$ of $M'$ into itself in the following way:

1. $\forall x \in X, \forall a \in a_x, \widetilde{x}(a) = -a$;
2. if $(x, y) \in X \times X$ is such that $xy$ is of order 2, then $\widetilde{x}(a_y) = a_y$;
3. if $\{x, y\}$ is an edge of $T$, then $\widetilde{x}(a_y) = \widetilde{y}(a_x) = a_x + a_y$;
4. if $\{x, y\}$ is in $J$, then $\widetilde{y}(a_x) = a_x + a_y l_{xy}$, $\widetilde{x}(a_y) = a_y + a_x l_{xy}^{-1}$ (for in this case $l_{xy} l_{yx} = Id$).

So, for each $x$ in $X$, we have defined an element $\widetilde{x}$ of $End(M)$. Put $\widetilde{X} = \{\widetilde{x} \mid x \in X\}$ and $\widetilde{G} =< \widetilde{X} >$. By construction $\widetilde{G}$ is a subgroup of $GL(M')$. It is not hard to see that the map $\varphi : X \to \widetilde{X} : x \longmapsto \widetilde{x}$ defines a homomorphism $\varphi : G \to \widetilde{G}$. As each $\widetilde{x}$ commutes with the action of $L$, $M'$ is an $L$–module and for each $\widetilde{x} \in \widetilde{X}$, $[M', \widetilde{x}] = N_x$ is a cyclic $L$–module, so $G$ acts, through $\varphi$, as a reflection group on $M'$.

## 5 Some examples

We keep the preceding notation and hypothesis. We suppose also that for all $x$ and $y$ in $X$, the order of $xy$ is $1, 2$ or $3$.

When $|J| = 1$, we have a unique circuit with pending leaves. Suppose that $\Gamma$ is a circuit: $\Gamma = \{x_0, x_1, \ldots, x_n\}$ with $E(\Gamma) = \{\{x_i, x_{i+1}\} \mid 0 \leq i \leq n-1\} \cup \{\{x_n, x_0\}\}$.

Let $M$ be a free left $K$–module of rank $(1+n)p$, $p$ integer $\geq 1$ and let $(b_{i,j} \mid 0 \leq i \leq n, 1 \leq j \leq p)$ be a basis of $M$. Put $b_i = (b_{i,j} \mid 1 \leq j \leq p)$ $(0 \leq i \leq n)$. Let $l \in GL_p(K)$. We define $n+1$ linear applications $x_i : M \to M$ by:

1. $x_i(b_i) = -b_i$ $(0 \leq i \leq n)$
2. if $x_i x_j$ is of order 2, $x_i(b_j) = b_j$ (condition (MG) for $n = 2$);
3. if $x_i x_j$ is of order 3, $\{i, j\} \neq \{0, n\}, x_i(b_j) = b_i + b_j$ (condition (MG) for $n = 3$);
4. $x_0(b_n) = b_n + b_0 l$, $x_n(b_0) = b_0 + b_n l^{-1}$ (condition (MG) for $n = 3$).

Then $G =< x_0, x_1, \ldots, x_n >$ is isomorphic to a quotient of $W(\widetilde{A_n})$.

More precisely, if $\Lambda$ is the translation subgroup of $W(\widetilde{A_n})$ and if $q\Lambda = \{t^q \mid t \in \Lambda\}$ where $q \in N$, then $q\Lambda$ is in the kernel of the representation if and only if $\sum_{0 \leq j \leq q-1} l^j = 0$. Furthermore, if the basis $b_0$ of $[M, x_0]$ is changed, then $l$ is changed into a conjugate in the group $GL_p(K)$.

- If $K = \mathbb{Z}$, $p = 1$ and $l = I$, the standard representation of $W(\widetilde{A_n})$ is obtained (as a reflection group).
- If $K = \mathbb{Z}, p = 1$ and $l = -I$, the standard representation of $W(D_{n+1})$ is obtained (as a reflection group).
- If $K = \mathbb{C}, p = 1$ and $l = \varsigma$ is a primitive $m$–root of unity, we obtain the standard representation of $G(m, m, n)$ (as a reflection group).
- If $K = \mathbb{Q}, p = 1$ and $l \neq I$ and $l \neq -I$, then we obtain infinitely many non equivalent representations of $W(\widetilde{A_n})$ (as a reflection group)

As the only condition on $l$ in this case is that it be invertible, by choosing $l$ in $GL_p(\mathbb{Q})$ in such a way that $\mathbb{Q}^p$ is an irreductible $\mathbb{Q} < l >$–module, we can obtain infinitely many representations of $W(\widetilde{A_n})$ as a reflection group on vector spaces over algebraic extensions of $\mathbb{Q}$.

Now we give an example where the algebra $L$ is isomorphic to $\mathbb{H}$, the field of quaternions. Let $G$ be the group $C_2 \times PSU(5,4)$ which has the presentation: generators: $X = \{x_1, x_2, x_3, x_4, x_5\}$, relations:

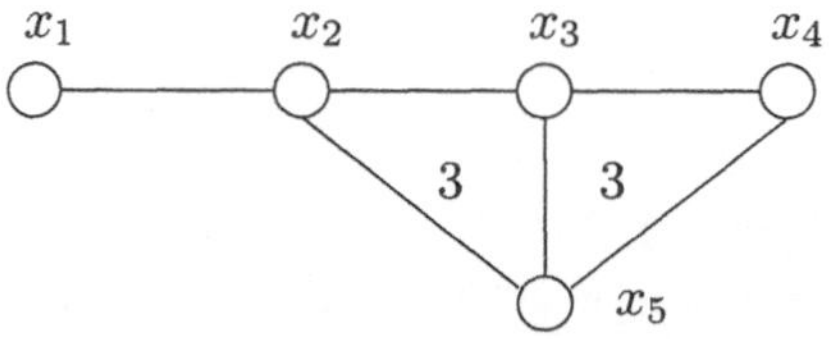


$$(x_3^{x_2} x_5)^3 = (x_3^{x_4} x_5)^3 = (x_3^{x_2 x_4} x_5)^3 = 1 \qquad (2)$$

We take as spanning tree $T$ the graph obtain from $\Gamma$ by removing the edges $\{x_2, x_5\}$ and $\{x_4, x_5\}$. The algebra $L$ is generated by $l = l_{x_2 x_5}$ and $m = l_{x_4 x_5}$ and we have the following relations:

$$1 + l + l^2 = 0 \qquad (3)$$

$$1 + m + m^2 = 0 \qquad (4)$$

$$lm^{-1} + ml^{-1} = 0 \qquad (5)$$

Then the group generated by $l$ and $m$ is isomorphic to $SL_2(3)$ and the algebra $L$ is isomorphic to the Hurwitz integers if $K = \mathbb{Z}$ and isomorphic to $\mathbb{H}$ if $K = \mathbb{R}$.

Others examples (all with the hypothesis: $\forall x, y \in X$, the order of $xy$ is $1, 2$ or $3$) can be found in Zara [7] and Zara [8].

# 6 Odds and ends. The future

For the future, one must first solve the spanning tree problem when $X$ is infinite: is the isomorphism type of the algebra $L$ an invariant of the generalized reflection group $(G, X)$ ? Another problem: is there any restriction on the algebra $L$ ?

The most important problem is to prove that $G$ can be realized as a reflection group in the general case and this I hope to do in the near future.

The axiom (MG) can be written multiplicatively so we can speak of the generalized reflection group $(G, X)$ acting on the group $M$. The results in [9] show that in this case the structure of $M$ is severely restricted.

We can also begin a study of general odd-transpositions groups in the spirit of the study of 3–transpositions groups.

**Acknowledgment.** I thank the members of the Amiens Seminar for many helpful conversations, particularly Serge Bouc, François Digne and Alexander Zimmermann. I thank also the (anonymous) Referee for his help in improving the paper. Finally, I thank warmly Antonio Pasini for his help and patience during the preparation of this manuscript.

# References


[1] Cohen A.M., Finite Quaternionic Reflection groups, Journal of Algebra 64 (1980) 293–324.

[2] Coxeter H.S.M., Discrete Groups generated by Reflections, Annals of Math. 35 (1934) 588–621.

[3] Ore O., Theory of graphs, Colloquium publications 38 (1962) American Math. Soc., Providence R.I..

[4] Shephard G.C. and Todd J.A., Finite Unitary Reflection groups Can. J. of Math. 6 (1954) 274–304.

[5] Zara F., Classification des couples fischeriens, Thèse, Amiens, 1985.

[6] Zara F., A first step toward the classification of Fischer groups, Geom. Ded.25 (1988) 503–510.

[7] Zara F., (G,D)-modules et algèbres de Clifford, Geom. Ded.35 (1990) 155–176.

[8] Zara F., Constructing new Fischer groups from old ones, Europ. J. Combin. 15 (1996) 87–104.

[9] Zara F., Action of Reflection Groups on Nilpotent Groups, Europ. J. Combin. (1997) 18, 231–242.



François Zara,
Université de Picardie Jules Verne,
Faculté de Mathématiques et Informatique,
33 rue Saint Leu, F-80033 Amiens Cedex, France
e-mail: `zara@mathinfo.u-picardie.fr`

## LM • Lectures in Mathematics, ETH Zürich

Department of Mathematics
Research Institute of Mathematics

*Each year the Eidgenössische Technische Hochschule (ETH) at Zürich invites a selected group of mathematicians to give postgraduate seminars in various areas of pure and applied mathematics. These seminars are directed to an audience of many levels and backgrounds. Now some of the most successful lectures are being published for a wider audience through the* ***Lectures in Mathematics, ETH Zürich*** *series. Lively and informal in style, moderate in size and price, these books will appeal to professionals and students alike, bringing a quick understanding of some important areas of current research.*

R.J. LeVeque
**Numerical Methods for Conservation Laws**
*2nd Edition, 3rd Printing 1994.*
1992. ISBN 3-7643-2723-5

R. Narasimhan
**Compact Riemann Surfaces**
1992. ISBN 3-7643-2742-1

A.J. Tromba
**Teichmüller Theory in Riemannian Geometry**
1992. ISBN 3-7643-2735-9

G. Baumslag
**Topics in Combinatorial Group Theory**
1993. ISBN 3-7643-2921-1

M. Giaquinta
**Introduction to Regularity Theory for Nonlinear Elliptic Systems**
1993. ISBN 3-7643-2879-7

O. Nevanlinna
**Convergence of Iterations for Linear Equations**
1993. ISBN 3-7643-2865-7

R.-P. Holzapfel
**The Ball and Some Hilbert Problems**
1995. ISBN 3-7643-2835-5

J.F. Carlson
**Modules and Group Algebras**
*Notes by Ruedi Suter*
1996. ISBN 3-7643-5389-9

L. Simon
**Theorems on Regularity and Singularity of Energy Minimizing Maps**
*based on lecture notes by Norbert Hungerbühler*
1996. ISBN 3-7643-5397-X

M. Freidlin
**Markov Processes and Differential Equations: Asymptotic Problems**
1996. ISBN 3-7643-5392-9

J. Jost
**Nonpositive Curvature**
**Geometric and Analytic Aspects**
1997. ISBN 3-7643-5736-3

Ch.M. Newman
**Topics in Disordered Systems**
1997. ISBN 3-7643-5777-0

M. Yor
**Some Aspects of Brownian Motion, Part I**
1992. ISBN 3-7643-2807-X
**Part II**
1997. ISBN 3-7643-5717-7

115 BABELON/CARTIER/KOSMANN-SCHWARZBACH (Eds). Integrable Systems. The Verdier Memorial Conference
116 DAVID (Ed.). Séminaire de Théorie des Nombres, Paris, 1991-1992
117 AUDIN/LAFONTAINE (Eds). Holomorphic Curves in Symplectic Geometry
118 VAISMAN. Lectures on the Geometry of Poisson Manifolds
119 JOSEPH/MIGNOT/MURAT/PRUM/RENTSCHLER (Eds) First European Congress of Mathematics (Paris, July 6-10, 1992). Volume I: Invited Lectures (Part 1)
120 JOSEPH/MIGNOT/MURAT/PRUM/RENTSCHLER (Eds). First European Congress of Mathematics (Paris, July 6-10, 1992). Volume II: Invited Lectures (Part 2)
121 JOSEPH/MIGNOT/MURAT/PRUM/RENTSCHLER (Eds). First European Congress of Mathematics (Paris, July 6-10, 1992). Volume III: Round Tables
122 GUILLEMIN. Moment Maps and Combina torial Invariants of Hamiltonian $T^n$-spaces
123 BRYLINSKI/BRYLINSKI/GUILLEMIN/KAC (Eds). Lie Theory and Geometry: In Honor of Bertram Kostant
124 AEBISCHER/BORER/KÄLIN/LEUENBERGER/REIMANN. Symplectic Geometry. An Introduction based on the Seminar in Bern, 1992
125 LUBOTZKY. Discrete Groups, Expanding Graphs and Invariant Measures
126 RIESEL. Prime Numbers and Computer Methods for Factorization
127 HÖRMANDER. Notions of Convexity
128 SCHMIDT. Dynamical Systems of Algebraic Origin
129 DIJKGRAAF/FABER/VAN DER GEER (Eds). The Moduli Space of Curves
130 DUISTERMAAT. Fourier Integral Operators
131 GINDIKIN/LEPOWSKY/WILSON. Functional Analysis on the Eve of the 21st Century in Honor of the 80th Birthday of I.M. Gelfand. Vol. I
132 GINDIKIN/LEPOWSKY/WILSON. Functional Analysis on the Eve of the 21st Century in Honor of the 80th Birthday of I.M. Gelfand. Vol. II
131/132 Set Vols I+II
133 HOFER/TAUBES/WEINSTEIN/ZEHNDER (Eds). The Floer Memorial Volume
134 CAMPILLO LÓPEZ/ NARVÁEZ MACARRO (Eds). Algebraic Geometry and Singularities
135 Amrein/Boutet de Monvel/Georgescu. $C_0$-Groups, Commutator Methods and Spectral Theory of $N$-Body Hamiltonians
136 BROTO/CASACUBERTA/MISLIN (Eds). Algebraic Topology. New Trends in Localization and Periodicity
137 VIGNÉRAS. Représentations $l$-modulaires d'un groupe réductif $p$-adique avec $l \neq p$
140 KNAPP. Lie Groups Beyond an Introduction
141 CABANES (Ed.). Finite Reductive Groups: Related Structures and Representations
142 MONK. Cardinal Invariants on Boolean Algebras
143 GONZÁLEZ-VEGA/RECIO (Eds). Algorithms in Algebraic Geometry and Applications
144 BELLAÏCHE / RISLER (Eds). Sub-Riemannian Geometry
145 ALBERT / BROUZET / DUFOUR (Eds). Integrable Systems and Foliations Feuilletages et Systemes Integrables
146 JARDINE. Generalized Etale Cohomology Theories
147 DI BIASE. Fatou Type Theorems Maximal Functions and Approach Regions
148 HUANG. Two-Dimensional Conformal Geometry and Vertex Operator Algebras
149 SOURIAU. Structure of Dynamical Systems. A Symplectic View of Physics
150 SHIOTA. Geometry of Subanalytic and Semialgebraic Sets
151 HUMMEL. Gromov's Compactness Theorem for Pseudo-holomorphic Curves
153 BUESCU. Exotic Attractors. From Liapunov Stability to Riddled Basins
154 BÖTTCHER/KARLOVICH. Carleson Curves, Muckenhoupt Weights, and Toeplitz Operators
155 DRAGOMIR/ORNEA. Locally Conformal Kähler Geometry
156 GUIVARC'H/JI/TAYLOR. Compactifications of Symmetric Spaces
157 MURTY/MURTY. Non-vanishing of $L$-functions and Applications
158 TIRAO/VOGAN/WOLF (Eds). Geometry and Representation Theory of Real and $p$-Adic Groups
159 THANGAVELU. Harmonic Analysis on the Heisenberg Group

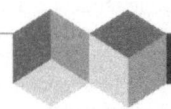

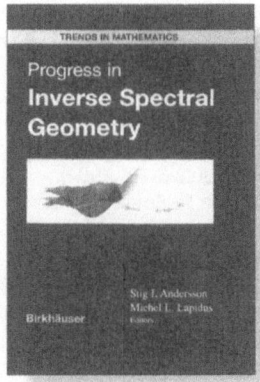